IET MATERIALS, CIRCUITS AND DEVICES SERIES 67

Frontiers in Securing IP Cores

Other volumes in this series:

Volume 1	**Advances in High-Power Fiber and Diode Laser Engineering** I. Divliansky (Editor)
Volume 2	**Analogue IC Design: The current-mode approach** C. Toumazou, F.J. Lidgey and D.G. Haigh (Editors)
Volume 3	**Analogue-Digital ASICs: Circuit techniques, design tools and applications** R.S. Soin, F. Maloberti and J. France (Editors)
Volume 4	**Algorithmic and Knowledge-Based CAD for VLSI** G.E. Taylor and G. Russell (Editors)
Volume 5	**Switched Currents: An analogue technique for digital technology** C. Toumazou, J.B.C. Hughes and N.C. Battersby (Editors)
Volume 6	**High-Frequency Circuit Engineering** F. Nibler *et al.*
Volume 8	**Low-power High-Frequency Microelectronics: A unified approach** G. Machado (Editor)
Volume 9	**VLSI Testing: Digital and mixed analogue/digital techniques** S.L. Hurst
Volume 10	**Distributed Feedback Semiconductor Lasers** J.E. Carroll, J.E.A. Whiteaway and R.G.S. Plumb
Volume 11	**Selected Topics in Advanced Solid State and Fibre Optic Sensors** S.M. Vaezi-Nejad (Editor)
Volume 12	**Strained Silicon Heterostructures: Materials and devices** C.K. Maiti, N.B. Chakrabarti and S.K. Ray
Volume 13	**RFIC and MMIC Design and Technology** I.D. Robertson and S. Lucyzyn (Editors)
Volume 14	**Design of High Frequency Integrated Analogue Filters** Y. Sun (Editor)
Volume 15	**Foundations of Digital Signal Processing: Theory, algorithms and hardware design** P. Gaydecki
Volume 16	**Wireless Communications Circuits and Systems** Y. Sun (Editor)
Volume 17	**The Switching Function: Analysis of power electronic circuits** C. Marouchos
Volume 18	**System on Chip: Next generation electronics** B. Al-Hashimi (Editor)
Volume 19	**Test and Diagnosis of Analogue, Mixed-Signal and RF Integrated Circuits: The system on chip approach** Y. Sun (Editor)
Volume 20	**Low Power and Low Voltage Circuit Design with the FGMOS Transistor** E. Rodriguez-Villegas
Volume 21	**Technology Computer Aided Design for Si, SiGe and GaAs Integrated Circuits** C.K. Maiti and G.A. Armstrong
Volume 22	**Nanotechnologies** M. Wautelet *et al.*
Volume 23	**Understandable Electric Circuits** M. Wang
Volume 24	**Fundamentals of Electromagnetic Levitation: Engineering sustainability through efficiency** A.J. Sangster
Volume 25	**Optical MEMS for Chemical Analysis and Biomedicine** H. Jiang (Editor)
Volume 26	**High Speed Data Converters** A.M.A. Ali
Volume 27	**Nano-Scaled Semiconductor Devices** E.A. Gutiérrez-D (Editor)
Volume 28	**Security and Privacy for Big Data, Cloud Computing and Applications** L. Wang, W. Ren, K.R. Choo and F. Xhafa (Editors)
Volume 29	**Nano-CMOS and Post-CMOS Electronics: Devices and modelling** S.P. Mohanty and A. Srivastava
Volume 30	**Nano-CMOS and Post-CMOS Electronics: Circuits and design** S.P. Mohanty and A. Srivastava
Volume 32	**Oscillator Circuits: Frontiers in design, analysis and applications** Y. Nishio (Editor)
Volume 33	**High Frequency MOSFET Gate Drivers** Z. Zhang and Y. Liu
Volume 34	**RF and Microwave Module Level Design and Integration** M. Almalkawi
Volume 35	**Design of Terahertz CMOS Integrated Circuits for High-Speed Wireless Communication** M. Fujishima and S. Amakawa
Volume 38	**System Design with Memristor Technologies** L. Guckert and E.E. Swartzlander Jr.
Volume 39	**Functionality-Enhanced Devices: An alternative to Moore's law** P.E. Gaillardon (Editor)

Volume 40	**Digitally Enhanced Mixed Signal Systems** C. Jabbour, P. Desgreys and D. Dallett (Editors)
Volume 43	**Negative Group Delay Devices: From concepts to applications** B. Ravelo (Editor)
Volume 45	**Characterisation and Control of Defects in Semiconductors** F. Tuomisto (Editor)
Volume 47	**Understandable Electric Circuits: Key concepts, 2nd Edition** M. Wang
Volume 53	**VLSI Architectures for Future Video Coding** M. Martina (Editor)
Volume 58	**Magnetorheological Materials and Their Applications** S. Choi and W. Li (Editors)
Volume 60	**IP Core Protection and Hardware-Assisted Security for Consumer Electronics** A. Sengupta and S. Mohanty
Volume 68	**High Quality Liquid Crystal Displays and Smart Devices: Vol. 1 and Vol. 2** S. Ishihara, S. Kobayashi and Y. Ukai (Editors)
Volume 69	**Fibre Bragg Gratings in Harsh and Space Environments: Principles and applications** B. Aïssa, E.I. Haddad, R.V. Kruzelecky and W.R. Jamroz
Volume 70	**Self-Healing Materials: From fundamental concepts to advanced space and electronics applications, 2nd Edition** B. Aïssa, E.I. Haddad, R.V. Kruzelecky and W.R. Jamroz
Volume 71	**Radio Frequency and Microwave Power Amplifiers: Vol. 1 and Vol. 2** A. Grebennikov (Editor)
Volume 73	**VLSI and Post-CMOS Electronics: Vol. 1: VLSI and post-CMOS electronics and Vol. 2: Materials, devices and interconnects** R. Dhiman and R. Chandel (Editors)

Frontiers in Securing IP Cores

Forensic detective control and obfuscation techniques

Anirban Sengupta

The Institution of Engineering and Technology

Published by The Institution of Engineering and Technology, London, United Kingdom

The Institution of Engineering and Technology is registered as a Charity in England & Wales (no. 211014) and Scotland (no. SC038698).

First published 2019

The Institution of Engineering and Technology
Michael Faraday House
Six Hills Way, Stevenage
Herts, SG1 2AY, United Kingdom

www.theiet.org

British Library Cataloguing in Publication Data
A catalogue record for this product is available from the British Library

ISBN 978-1-83953-031-9 (hardback)
ISBN 978-1-83953-032-6 (PDF)

Typeset in India by MPS Limited
Printed in the UK by CPI Group (UK) Ltd, Croydon

Contents

Preface

This book on *Frontiers in Securing IP Cores: Forensic detective control and obfuscation techniques* presents state-of-the-art technological solutions for securing and protecting digital signal processing (DSP) circuits used in consumer electronics (CE) systems. DSP cores, or integrated circuits, are an integral component of CE systems and serve as the work-horse or engine in smart electronics for computationally intensive, data-crunching and control-intensive applications such as digital filters, discrete cosine transformation, fast-Fourier transformation, JPEG, Haar-wavelet transformation, etc. Thus security or protection of these DSP integrated circuits or IP cores against standard threats, IP abuse or misuse, etc., becomes highly essential for modern CE devices. This book, for the first time, covers frontiers of securing hardware IP (for DSP applications) using forensic detective control mechanisms and obfuscation under a single umbrella.

Broadly the themes of this book include:

1. Introduction to hardware (IP) security: forensic detective control and obfuscation of DSP cores
2. Forensic detective control using hardware steganography for IP core protection
3. Forensic detective control using digital signature–based watermark for IP core protection
4. Protection of fault-secured IP core using digital signature based watermarking
5. Multi-level watermarking for IP protection
6. Protecting the rights of an IP buyer using cryptosystem–based multi-variable fingerprinting
7. Multi-phase obfuscation for fault-secured DSP circuits
8. Hologram-based structural obfuscation for securing DSP cores
9. Security of functionally obfuscated DSP cores

Chapter 1 presents an *Introduction hardware security and IP protection*. The prominent features of this chapter include providing a security perspective, outlining hardware security, presenting background on IP cores and their importance, background on why DSP cores are vital for modern CE systems – with an overview of the high-level synthesis framework.

Chapter 2 presents *Forensic detective control using hardware steganography for IP core protection*. The prominent features of this chapter include defining hardware steganography, the differences between hardware steganography and hardware watermarking, advantages of hardware steganography and forensic detective control using hardware steganography for IP core protection.

Chapter 3 presents *Forensic detective control using digital signature–based watermarks for IP core protection*. The prominent features of this chapter include defining digital signatures, the difference between a digital-based watermark and a normal watermark, advantages of digital signature–based watermarks and forensic detective control using digital signature–based watermarks for IP core protection.

Chapter 4 presents *Protection of fault secured IP core using digital signature–based watermark*. The prominent features of this chapter include an introduction to fault security, background on transient faults, applying digital signatures on fault-secured IP cores, and analysis of digital signature–based IP protection for fault-secured DSP circuits.

Chapter 5 presents *Multi-level watermarks for IP protection*. The prominent features of this chapter include defining multi-level watermarking, the importance and utility of multi-level watermarks for DSP cores, design process of multi-level watermarking for IP core protection of DSP applications as well as a discussion and comparative analysis with contemporary watermarking approaches.

Chapter 6 presents *Protecting the rights of an IP buyer using cryptosystem-based multi-variable fingerprinting*. The prominent features of this chapter include an overview on protecting the IP buyer's rights, difference between IP buyer's right and IP vendor's right, details on crypto-based fingerprinting and the design process of a fingerprinted DSP core.

Chapter 7 presents *Multi-phase obfuscation for fault-secured DSP circuits*. The prominent features of this chapter include an introduction to structural obfuscation, the difference between structural and functional obfuscation, definition of multi-phase obfuscation, algorithms for transient-fault security, significance of multi-phase obfuscation for fault-secured DSP circuits and case studies on DSP applications.

Chapter 8 presents *Hologram-based structural obfuscation for securing DSP cores*. The prominent features of this chapter include a background on hologram-based obfuscation, algorithms for performing hologram-based obfuscation, design process of generating obfuscated DSP circuits and analyses of several case studies.

Chapter 9 presents *Security of functionally obfuscated DSP cores*. The prominent features of this chapter include an overview on functional obfuscation of DSP circuits, using crypto-logics for securing functionally obfuscated DSP circuits against removal attacks, comparative view of different crypto-logics used for securing IP cores in obfuscation and a discussion on alternative, lightweight approaches.

The authors believe that there is no book that presents the frontiers of securing hardware IP (especially for DSP applications) using a forensic detective control mechanism and obfuscation under one canopy. This central theme is important to researchers and graduate students in different areas of specialisation, as it encompasses overlapping contents of EDA–CAD, hardware design security, VLSI design (high-level synthesis, RTL, gate-level synthesis, etc.), IP core protection, system-on-chip design and, finally, application of a specific processor design for CE applications. The book offers theoretical exercises for each chapter. This presents abundant prospects to test the understanding for each security topic and hone readers' understanding. The book is designed keeping in mind the possibility of seamless integration with any Masters and Ph.D. level course as well as senior Bachelors course. Furthermore, it also serves as a designer's handbook for those who are eager to learn designing secured hardware, especially IP cores for DSP kernels.

Dr Anirban Sengupta
Fellow of IET, Fellow of British Computer Society (BCS),
Senior Member of IEEE
IEEE Distinguished Lecturer (IEEE CE Society)
IEEE Distinguished Visitor (IEEE Computer Society)
Chair, IEEE Computer Society Technical Committee on VLSI
Founder and Chair, IEEE Consumer Electronics Society Bombay Chapter
Deputy EiC, IET Computers and Digital Techniques, EiC, IEEE VLSI Circuits
and Systems Letter
Associate Professor, Computer Science and Engineering,
Indian Institute of Technology Indore, Indore, India
Email: asengupt@iiti.ac.in
Web: http://www.anirban-sengupta.com

Acknowledgements

I would like to thank my family and friends for the support and encouragement throughout the execution of the book project. I would also like to thank the Indian Institute of Technology (IIT) Indore for the support in this work.

About the author

Dr Anirban Sengupta is Associate Professor in Computer Science and Engineering at the Indian Institute of Technology (IIT) Indore, where he directs the research lab on 'CAD for Consumer Electronics Hardware Device Security and Reliability'. He is an elected *Fellow of IET and Fellow of British Computer Society (FBCS), United Kingdom*. He holds a Ph.D. and M.Sc. in Electrical and Computer Engineering from Ryerson University, Toronto (Canada) and is a registered Professional Engineer of Ontario (P.Eng.). He has been an active researcher in the emerging areas of 'Hardware Security', 'IP Core Protection' and 'Digital Rights Management for Electronics Devices'. He was awarded with the prestigious IEEE Distinguished Lecturer by IEEE Consumer Electronics Society in 2017 and IEEE Distinguished Visitor by IEEE Computer Society in 2019.

He has more than *210 publications including 11 patents and 3 International Books*. He is an author of a *Book from IET on 'IP Core Protection and Hardware-Assisted Security for Consumer Electronics'* published in 2019 in the United Kingdom. He is currently *Deputy Editor-in-Chief* of IET Computers and Digital Techniques Journal that has a publishing history of over 40 years and *Editor-in-Chief* of IEEE VLSI Circuits and Systems Letter of IEEE Computer Society TCVLSI. He is currently the *Chairman* of IEEE Computer Society TCVLSI. He also serves on several editorial positions as senior editor, associate editor, editor and guest editor of several IEEE transactions/journals, IET and Elsevier journals including *IEEE Transactions on Aerospace and Electronic Systems (TAES), IEEE Transactions on VLSI Systems, IEEE Transactions on Consumer Electronics, IEEE Access Journal, IET Journal on Computer and Digital Techniques, IEEE Consumer Electronics, IEEE Canadian Journal of Electrical and Computer Engineering, IEEE VLSI Circuits and Systems Letter* and *Elsevier Microelectronics Journal*. He further serves as guest editor of IEEE Transactions on VLSI Systems, IEEE Access and IET Computers and Digital Techniques. He was the general/conference chair of the 37th IEEE International Symposium on Consumer Electronics (ICCE) 2019, Las Vegas and Technical Program Chair of 36th IEEE International Conference on Consumer Electronics (ICCE) 2018 in Las Vegas, 9th IEEE International Conference on Consumer Electronics (ICCE) – Berlin 2019, 15th IEEE International Conference on Information Technology (ICIT) 2016, 3rd IEEE International Symposium on Nanoelectronic and Information Systems (iNIS) 2017.

More than a dozen of his IEEE publications have appeared in 'Top 50 Most Popular Articles' with few in 'Top 5 Most Popular Articles' from IEEE periodicals. His patents have been cited in industry patents of *IBM Corporation, Siemens*

Corporation, Qualcomm, Amazon Technologies, Siemens (Germany), MATHWORKS Inc., Ryerson University and STC University of Mexico multiple times. His professional work has received wide media coverage nationally and internationally such as in *IET International News (UK), Times of India, Central Chronicle, DBPOST News, Free Press Journal, Dainik Bhaskar*, etc. He has supervised more than 30 candidates including several graduated Ph.D. candidates, research assistants, associates and B.Tech students, all of whom were placed in academia and industry. He has successfully commissioned special issues in IEEE TVLSI, IEEE TCAD, IET CDT, IEEE Access as well as IEEE CEM. He has been awarded the highest rating (excellent) by the Department of Science and Technology (DST), India based on the performance of funded projects in 2017. His ideas have been awarded funding from the DST, Council of Scientific and Industrial Research (CSIR), India and Department of Electronics and IT (DEITY), India.

Foreword of the IET book 'Frontiers in Securing IP Cores: Forensic detective control and obfuscation techniques'

- The first book to present frontiers of securing hardware for intellectual property (especially for Digital signal processing (DSP) circuits) using forensic detective control mechanisms and obfuscation under one canopy. DSP circuits and applications are key components of consumer electronics (CE) systems and the Internet of things.
- The book offers theoretical exercises at the end of each chapter. This presents abundant opportunities to test and refine the readers' understanding of all the topics covered in the book.
- The publication serves as a designers' handbook for those eager to learn how to design secured hardware, especially for IP cores with DSP kernels.
- A comprehensive coverage of DSP and hardware security concepts with relevant background material and descriptions of current industrial practices are key distinctive features of this textbook. Further, some advanced, exotic, topics for future hardware IP protection are also presented.
- It can be an essential read for a hardware designer, FPGA programmer, CAD tool designer, as well as researchers and academics involved in the design of secure CE devices.
- The book offers state-of-the-art knowledge on providing security for DSP circuits and is appearing at a very timely juncture when there is significant demand for securing hardware IPs for CE systems.
- Dr Anirban Sengupta is an extraordinary professor and author in the area of hardware security, IP protection and design optimization with over 200 technical papers published and 11 patents. Dr Sengupta is a Fellow of The Institution of Engineering and Technology and Fellow of British Computer Society. He is a renowned authority in emerging technologies in forensic science security. He has received numerous IEEE awards, is an accomplished author and also the chair of a number of IEEE CE conferences.

Stephen Dukes, IEEE Life Fellow
Former President, IEEE Consumer Electronics Society
Former VP of Conferences, IEEE Consumer Electronics Society
Former Board of Governors, IEEE Consumer Electronics Society

Foreword for IET book 'Frontiers in Securing IP Cores: Forensic detective control and obfuscation techniques'

Frontiers in Securing Hardware IP: Forensic detective control and obfuscation is a timely and very important book. Digital signal processing (DSP) is widely spread in consumer technology and in many other fields as well. This book provides a solid overview of security issues related to hardware for DSP applications. The way this book is structured makes it useful for teaching purposes as well as self-study. Each of the chapters contains case studies and review questions and exercises.

The book introduces the reader to important fundamental concepts and the need for security in IP cores, and then leads on to forensic detective control using hardware steganography for IP core protection. Then the book discusses watermarking as a protection tool, followed by cryptosystems and buyer's and vendor's rights. The next topic is the obfuscation of fault-secured DSP circuits, followed by hologram-based obfuscation. The last chapter covers functional obfuscation.

In today's digitally connected environment, where security is a prime concern, this book fills a very important gap. I am confident that readers will find *Frontiers in Securing Hardware IP: Forensic detective control and obfuscation* to be a valuable resource.

Stefan Mozar,
PhD, Fellow IEEE, FIEAust, CPEng, IntPE(AUS)
Past President, IEEE CE Society

List of acronyms

Chapter 2

$N_i(R_j)$	resource configuration, where N_i is the number of resources of type R_j
(Si, Sj)	a node pair
S	a set of edges (in a CIG) between nodes of same colour
E_{s_i,s_j}	hardware entropy of a swapping pair of an edge between the node pair (Si, Sj)
$E^m_{s_i,s_j}$	maximum entropy of an edge between the node pair (Si, Sj)
E^t	entropy threshold
S^*	a subset of set S containing only those edges that qualify for insertion in the CIG based on maximum entropy $E^m_{s_i,s_j}$
T	control step
P^c	probability of coincidence
n	number of colours used in the CIG before implanting steganography
k	number of effective constraints (hidden edges) embedded into the CIG
$C_d(R_i)$	the design cost post-embedding steganography with vendor-specified resource configuration R_i
L_d	design latency
A_d	hardware area
$L_{\max}$	maximum execution latency
$A_{\max}$	maximum hardware area
w_1	user-specified weight for execution latency
w_2	user-specified weight for hardware area
R^e	number of extra registers required after implanting steganography

Chapter 3

α, β', γ, i, I, T, !	signature digits of a triple-phase watermark
W_t	64-bit word used in the SHA-512 algorithm
K_t	64-bit constant used in the SHA-512 algorithm
C	cipher-data of RSA encryption algorithm

M	decimal value (corresponding to 8-bit) input to RSA encryption algorithm
d	owner's private key
n	multiplication of two distinct prime numbers (p_1 and p_2)
R_i	user-specified resource configuration
T	control step
Q	digital signature size chosen by the vendor
$\langle Si, Sj \rangle$	encoded edges as digital-signature constraints
P^c	probability of coincidence
£	number of colours used in the CIG before implanting digital signature
k	number of digits in a digital signature, i.e., the digital-signature size
$C_d(R_i)$	the design cost post-embedding a digital signature with vendor-specified resource configuration R_i
L_d	design latency
A_d	hardware area
$L_{\max}$	maximum execution latency
$A_{\max}$	maximum hardware area
w_1	user-specified weight for execution latency
w_2	user-specified weight for hardware area
R^e	number of extra registers required after implanting a digital signature

Chapter 4

K_c	strength of transient fault
UF	loop-unrolling factor
X	user-specified hardware resources set
$N(R_D)$	the number of hardware resources of type R_D
SDFG^{DMR}	scheduled data flow graph of DMR design
ps	pico-seconds
CS	control step
E	encrypted decimal value obtained using RSA encryption
H	decimal equivalent of hashed bitstream block (8-bit)
k	vendor private key
n	multiplication of two distinct prime numbers (p and q)
$y[n]$	output of FIR filter
$x[n-i]$	input 'X' to FIR filter at time instant '$n-i$'

$h[i]$	constant value input to FIR filter
i	variable iterating from 0 to 8
S	digital signature length
$\langle V_i, V_j \rangle$	encoded edges as digital-signature constraints
P^c	probability of coincidence
c	number of colours used in the baseline design
$C_f(X_i)$	the design cost post-embedding a digital signature in a transient, fault-secured design with vendor-specified resource configuration X_i
L_T	design latency
A_T	hardware area
$L_{\max}$	maximum execution latency
$A_{\max}$	maximum hardware area
ϕ_1	user-specified weight for execution latency
ϕ_2	user-specified weight for hardware area
£	number of extra registers required after implanting a digital signature

Chapter 5

X	resource configuration
$N_m(R_m)$	number of resource type R_m
'ω', 'θ', 'φ'	variables of a multi-level watermark
u, v	number of adders and multipliers, respectively, in the RTL design
T	control step
T_t	tamper tolerance
K	number of encoding variables
S	signature size

Chapter 6

x	force even operation to execute in odd control steps
y	force odd operation to execute in even control steps
z	encoded value of edge with node pair as (odd, odd) in coloured interval graph
i	encoded value of edge with node pair as (prime, prime)
I	encoded value of edge with node pair as (even, even)
T	encoded value of edge with node pair as (odd, even)
$!$	encoded value of edge with node pair as (0, any integer)
A^t_{fp}	total area
$N(R_k)$	number of kth functional unit

$N(mux)$	number of multiplexers
$N(reg)$	number of registers
$A(R_k)$	area of the kth functional unit
$A(mux)$	area of multiplexers
$A(reg)$	area of registers
L_{fp}^{t}	total execution latency
$d(op_k)$	delay of kth operation in a given control step
T	control step
$d(reg)$	delay of a register
C_{fp}	the design cost post-embedding a fingerprint
L_{fp}^{t}	design latency
A_{fp}^{t}	hardware area
L_m	maximum execution latency
A_m	maximum hardware area
w_1	user-specified weight for execution latency
w_2	user-specified weight for hardware area
OPN	operation number
(Si, Sj)	a node pair in the CIG
P^c	probability of coincidence
n	number of colours used in the CIG before implanting fingerprint constraints
k	number of additional constraints-edges added or effective fingerprint size

Chapter 7

K^c	strength of transient fault
A_{Ob+FS}^{T}	total area of obfuscated, fault-secured design
$N(R_k)$	number of kth functional unit
$N(mux)$	number of multiplexers
$N(reg)$	number of registers
$A(R_k)$	area of the kth functional unit
$A(mux)$	area of multiplexers
$A(reg)$	area of registers
L_{Ob+FS}^{T}	total execution latency of obfuscated, fault-secured design
ns	nano-second
T	control step
$d(reg)$	delay of a register

C_{Ob+FS}	the design cost of the obfuscated, fault-secured design
L^m_{Ob+FS}	maximum execution latency
A^m_{Ob+FS}	maximum hardware area
w_1	user-specified weight for execution latency
w_2	user-specified weight for hardware area
$\$$	custom functional unit
opn	operation number
S^{MP}_{ob}	strength of multi-phase obfuscation
T	number of transformations used in obfuscation
μm	micrometre

Chapter 8

H	select line for switching muxes of hologram
$Y[n]$	output of IIR filter
$Y'[n]$	output of FIR filter
$\{a1, a2, a3, b0, b1, b2, b3\}$	set of input coefficients of IIR filter
$\{h0, h1, h2, h3\}$	set of input coefficients of FIR filter
$\{X[n], X[n-1], X[n-2], X[n-3]\}$	set of previous values of input of IIR filter
$\{Y[n-1], Y[n-2], Y[n-3]\}$	set of previous values of output of IIR filter
$\{X'[n-1], X'[n-2], X'[n-3]\}$	set of previous values of input of FIR filter
$X[0]$	first output value of 4-point DCT
$x'[0]$	first output value of 4-point IDCT
$\{k1, k2, k3, k4\}$	set of generic values of input coefficients of 4-point DCT
$\{k'1, k'2, k'3, k'4\}$	set of generic values of input coefficients of 4-point IDCT
$\{x[0], x[1], x[2], x[3]\}$	set of input values of 4-point DCT
$\{X'[0], X'[1], X'[2], X'[3]\}$	set of input values of 4-point IDCT
G^a_c	number of gates affected due to obfuscation
ΔG^b_c	difference in gate count between obfuscated design and baseline
G^i_c	number of gates modified in terms of input connectivity after obfuscation
ps	pico-seconds
S_{ob}	strength of obfuscation

G_c^T	total number of gates in a baseline (un-obfuscated) design
T	control step

Chapter 9

T_{ilb}	total number of ILBs (T_{ilb}) inserted into a DSP core
Bn	registers $B0$–$B15$ used in AES
$A0, A1, \ldots, A15$	input registers of AES (each of 8-bit size)
Str_Reg_ip	control signal of input registers
Sel_mux($S0, \ldots, S3$)	select line of 16:1 mux used in AES
Str_RegA	control signal of register A
Str_RegSB	control signal of S-Box
Str_RegB	control signal of register B
Desel_demux ($S'0, \ldots, S'3$)	select line of 1:16 demux used in AES
Str_RegSRI	control signal of the registers ($B0$–$B15$) before shift-row operation
Str_RegSRO	control signal of the registers ($B0$–$B15$) after shift-row operation
Str_Reg2Bn	control signal for the $2\times Bn$ operation in the GF(2^8)
Sel_2Bn	select line of 2:1 mux used in $2\times Bn$ operation
Str_Vip	control signal of the registers ($V0$–$V3$) before circular-shift operation (in key scheduler)
Str_Vop	control signal of the registers ($V0$–$V3$) after circular-shift operation (in key scheduler)
Sel_SB	select line of 4:1 mux used in AES key-scheduler
Str_RegC	control signal of register C
Str_RegD	control signal of register D
Desel_SB	select line of 1:4 demux used in AES key-scheduler
RC(i)	round constant
Str_Reggf	control signal of registers ($g'1$, $g'2$, $g'3$ and $g'4$)
$K0, K1, \ldots, K15$	a 128-bit for round-0 of AES
$K0$	first byte of 128-bit key
$K1$	second byte of 128-bit key
$K15$	sixteenth byte of 128-bit key
$K^10, K^11, \ldots, K^115$	a 128-bit key generated from key-scheduler for round-1 of AES
a, **b**, **c**, **d**, **e**, **f**, **g**, **h**	eight hash buffers used in the SHA-512 algorithm

W_r	word used in Rth round of the SHA-512–based key generation hardware
K_r	constant used in Rth round of the SHA-512–based key generation hardware
S, C	two sets of challenge bits used in an anti-removal logic (ARL) unit
N	size of the challenge-bits in set S
M	size of the challenge-bits in set C
$C_d(R_i)$	the design cost the integrated SHA-512–based key generation hardware with a functionally obfuscated DSP core
L_d	design latency
P_d	power
$L_{\max}$	maximum execution latency
$P_{\max}$	maximum power
w_1	user-specified weight for execution latency
w_2	user-specified weight for hardware area

Chapter 1

Introduction to hardware (IP) security: forensic detective control and obfuscation of digital signal processing (DSP) cores

Anirban Sengupta[1]

This chapter introduces the reader to the concepts of hardware security as well as exposing them to the new frontiers on hardware (intellectual property (IP) core) security from the perspective of forensic detective control mechanisms and obfuscation techniques. The book explicitly emphasizes the emerging solutions in the literature on hardware security of digital signal processing (DSP) cores. This has not been covered in any book so far.

The reader is expected to have some foundational knowledge of very-large-scale integration (VLSI), digital hardware design flow, high-level synthesis (HLS) and optimization. The readers can use *IP Core Protection and Hardware-Assisted Security for Consumer Electronics* to gain some pre-knowledge in the field.

1.1 Hardware security: an introduction

Hardware security refers to the branch of study that deals with security and trust from the hardware perspective. Hardware security and IP core protection is an emerging area of research in the semiconductor community that focusses on protecting designs against standard threats such as reverse engineering (RE), counterfeit, forgery, malicious hardware modification, etc. The frameworks to formulate a hardware security methodology depend on the threat model and the target application (hardware core). Although hardware security algorithms can be applied at any design abstraction level of hardware design flow, sometimes it is advantageous to employ security constraints (or protocol) at the early stages of hardware design flow in order to combat the implementation complexity, design overhead and covertness. More specifically, this is truer for complex applications such as DSP or multimedia cores which have extremely large gate counts and involve design trade-offs. However, it also depends on the technique, specifications and requirements (Sengupta, 2017a, 2017b; Sengupta and Kundu, 2017; Sengupta *et al.*, 2018).

[1]Computer Science and Engineering, Indian Institute of Technology Indore, Indore, India

In a typical system-on-chip life cycle, every stage of the design flow is vulnerable to threats such as RE, backdoor Trojan insertion in third-party IPs, IP piracy, IP cloning, Trojan insertion at foundry, overproduction, leakage of secret information, etc. Security solutions that can handle threats for each abstraction level is crucial (Sengupta, 2015a, 2015b). Hardware security can be broadly classified into two categories, namely (1) forensic detective control mechanisms and (2) preventive control mechanisms. Forensic detective control mechanisms refer to those classes of methodologies that are employed to detect and control pirated, cloned and forged IPs (hardware) in the design chain. These methods are capable to resolve fraudulent IP ownership through forensic analysis and detect IP abuse or infringement. Some examples include hardware watermarking – single-phase watermarking, multi-phase watermarking and multi-level watermarking, hardware steganography, IP fingerprinting, digital signature for hardware IP protection, etc. (see Figure 1.1). In the context of IP protection, hardware steganography is conceptually different from hardware watermarking. Figure 1.2 shows the concept behind hardware steganography for IP core protection. In the generic model of hardware steganography, the larger the cover data of the design is the easier it is to hide the embedded data in the design. Thus cover data is the original design with no hidden secret information (of the IP vendor), the resultant design with encoded secret information is referred as stego design and stego key is a secret key utilized in the embedding process, thereby increasing security. These marks in hardware steganography and watermarking are covert by nature (e.g., see Figure 1.3 which shows a comparison between non-watermarked vs. watermarked finite impulse response (FIR) filters) and inclusion of these hidden covert marks in the hardware design also acts as a deterrent for attackers. As evident from Figure 1.3, it is very difficult to differentiate (detect) a watermarked design from its non-watermarked counterpart for an attacker, thereby ensuring its covert

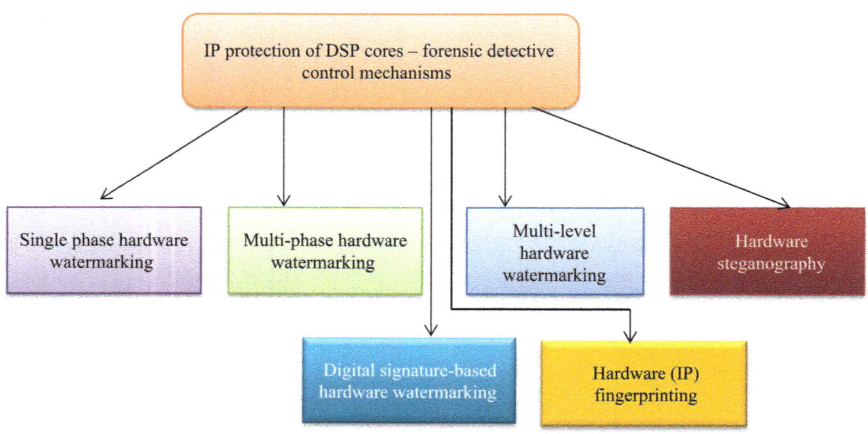

Figure 1.1 IP protection of DSP cores – forensic detective control mechanisms

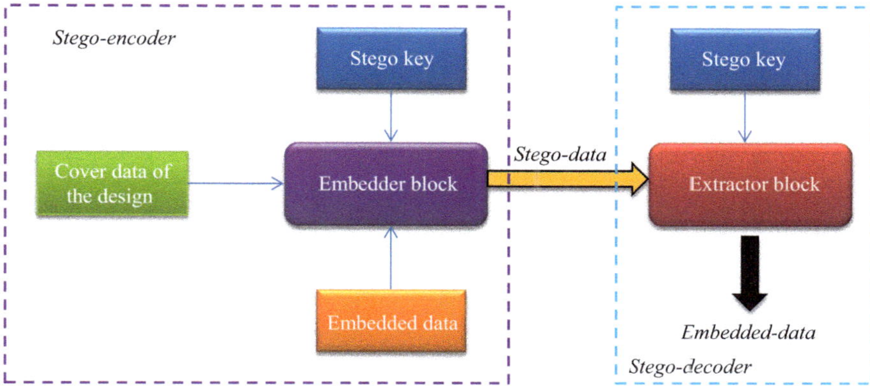

Figure 1.2 Basic steganography model

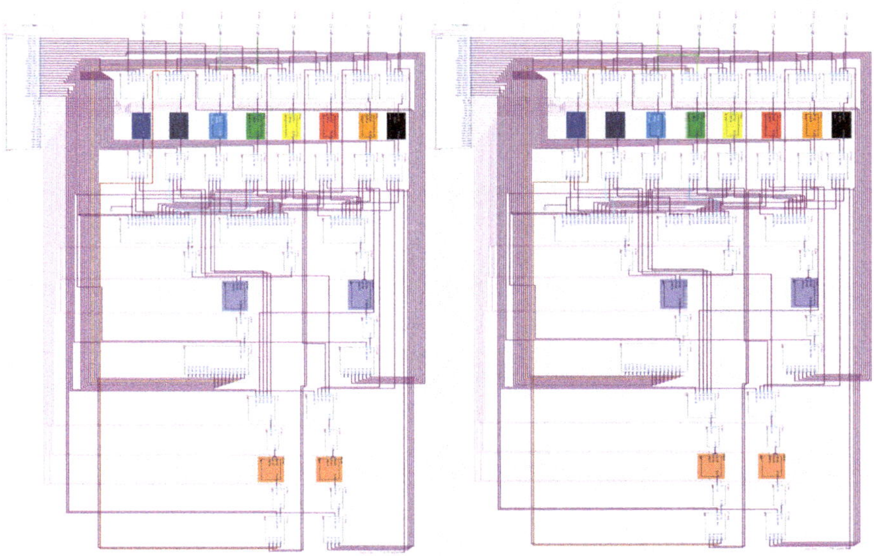

Figure 1.3 Watermarked FIR vs. non-watermarked FIR at RTL

characteristics. However, it is straightforward for an original IP designer (vendor). Further, through these forensic detective control algorithms, a hardware designer or company can protect their design IP from misuse by attackers (present in a competitor's firm, foundry, etc.) by providing evidence during law enforcement protection. On the other hand, preventive control mechanisms refer to those classes of techniques that can prevent IP misuse or infringement. These methods are also sometimes called active protection techniques. Some examples include structural

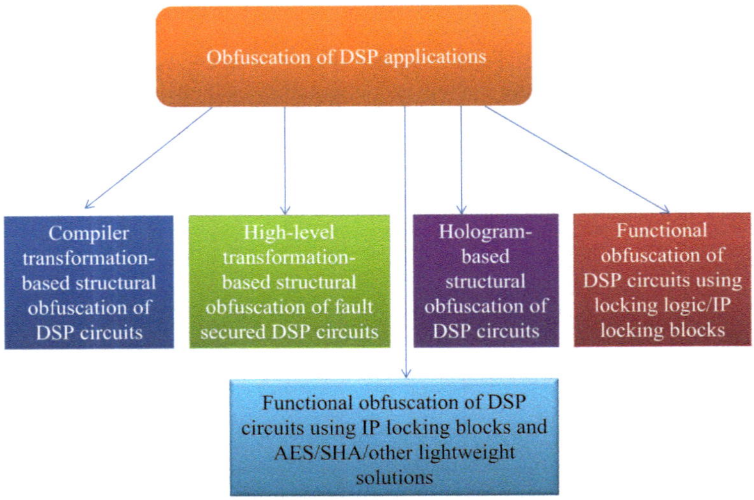

Figure 1.4 Hardware security of DSP applications using obfuscation

obfuscation and functional obfuscation. Structural obfuscation transforms a design into one that is functionally equivalent to the original, but is significantly more difficult to reverse engineer, while the second is an active protection type that locks the design through a secret key. Some examples of structural obfuscation methods used for DSP core protection are compiler transformation-based structural obfuscation, hologram-based structural obfuscation, high-level transformation-based structural obfuscation, etc. The structural obfuscation methods can be applied for normal DSP circuits or more complex fault-secured DSP circuits. Some examples of functional obfuscation methods used for DSP cores include usage of locking gates and IP locking blocks (see Figure 1.4) (Sengupta and Kundu, 2017; Sengupta and Roy, 2018a, 2018b; Alkabani and Koushanfar, 2007; Roy *et al.*, 2008; Sengupta and Mohanty, 2019).

1.2 Hardware security of DSP cores: using high-level synthesis (HLS) framework or ESL

At the heart of smart consumer electronic systems, there are several DSP blocks operating either as a standalone application-specific integrated circuit (ASIC) or embedded as a part of a system-on-chip platform. These DSP kernels are computation intensive cores that are either data intensive or control intensive. They mainly process huge mathematical calculations or data crunching algorithms and consume a huge amount of resources, power, energy, latency, etc. These DSP or multimedia cores are an integral part of modern consumer electronic systems. Some examples include digital filters (finite impulse response, infinite impulse response, etc.), discrete cosine transformation (DCT), Haar wavelet transformation (HWT), fast Fourier transformation (FFT), joint photographic experts group (JPEG), etc. Some of the applications of these DSP blocks to modern consumer electronics devices include FIR filters used in audio devices and high definition TVs; HWT used in lossless image compression of camera systems; JPEG compression/decompression (CODEC) process used in image and video compression of camera systems; DCT module used in audio and image compression; inverse DCT used in image and audio decompression; low-pass filter, FFT used as anti-aliasing filter in a spectrum analyser as well as MP3 audio encoder chains; IIR filters used in audio processors for equalization of speakers, etc. As is evident, these cores are highly critical for the successful functioning of DSP chips or system on chip (SoC platforms used in consumer applications. Let us take a look at a simple application where in a user's smartphone the routine tasks such as handling incoming calls, wireless services, etc., are being take care of by the main microprocessor. While the application-specific tasks such as handling image or video are not being done through the main processor but a co-processor called a 'DSP chip'. This is because computing these images or videos involves huge mathematical calculations thus overloading the main processor to do these would result into unacceptable delays and affect the other routine functions. The user experience would get adversely affected. Therefore these DSP cores (sitting as co-processor as ASIC or integrated in an SoC) are responsible for performing the heavy computations. Refer to Figure 1.5 for a practical application of DSP cores in an MP3 audio encoder (Sengupta and Mohanty, 2019).

As discussed, since these DSP or multimedia cores are either used as a standalone ASIC or integrated as an IP core in a SoC, security of these cores against standard threats such as IP piracy, cloning, RE, Trojan insertion, etc., becomes extremely crucial. Use of forensic detective control mechanisms such as hardware watermarking and hardware steganography can significantly bolster the security of such IP cores against piracy through robust detection. Detection of counterfeited

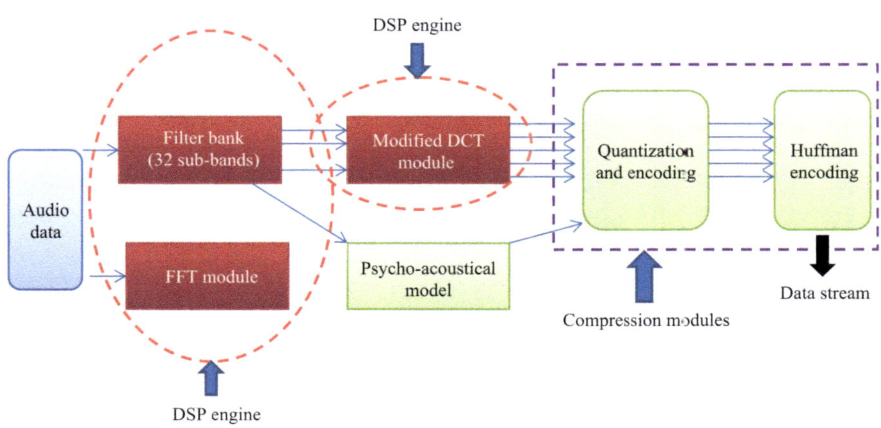

Figure 1.5 MP3 encoder chain for audio compression

and pirated CE devices is significant from the perspective of consumer safety, protecting IP/brand value, returning revenue and resolving traffic bleed. Further, use of obfuscation techniques can provide an active mode of protection by preventing an attacker from misusing the IP or inserting malicious logic inside it. However, securing complex DSP cores (comprising of more than 20k gates) is not trivial. Selecting the optimal level of abstraction for embedding security constraints is crucial. As these designs are complex gate structures, the design process of these applications typically starts from the highest abstraction level (such as at the system or algorithmic levels). It is subsequently converted into its register transfer level (RTL) counterpart using electronic system level (ESL) or HLS frameworks. It is during this synthesis process that the security constraints are implanted into the design flow. Thus watermarking and steganography constraints are mostly embedded during HLS flow. Even structural obfuscation uses an HLS framework to perform structural obfuscation of the DSP designs. Nevertheless, functional obfuscation relies on gate-level structures obtained post-HLS and RTL synthesis to encrypt the design using locking logic. More details are provided in Chapters 7–9. Another dimension using an HLS framework to incorporate security is the design space exploration (DSE) of the architecture as it involves design trade-offs between security, energy, area, delay, etc. (Holzer *et al.*, 2007; McFarland *et al.*, 1988; Paulin and Knight, 1989; Blickle and Thiele, 1995; Sengupta, 2015; Sengupta and Mohanty, 2016; Jain and Panda, 2007; Jha and Zhong, 2002; Katkoori and Vemuri, 2000; Coussy *et al.*, 2009; Coussy and Morawiec, 2008). This is discussed in more detail in Section 1.3.

1.3 Security: energy trade-off in high-level synthesis for hardware security of DSP cores

Design space of DSP cores contains innumerable design alternatives of the same functionality. Exploring an optimal design alternative from several design variances is a challenging task and involves trade-offs. Trade-offs can be between security and energy, security and delay, security and power, etc. Incorporating security algorithms into the design often incurs overheads in the form of latency or area/power. These design overheads need to be accounted for while exploring an optimal design architecture that can implement the desired functionality. During the design of a DSP-based IP core using an ESL (or HLS) framework, DSE is an integral component that explores optimal design architecture. However, during DSE it becomes critical to consider the trade-offs of area/energy/delay with security in order to obtain a low overhead, low-cost solution. The DSE system within the HLS framework needs to be tightly coupled with hardware security algorithm (such as watermarking/steganography/obfuscation process) such that after performing (or embedding) the security algorithm the resultant design latency and area/power is fed into the exploration process. This iteratively continues to explore a low-cost, secured DSP design. The DSE system is based on mostly evolutionary algorithms or heuristics that have innate capabilities to maintain exploration–exploitation trade-off during the search for an optimal solution. Some of the most widely used DSE systems are based on a genetic algorithm (GA), particle swarm optimization (PSO), bacterial foraging optimization algorithm (BFOA), firefly algorithm (FA), fuzzy logic, etc. These DSE systems in HLS frameworks are also capable to explore the security–energy trade-off for fault-secured DSP circuits besides regular DSP circuits. A pictorial overview of HLS and logic synthesis frameworks coupled with hardware security algorithms is shown in Figure 1.6. As evident from the diagram, the security algorithms for DSP circuits can be employed at architecture-level/ESL (during HLS), RT-level or at gate-level design, depending on the technique used and threat model. For example, hardware watermarking can be employed at ESL and RTL, while hardware fingerprinting and steganography can be employed at ESL (during HLS) and structural obfuscation can be employed during HLS, while, finally, functional obfuscation is employed at gate-level representation of the design (Sengupta and Mohanty, 2019). Each of these security algorithms in the context of different threat models for DSP cores will be explained and demonstrated in Chapters 2–9.

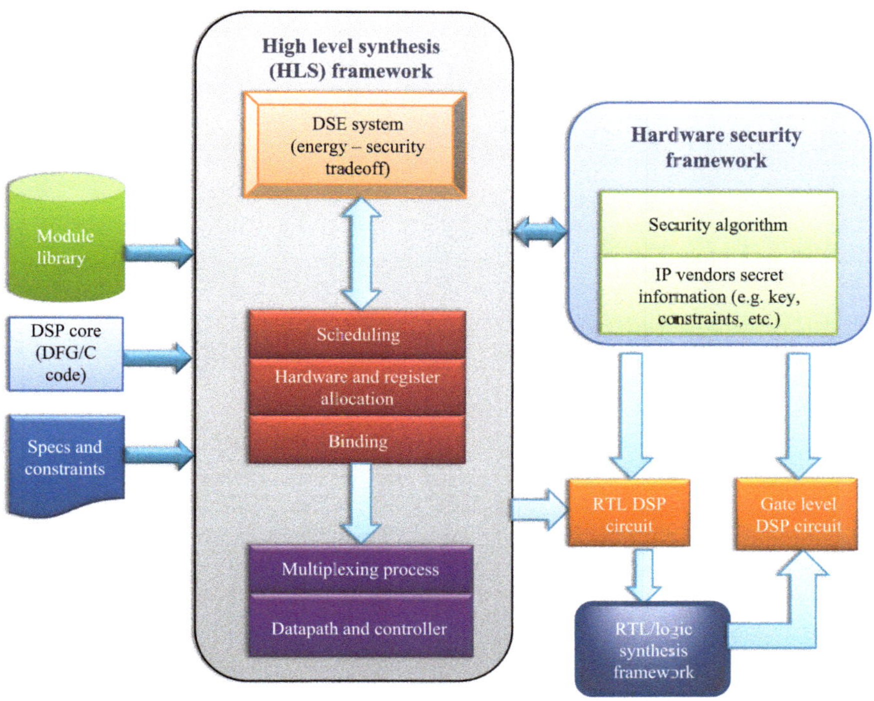

Figure 1.6 Hardware security algorithms integrated with HLS and logic synthesis phases

1.4 Design space exploration systems used in HLS: utility in performing energy or area–security trade-offs

The DSE process in HLS can be formulated using evolutionary algorithms such as a GA (Gallagher *et al.*, 2004; Krishnan and Katkoori, 2006; Torbey and Knight, 1988a, 1988b; Mandal *et al.*, 2000; Heijlingers *et al.*, 1995; Dhodhi *et al.*, 1995), PSO, bacterial foraging optimization, etc. In this section, we discuss how to map an evolutionary algorithm to DSE such that energy–security trade-off can be performed. The section is divided into several sub-sections based on the evolutionary or nature-inspired framework used for DSE in the literature.

1.4.1 Genetic algorithm (GA)-based design space exploration framework

The work by Sengupta *et al.* (2012) proposed a DSE system using a multi-structure GA for DSP cores. This framework is capable of being integrated into any existing security framework such as a watermarking algorithm, fingerprinting algorithm, obfuscation, etc. The work (Sengupta *et al.*, 2012) is an extension of another similar work (Krishnan and Katkoori, 2006) based on a GA. Let us first discuss the DSE framework proposed by Sengupta *et al.* (2012). In this DSE system, the algorithm performs a trade-off between power and execution time, where execution time is a function of latency, number of pipelined data sets and cycle time. This is more applicable for data-intensive DSP applications than its control-intensive counterpart. The multi-structure GA topology reported by the authors used a different topology where each functional unit (FU) type (available in the module library) of a DSP core design is represented by an independent chromosomal structure. The encoded value of each gene in the chromosome includes the datapath nodes used in the scheduling process. In other words, each operation type is encoded through a unique chromosome structure. Furthermore, the seeding process of this approach used special chromosomal structures that guide the exploration search towards an optimal solution. On the other hand, Krishnan and Katkoori (2006) suggested the use of a single-chromosomal structure for encoding the datapath schedule of all the available resources. There is also a significant difference between the fitness function used in the cost evaluation of both approaches. In Krishnan and Katkoori (2006) the cost function included normalized latency and the area of the design. However for Sengupta *et al.* (2012) the cost function included normalized execution time and power of the design. This indicates that the cost function used in Sengupta *et al.* (2012) can consider data-pipelined DSP designs as it incorporates features of latency, number of pipelined data and cycle time (which is the difference between the end times of two consecutive data elements). Both approaches, however, use the same control parameters such as mutation probability, crossover probability, etc. The module library of Sengupta *et al.* (2012) comprises of available versions of resources, maximum available resources, clock cycle, area occupied, etc. The approach in Sengupta *et al.* (2012) is called a multi-structure GA (MSGA) and includes two types of chromosomes, namely (1) nodal string and (2) resource allocation string. The nodal string further comprises of two special chromosomes: (i) P1 which is encoded based on the load factor (α) of each resource from the as soon as possible (ASAP) scheduling; and (ii) P2 which is encoded based on the difference of the latency obtained by using ASAP scheduling with maximum resources (L^{ASAP}) and the load factor (α) for each operation (o^i) obtained for the P1 chromosomes. There are also additional nodal strings besides these special ones that are encoded based on a perturbation function lying between P1 and P2. This is performed because an optimal solution to the DSE problem is expected to lie between parallel and serial implementation. The resource allocation string contains two special individuals; one that encodes the value of the maximum resources and the other that encodes the value of minimum resources. Overall the process of performing GA-based DSE follows 12 steps:

1. Feed module library and data flow graph (DFG) of the DSP core.
2. Schedule the DFG using the ASAP algorithm and evaluate the latency of the design.
3. Generate the initial population of chromosomes based on encoded values such as load factor (which is the summation of the delays of the successor operations and the operation itself).
4. Performing crossover for two independent strings (or chromosomes), i.e., nodal string and resource allocation string. Crossover can be of several different types such as direct crossover, multi-point crossover, etc. Before crossover is performed, it is often a choice to employ a binary tournament selection method to select the 'parents' that will be used for crossover.
5. Decode the chromosome using heuristic (e.g. in Sengupta *et al.*, 2012), load factor heuristic was used. The output of this is a valid scheduling.
6. In case of multiple versions of the same FU type in the library, invoke another function to choose the preferred version. The output of this is a valid hardware assignment in the scheduling graph. An example of such a function is shown as:
   The function ( ) may be composed of the following modes:
   (i) Each operation is allocated to the FU of a particular kind with the lowest cost.
   (ii) All operations are allocated to the fast FU, including operations on critical and non-critical paths.
   (iii) Operations are randomly allocated to the FU of that kind.
   (iv) All operations are allocated to the FU with minimum area overhead of that kind.
7. Then employ cost function on the decoded chromosome to evaluate its fitness. In this process, first the power of the design is determined and then the execution time is computed. Then the design cost is subsequently evaluated.
8. Mutation is performed next based on the mutation probability on the least-fit nodal string. This generates a new chromosome structure.

9. The new mutated chromosome structure is passed through the fitness evaluation function.
10. This is done for all the individuals in the population of generation 1 (in iteration 1) and the best population thus far is forwarded to the next generation.
11. This process is continued until the maximum number of generations has been reached or the terminating condition is met.
12. The output of this process is expected to generate an optimal design solution.

This process can be integrated with a security framework for DSP cores to optimize the cost of the design or to perform security–cost trade-off. An overview of GA-DSE used in an HLS is shown in Figure 1.7.

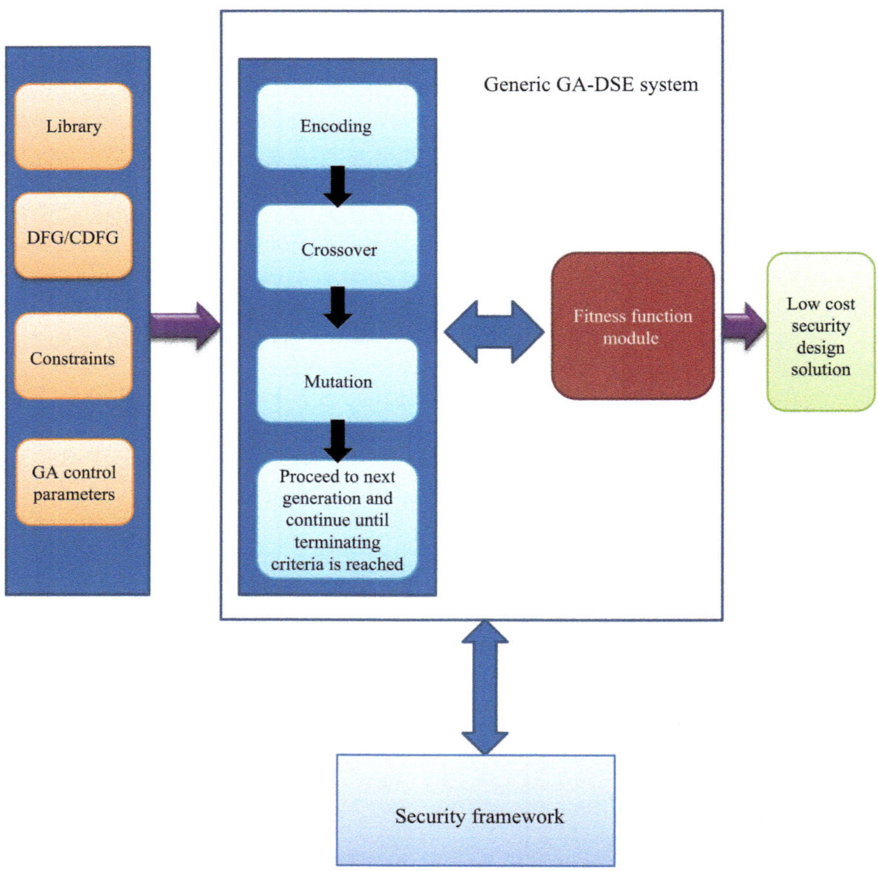

Figure 1.7 A generic GA-DSE framework that can be integrated with a security framework for obtaining low-cost security solutions

1.4.2 Particle swarm optimization (PSO)-based design space exploration framework

The PSO-driven DSE system (Mishra and Sengupta, 2014) is a very powerful tool in exploring trade-offs with respect to area–delay, power–delay, energy–security, etc. These have become popular owing to their several advantages compared to the GA counterpart. Some of the special advantages of PSO-DSE are listed here (Kennedy and Eberhart, 1995; Engelbrecht, 2005; Trelea, 2003):

1. Capable of providing high stochasticity using parameters such as inertia weight, acceleration coefficient and random variables.
2. Capable of preserving exploration–exploitation balance during DSE.
3. Ability to escape local minima.
4. Ability of faster convergence to the optimal solution.

Mishra and Sengupta (2014) proposed a PSO-DSE system that is integrated with HLS for DSP cores. Some of the attributes of this approach are: (1) simultaneous exploration of data path and loop unrolling factor through a multi-dimensional PSO; (2) use of a predictive model for latency determination of a loop unrolled control DFG (CDFG); (3) ability to perform trade-off between power–execution time as well as control states–latency; and (4) analysis of the effect of swarm size on the quality of the final solution explored. Besides these two options, there has been another approach based on integrated GA and weighted sum PSO (WSPSO) by Harish Ram *et al.* (2012). The authors employed a non-dominated sorting GA II in this work. Furthermore the fundamental structure employed for executing the GA has been tailored for applying WSPSO to the multi-objective, HLS. In this work a particle is crossed over with the current particle that represents the global best (or particle best) to mimic the velocity function used in PSO. On the other hand, in Mishra and Sengupta (2014) the DSE framework accepts a list of pre-processed unrolling factors that filters out the unfit, unrolling factors from participating in the exploration process. Here unrolling factors are those that either result into a large remainder (sequential loops) or do not impact improvement of performance significantly. Some unrolling factors that may marginally improve the performance, however, may increase the code density too much as well as increase the control states significantly. These unrolling factors do not play a vital role in reaching a global optimal solution. The process of PSO-DSE can be explained in 11 steps:

1. Feed module library, user constraints and CDFG of DSP core.
2. Feed the control parameters such as swarm size, maximum iteration count, acceleration coefficient, etc.
3. Perform pre-processing of unrolling factors.
4. Perform particle encoding.
5. Determination of velocity (apply velocity clamping if necessary). Velocity clamping is sometimes necessary to control unwarranted exploration burst.
6. Evaluate the new particle location (apply adaptive perturbation if necessary). Adaptive perturbation may be necessary to control a boundary outreach problem.

7. Upgradation of the global particle best and local particle best.
8. Apply mutation (optional step) to find the new particle location.
9. In steps 6, 7 and 8, determine the fitness function.
10. Continue the process until the terminating criteria are reached.
11. The output is an explored optimal combination of the datapath resource configuration and loop unrolling factor.

As evident from this process, the DSE framework evolves through a concept of particle (local) best and global best. The particle best is the best fit solution thus far for that specific particle, while the global best is the best fit solution thus far among all particles in the search space. The work also discussed the mathematical model for estimating execution time of loop unrolled DSP applications. For example, some of the conditions considered by the authors were:

1. When the unrolling factor is equal to one (indicates no unrolling).
2. When unrolling factor evenly divides the loop count (I).
3. When unrolling factor unevenly divides loop count.

This is the only work in the literature that is capable of performing simultaneous exploration of optimal datapath resources and loop unrolling factors for loop-based CDFGs. The comparative analysis with GA-based frameworks (Sengupta *et al.*, 2012; Krishnan and Katkoori, 2006) suggested a reduction in exploration runtime of more than 90% and reduction in the final design cost of more than 28%. An overview of PSO-DSE used in an HLS is shown in Figure 1.8.

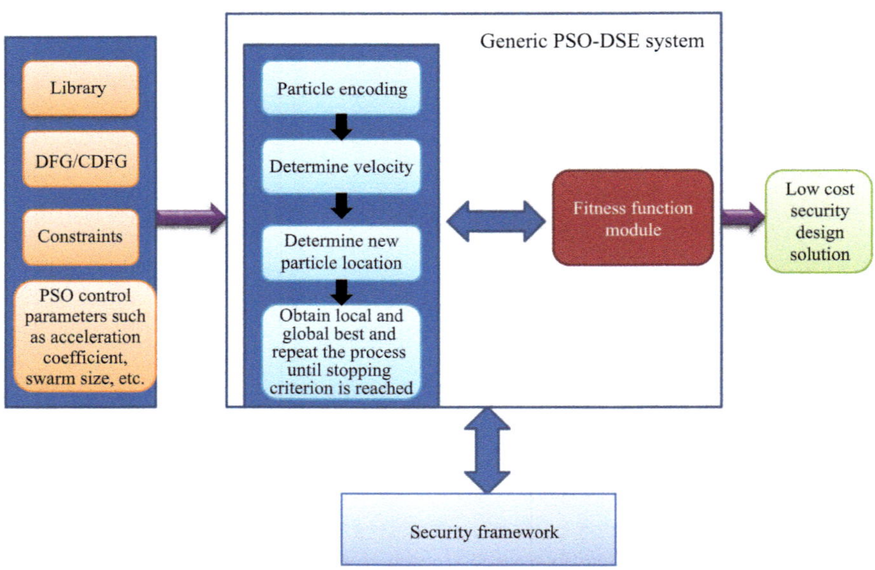

Figure 1.8 A generic PSO-DSE framework that can be integrated with a security framework for obtaining a low-cost security solution

1.4.3 Bacterial foraging optimization algorithm (BFOA)-based design space exploration framework

The BFOA-driven DSE system (Sengupta and Bhadauria, 2015) is a very powerful mechanism of exploring trade-offs with respect to area–delay, power–delay, energy–security, etc. These have also become popular like PSO-DSE in an HLS owing to their several advantages compared to the GA counterpart. Some of the special advantages of BFOA-DSE are (Das *et al.*, 2009; Sengupta and Bhadauria, 2015):

1. The real bacterial system involves four basic mechanisms, viz. chemotaxis, swarming, replication and elimination dispersal (ED).
2. The framework of this exploration process is motivated from the social foraging behaviour of *Escherichia coli*.
3. The exploration process provides enough exploration–exploitation balance like in PSO-DSE.
4. There are several variables that provide stochasticity such as step length ($C(i)$) and random number ($\Delta(i)$).
5. Phases such as ED are capable of exploring new solutions in the search space by deleting unfit solutions.

This exploration framework is iterated by mimicking the valid temperature range of $t_{\min} = 25$ °C and $t_{\max} = 45$ °C at which *E. coli* survives. However, elimination of a bacterium may happen at higher temperatures beyond 40 °C. Within this motility temperature interval through chemotactic movement in every step (j) of each bacterium, the BFOA-DSE explores new feasible solutions in the design space. Furthermore, the process of ED and replication are repeated after certain fixed intervals. Yet the main step in BFOA-DSE is chemotaxis that involves movement of a bacterium either for a certain period of time in the same direction or tumbling in the design space; the bacterium alternatingly moves between these locomotive movements. A small step size ($C(i)$) is not sufficient for exploring new design solutions in the context of BFOA-DSE in HLS. Thus a large step size is necessary. The BFOA-DSE approach can be summarized in ten steps:

1. Feed the module library, user constraints and DFG into the DSE system.
2. Feed the control parameters of BFOA into the system such as step size, population size, terminating criteria, maximum chemotaxis count, etc.
3. Feed a valid temperature range to govern the iteration process.
4. Check whether the current iteration count is below the maximum chemotaxis count specified. If 'yes' then move to step 5.
5. Check whether replication has been performed in the last iteration. If not, perform replication. Corresponding arrays ((*Rep* [j–]) and (*Ed* [j–])) are created for replication and the ED process to store the outcome, checking whether replication and ED have been performed in the last iterative step.
6. Similarly, check whether ED has been performed in the last iteration. If not, perform ED.

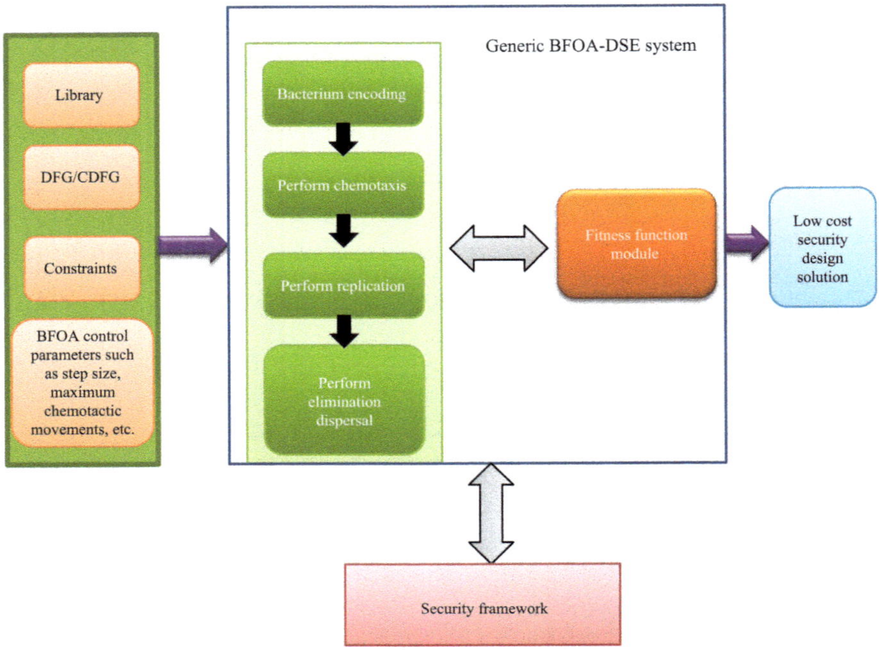

Figure 1.9 A generic BFOA-DSE framework that can be integrated with a security framework for obtaining a low-cost security solution

7. Continue with chemotaxis.
8. Evaluate cost of each new solution encountered.
9. Continue until the terminating criteria have been reached.
10. An optimal solution is obtained at the end of the stopping criterion.

The authors in this work have performed analysis on several aspects (Sengupta and Bhadauria, 2015). Some of these are: (1) analysis of variation of BFOA parameters which further includes sensitivity analysis on bacterium size, step size with respect to convergence time, exploration time and design cost; and (2) comparative analysis with other DSE approaches in terms of quality of results, exploration time, etc. An overview of the BFOA-DSE used in an HLS is shown in Figure 1.9.

1.4.4 Firefly algorithm (FA)-based design space exploration framework

FA-driven DSE systems are one of the most recent approaches reviewed in the literature. Even though it has not been popular in the context of HLS problems, we feel that it has huge potential in this area. FA is a special class of PSO algorithm that is also capable of performing security trade-offs with energy, area, delay, etc., during HLS. Some of the unique features of FA-DSE are (Sarkar *et al.*, 2017; Yang and He, 2013; Yang, 2009):

1. FA contains control parameters such as an absorption coefficient (inverse of attractiveness or brightness) and controls the step size parameter, which allows controlling the randomness as iterations proceed so that convergence can also be sped up by tuning these parameters.
2. FA is based on attraction and attractiveness decreases with distance. This leads to the fact that the whole population can be sub-divided into sub-groups and each group can swarm around each local optimum.
3. FA is capable of achieving exploration–exploitation balance during DSE.

The FA is another class of nature-inspired algorithms that communicate between other fireflies in the search space using generated flashes. Every firefly has a unique feature called 'attractiveness' that is proportional to its brightness (which is the inverse of absorption coefficient). This is because as the FA-DSE evolves with iteration count, the brightness increases (as the fireflies are moving towards the global best firefly), thus absorption coefficient decreases. In the case that the brightest firefly is not found in the design space, then the remaining fireflies move randomly in the search space. The movement of a firefly in the design space is governed by adding the drift factor to the current position of the firefly. Every firefly encodes the resource configuration of a DSP design. The drift factor comprises two components, namely, (1) attractiveness and (2) step-size parameter used for guided searching. The concept of encoding fireflies are: (1) the first firefly x_1: (representing the first seed) is constructed by mapping the hardware resources with minimum value, (2) the second firefly x_2 (representing the second seed) is constructed by mapping the hardware resources with maximum value and (3) the third firefly x_3 is constructed by mapping the average value of minimum and maximum hardware resource. An overview of FA-DSE used in an HLS is shown in Figure 1.10.

1.5 Comparative perspective of different security techniques used for DSP cores

Let us first understand why traditional forms of security and protection for IP cores are not sufficient. Conventional forms of protection include patents, copyrights, trademarking, etc. However, these are not applicable for reusable IP cores. This is primarily due to three reasons:

1. Design cost or implementation cost.
2. The obligation of proof to exhibit infringement is too costly.
3. Occasionally some protection techniques are not applicable for securing hardware IP.

Thus other hardware security techniques such as hardware watermarking, hardware fingerprinting, hardware obfuscation, digital signature–based IP protection and IP metering, etc., have become popular. In Section 1.5.1 we provide a comparative

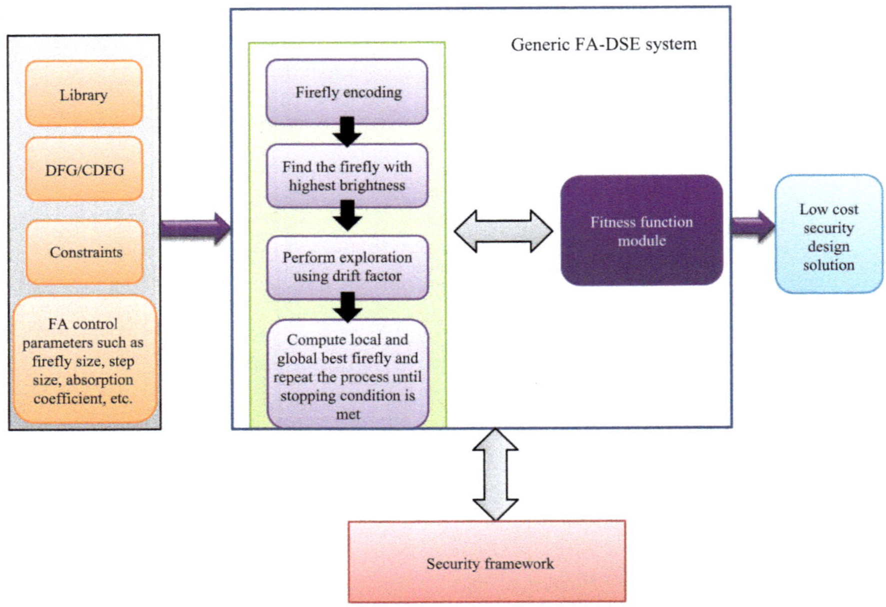

Figure 1.10 A generic FA-DSE framework that can be integrated with a security framework for obtaining a low-cost security solution

perspective of different hardware security techniques from the point of view of threat model, target objective and complexity.

1.5.1 Hardware watermarking versus hardware steganography

Hardware watermarking is useful to combat threat models such as IP piracy, IP cloning and fraudulent claims of ownership. Hardware watermarking carefully embeds a secret mark into the design based on the IP vendor's choice. This secret mark is determined through a vendor signature that is supposed to be robust and strong. Hardware watermarking is highly capable to detect a pirated, cloned IP and also capable to nullify false claims of ownership. Thus watermarking provides forensic detective control of fake IPs from the design chain and market. However, watermarking is not capable of preventing fake IPs being designed and manufactured. Hardware steganography, on the other hand, is a new concept that does not use an external signature for resolving piracy, cloning or fraudulent claims of ownership. Hardware steganography mixes secret vendor information within the design itself based on a stego key. Since there is no external signature (unlike watermarking), the probability of tampering with the signature is nil. Therefore hardware steganography is a broader concept and encapsulates the concept of hardware watermarking theoretically. However, the practical design process (encoding) of hardware steganography is different from watermarking. Both processes include a decoder system that is essential to provide ownership or detect clones/pirated IPs in court through forensic analysis. The complexity of steganography is lower than watermarking because in watermarking the signature size needs to be carefully chosen. The underlying reason is the careful selection of signature size that maintains a balance between the lower probability of coincidence and the lower probability of tampering. A larger size signature provides lower probability of coincidence, but higher probability of tampering. On the contrary, a smaller size signature provides higher probability of coincidence, but lower probability of tampering. Thus there is a trade-off involved in watermarking. In contrast, hardware steganography does not include this trade-off due to no presence of signature. More details are provided in Chapters 2 and 5.

1.5.2 Hardware watermarking versus digital signature

A digital signature–based watermark for IP protection includes: (1) extraction of bit stream representation of the DSP design though some encoding; (2) generation of the design digest of the extracted bit stream; (3) creation of the digital signature using encryption from the vendor's private key; and (4) conversion of the digital signature into equivalent watermarking constraints using further encoding. The final hidden constraints generated are implanted into the design covertly without affecting the functionality of the design. The process does not incur much design overhead and is robust against tampering and other external attacks. In a digital signature scheme, the signature is not provided by the vendor directly. Instead the signature is generated using multi-layered encoding, hashing and encryption. This provides the IP vendor more evidence of ownership in case of ownership conflict. Furthermore, ownership validation can be achieved using a decryption process as well. This technique ensures that the signature does not need to be remembered and the vendor's private key can be used to prove ownership in case the signature is compromised. On the other hand, regular hardware watermarking accepts a vendor signature and converts it into constraints for insertion into the design. In case the signature gets stolen or is compromised, it becomes very difficult for the genuine IP owner to prove ownership. However, implementation complexity of a digital signature scheme is higher than watermarking. Thus although the threat model and target objective of both types of approaches are the same, a digital signature is more robust and stronger than watermarking in terms of providing more digital evidence for ownership resolution as well as providing forensic detective control of pirated/cloned IPs. More details are provided in Chapters 3 and 4, respectively.

1.5.3 Hardware fingerprinting versus hardware steganography

Hardware fingerprinting is a method for implanting an IP buyer's secret mark into the design. Hardware fingerprinting is generated through encoding and crypto function, respectively. The threat model is to combat a dishonest IP seller from selling unauthorized copies of the IP or illegal redistribution of extra copies without the knowledge of the IP buyer. In other words, the threat model includes unlawful redistribution/resell of IP core copies in the market by a dishonest IP seller. It is a forensic detective control of illegal IPs from the chain. As is evident, the threat model of hardware fingerprinting is entirely different from hardware steganography. The latter is used for protecting IP vendors' rights, while the former is for protecting IP buyers' rights. The complexity of crypto-based hardware fingerprinting is higher than hardware steganography. More details on hardware fingerprinting are provided in Chapter 6.

1.5.4 Hardware obfuscation versus digital signature

Hardware obfuscation is employed when the hardware threat anticipated is an IP netlist theft/counterfeit/piracy through RE. There are two types of hardware obfuscation, namely, (1) structural obfuscation and (2) functional obfuscation. The threat model of hardware obfuscation is significantly different from the digital signature scheme. Hardware obfuscation is an active mode of protection that tries to prevent an attacker from using the design netlist by making the RE process harder. Structural obfuscation is done by making the design architecture unobvious to an attacker while functional obfuscation is done by locking the design netlist using locking key-gates.

1.6 Conclusion

This chapter introduced concepts of hardware security, background of forensic detective control mechanisms, types of obfuscation, importance of security of DSP cores, importance of DSP cores in the context of CE systems and how HLS and DSE frameworks play a major role in integrating security algorithms. A summary of major threats for DSP cores were also highlighted. The chapter presented a real-life case study of an MP3 encoder for audio compression emphasizing the DSP blocks used. The significance of security–energy or security–delay trade-offs in the context of DSP IP design was also explained. How these trade-offs can be integrated with HLS and logic synthesis frameworks was discussed. Through this chapter the foundation of DSP design, importance of its security and design frameworks (abstraction levels) that can be exploited are motivated. The reader should be clear about these fundamental topics before diving into the hardware security and IP protection mechanisms discussed in subsequent chapters.

1.7 Questions and exercises

What is hardware security?

1. Name some standard threats to semiconductor IPs.
2. What is forensic detective control?
3. How are preventive control mechanisms employed?

4. State the difference between IP fingerprinting and IP watermarking in terms of a threat model.
5. What is hardware steganography?
6. What is cover data of the design in the context of hardware steganography?
7. What is HLS?
8. Name some major DSP blocks that are found in everyday CE systems.
9. What are some of the structural obfuscation techniques used for DSP core security?
10. What are the DSP modules used in an MP3 encoder for audio compression?
11. What is the use of psychoacoustic model in an MP3 encoder?
12. Explain the significance of security–energy trade-off during DSP design.
13. Explain the design flow of high level synthesis.
14. What are the benefits of implanting security constraints at ESL compared to lower levels?
15. What security algorithms are employed at gate-level design?
16. What are some of the common DSE systems used in HLS frameworks?
17. Where is hardware steganography employed for IP protection?
18. What are scheduling, allocation and binding? Explain with examples.
19. What is the input form for a DSP core in an HLS framework?
20. How is a datapath different from a controller structure?
21. Can security constraints be implanted at RTL representation of a DSP circuit?
22. How is fault secured DSP circuit different from a normal DSP circuit?
23. What is the threat model for digital signature-based watermark?
24. Describe any real-life example where DCT and JPEG modules play an important role.
25. What is a genetic algorithm?
26. How is a genetic algorithm mapped to an HLS problem?
27. What is crossover?
28. What is the difference between direct crossover and partially matched crossover?
29. Explain the encoding algorithm of multi-structure genetic algorithm-based DSE.
30. Explain the decoding algorithm of multi-structure genetic algorithm-based DSE.
31. What are the different types of fitness functions used in genetic algorithm-based DSE?
32. What is data pipelining of DSP cores?
33. What is latency? Explain the difference between latency and life cycle time.
34. What are the control parameters used in GA-based DSE?
35. What is present in the module library of a DSE system?
36. Explain the encoding rule of the special chromosomes used in MSGA-DSE.
37. How is a nodal string different from a resource allocation string?
38. How is an FU version chosen in GA-DSE?
39. Explain the modes of the function used to select a desired FU version during DSE.
40. What is PSO?
41. Explain the control parameters used in PSO-DSE.
42. What is the utility of velocity function in DSE?

43. What are inertia weight and acceleration coefficients? Explain their significance.
44. What is stochasticity in PSO-DSE?
45. What is exploration–exploitation balance in PSO-DSE?
46. How is pre-processing of loop unrolling factor performed?
47. What is the advantage of pre-processing loop unrolling factor before exploration initiates?
48. Explain the WSPSO method.
49. What is the difference between global best and local best in PSO?
50. What are remainder loops in CDFG?
51. What are velocity clamping and adaptive perturbations?
52. What are the steps used in BFOA-DSE when searching for an optimal design solution?
53. What is the significance of elimination-dispersal?
54. Explain the locomotive operations used in chemotaxis.
55. What is the valid temperature range at which BFOA-DSE is mimicked?
56. What is FA-DSE and how it is different from PSO-DSE?
57. What are the control parameters used in FA-DSE?
58. How is attractiveness related to absorption coefficient?
59. What is the drift factor in FA-DSE?
60. What are inputs to an FA-DSE system?
61. Which can provide more stochasticity, FA or PSO?
62. Describe the encoding process of an FA-DSE system.

References

Y. M. Alkabani and F. Koushanfar, "Active hardware metering for intellectual property protection and security," in Proceedings of 16th USENIX Security Symposium on USENIX Security Symposium (SS'07), Niels Provos (Ed.). USENIX Association, Berkeley, CA, USA, Article 20, 16 pages, 2007.

T. Blickle and L. Thiele, "A mathematical analysis of tournament selection," in Proceedings of 6th International Conference on Genetic Algorithms, Pages: 9–16, 1995.

P. Coussy and A. Morawiec, "High-Level Synthesis: From Algorithm to Digital Circuit," Springer, Berlin, Germany, 2008.

P. Coussy, G. Gajski, A. Takach, and M. Meredith "An introduction to high-level synthesis, special issue on high-level synthesis," IEEE Design and Test of Computers, vol. 26, no. 4, Pages: 8–17, 2009.

S. Das, A. Biswas, S. Dasgupta, and A. Abraham, "Bacterial foraging optimization algorithm: theoretical foundations, analysis, and applications," Studies in Computational Intelligence, vol. 203, Pages: 23–55, 2009.

M. K. Dhodhi, F. H. Hielscher, R. H. Storer, and J. Bhasker, "Datapath synthesis using a problem-space genetic algorithm," IEEE Transactions on Computer Aided Design of Integrated Circuits and Systems, vol. 14, Pages: 934–944, 1995.

A. P. Engelbrecht, "Fundamental of Computational Swarm Intelligence," John Wiley and Sons Limited, England, 2005.

J. C. Gallagher, S. Vigraham, and G. Kramer, "A family of compact genetic algorithms for intrinsic evolvable hardware," IEEE Transactions on Evolutionary Computation, vol. 8, no. 2, Pages: 1–126, 2004.

D. S. Harish Ram, M. C. Bhuvaneswari, and S. S. Prabhu, "A novel framework for applying multiobjective GA and PSO based approaches for simultaneous area, delay, and power optimization in high level synthesis of datapaths," VLSI Design, 12 pages, 2012, Article ID 273276.

M. J. M. Heijlingers, L. J. M. Cluitmans, and J. A. G. Jess, "High-level synthesis scheduling and allocation using genetic algorithms," in Proceedings of Asia South Pacific Design Automation Conference, Pages: 61–66, 1995.

M. Holzer, B. Knerr, and M. Rupp, "Design space exploration with evolutionary multi-objective optimisation," in International Symposium on Industrial Embedded Systems, 2007.

R. Jain and P. R. Panda, "An efficient pipelined VLSI architecture for lifting-based 2d-discrete wavelet transform," in Proceedings of the International Symposium on Circuits and Systems (ISCAS), Pages: 1377–1380, 2007.

N. K. Jha and L. Zhong, "Interconnect-aware high-level synthesis for low power," in 2002 International Conference on Computer-Aided Design (ICCAD '02), Pages: 110–117, 2002.

S. Katkoori and R. Vemuri, "Scheduling for low power under resource and latency constraints," in Proceedings of IEEE International Symposium on Circuits and Systems (ISCAS), Pages: 53–56, May 2000.

J. Kennedy and R. C. Eberhart, "Particle swarm optimization," in Proceedings of the 1995 IEEE International Conference on Neural Networks, Pages: 1942–1948, 1995.

V. Krishnan and S. Katkoori, "A genetic algorithm for the design space exploration of datapaths during high-level synthesis," IEEE Transactions on Evolutionary Computation, vol. 10, no. 3, Pages: 213–229, 2006.

C. Mandal, P. P. Chakrabarti, and S. Ghose, "GABIND: a GA approach to allocation and binding for the high-level synthesis of data paths," IEEE Transactions on VLSI, vol. 8, no. 5, Pages: 747–750, 2000.

M. C. McFarland, A. Parker, and R. Camposano, "Tutorial on high-level synthesis," in Proceedings of 25th Design Automation Conference, Pages: 330–336, 1988.

V. K. Mishra and A. Sengupta, "MO-PSE: adaptive multi-objective particle swarm optimization based design space exploration in architectural synthesis for application specific processor design," Elsevier Journal on Advances in Engineering Software, vol. 67, Pages: 111–124, 2014.

G. Paulin and J. P. Knight, "Force-directed scheduling for the behavioral synthesis of ASICs," IEEE Transactions on Computer Aided Design of Integrated Circuits and Systems, vol. 8, no. 6, Pages: 661–679, 1989.

J. A. Roy, F. Koushanfar, and I. L. Markov, "EPIC: ending piracy of integrated circuits," in 2008 Design, Automation and Test in Europe, Munich, Pages: 1069–1074, 2008.

P. Sarkar, A. Sengupta, S. Rathlavat, and M. K. Naskar, "Designing low-cost hardware accelerators for CE devices," IEEE Consumer Electronics Magazine, vol. 6, no. 4, Pages: 140–149, 2017.

A. Sengupta, "Protection of IP-core designs for CE products," IEEE Consumer Electronics Magazine, vol. 5, Pages: 83–89, 2015a.

A. Sengupta, "Protection of reusable IP core at architectural level," IEEE VLSI Circuits & Systems Letter, vol. 1, no. 2, Pages: 14–17, 2015b.

A. Sengupta, "Hardware security of CE devices: threat models and defence against IP Trojans and IP piracy," IEEE Consumer Electronics Magazine, vol. 6, no. 1, Pages: 130–133, 2017a.

A. Sengupta, "Hardware vulnerabilities and its effect on CE devices: design-for-security against Trojan," IEEE Consumer Electronics Magazine, vol. 6, no. 3, Pages: 126–133, 2017b.

A. Sengupta and S. Bhadauria, "Adaptive bacterial foraging driven datapath optimization: exploring power-performance tradeoff in high level synthesis," Applied Mathematics and Computation, vol. 269, Pages: 265–278, 2015.

A. Sengupta and S. Kundu, "Securing IoT hardware: threat models and reliable, low-power design solutions," IEEE Transactions on Very Large Scale Integration (VLSI) Systems, vol. 25, no. 12, Pages: 3265–3267, 2017.

A. Sengupta and S. P. Mohanty, "High-Level Synthesis of Digital Circuits in the Nanoscale, Mobile Electronics Era," Nano-CMOS and Post-CMOS Electronics: Circuits and Design (Eds: Saraju P. Mohanty & Ashok Srivastava), IET, Stevenage, Pages: 219–261, June 2016, e-ISBN: 9781785610004.

A. Sengupta and S. P. Mohanty, "IP Core Protection and Hardware-Assisted Security for Consumer Electronics," The Institute of Engineering and Technology (IET), Stevenage, 2019, ISBN: 978-1-78561-799-7, e-ISBN: 978-1-78561-800-0.

A. Sengupta and D. Roy, "Reusable intellectual property core protection for both buyer and seller," in 2018 IEEE International Conference on Consumer Electronics (ICCE), Las Vegas, NV, Pages: 1–3, 2018a.

A. Sengupta and D. Roy, "Multi-phase watermark for IP core protection," in 2018 IEEE International Conference on Consumer Electronics (ICCE), Las Vegas, NV, Pages: 1–3, 2018b.

A. Sengupta, S. P. Mohanty, and G. Rose, "Hardware–assisted design for security and protection of consumer electronics," IET Computers and Digital Techniques, vol. 12, no. 6, Pages: 249–250, 2018.

A. Sengupta, R. Sedaghat, and P. Sarkar, "A multi structure genetic algorithm for integrated design space exploration of scheduling and allocation in high level synthesis for DSP kernels," Swarm and Evolutionary Computation, vol. 7, Pages: 35–46, 2012.

E. Torbey and J. Knight, "Performing scheduling and storage optimization simultaneously using genetic algorithms," in Proceedings of IEEE Midwest Symposium on Circuits and Systems, Pages: 284–287, 1998b.

E. Torbey and J. Knight, "Performing scheduling and storage optimization simultaneously using genetic algorithms," in Proceedings of IEEE Midwest Symposium on Circuits and Systems, Pages: 284–287, 1998b.

I. C. Trelea, "The particle swarm optimization algorithm: convergence analysis and parameter selection," Information Processing Letters, vol. 85, no. 6, Pages: 317–325, 2003.

X.-S. Yang, "Firefly Algorithms for Multimodal Optimization," Stochastic Algorithms: Foundations and Applications, SAGA Lecture Notes in Computer Sciences, vol. 5792, Springer, Berlin, Heidelberg, Pages: 169–178, 2009.

X.-S. Yang and X. He, "Firefly algorithm: recent advances and applications," International Journal of Swarm Intelligence, vol. 1, no. 1, Pages: 36–50, 2013. DOI: 10.1504/IJSI.2013.055801.

Chapter 2

Forensic detective control using hardware steganography for IP core protection

Anirban Sengupta[1]

This chapter discusses forensic detective control using hardware steganography for intellectual property (IP) cores or integrated circuits. The chapter is organized as follows: Section 2.1 introduces the utility or applications of hardware steganography; Section 2.2 discusses the threat model for which hardware steganography is applicable; Section 2.3 presents comparative study on contemporary approaches; Section 2.4 explains the IP core steganography model; Section 2.5 discusses forensic detective control using hardware steganography for digital signal processing (DSP) cores; Section 2.6 presents the design process of the hardware steganography process for DSP cores; Section 2.7 analyses security properties of hardware steganography; Section 2.8 presents an analysis and comparison of hardware steganography with watermarking for different DSP IP cores; finally, Section 2.9 concludes the chapter.

2.1 Introduction

An IP core can be designed at different abstraction levels of design flow. The IP core design in the form of register transfer level (RTL) schematic or Very High Speed Integrated Circuit Hardware Description Language (VHDL)/Verilog description is termed a 'soft IP' core. In the case of more complex designs such as DSP kernels, RTL description is obtained through electronic system level (ESL) (high-level) synthesis. This is because ESL synthesis (1) reduces the design cycle due to automation of design process and (2) provides viability of design space exploration (DSE). The IP core design at other abstraction levels such as the netlist level is known as a 'firm IP' core. Further, the IP core design at the layout level in the form of GDS-II (layout) description is termed as a 'hard IP' core (Castillo *et al.*, 2007; Chakraborty and Bhunia, 2009; McFarland *et al.*, 1988).

A number of IP cores are integrated into a system-on-chip (SoC) design to realize a complex electronic system. Additionally, this integration of IP cores into

[1]Computer Science and Engineering, Indian Institute of Technology Indore, Indore, India

an SoC design is essentially employed to meet objectives such as reduction in design time and cost savings. The IP cores could be sold to SoC integrators or IC manufacturers to meet these objectives (Plaza and Markov, 2015). However, in the design cycle, these manufacturers or integrators may not be trustworthy and may invite some potential threats. Adversaries in the foundry or manufacturing house may realize their malicious intents such as piracy, counterfeiting or cloning of an IP core and also claim ownership (Sengupta, 2016, 2017).

Therefore forensic detective approaches can be employed to detect these threats. Hardware watermarking is one of the approaches in which an IP core designers/owners implant their signature into the design in the form of secret mark (Sengupta and Bhadauria, 2016). The implanted signature can be identified in the IP/IC to detect piracy, counterfeiting and cloning as well as the actual owner (Sengupta and Mohanty, 2019a, 2019b; Roy and Sengupta, 2019). Further, the signature can be implanted into the design at lower abstraction level such as gate-level or higher-abstraction level as in the case of an ESL. The owner's signature is implanted at the gate level during in-synthesis process or in the form of netlist/bit stream of an IP core design (Ziener and Teich, 2008; Le Gal and Bossuet, 2012). Additionally, the owner's signature is implanted at the ESL by modifying the register's allocation according to the signature combination during the register-allocation phase (Sengupta and Bhadauria, 2016; Sengupta and Roy, 2017, 2018; Sengupta *et al.*, 2018). However, the effect of the signature combination (which is not under full control of the designer) dictates security and design overheads. Further, to decide on an appropriate signature combination for a hardware watermark is considered a challenge to the designer (Sengupta and Rathor, 2019). The limitations of hardware watermarking and their key reasons are described in Section 2.3.

An alternative approach to hardware watermarking is '*hardware steganography*'. Steganography is a process of concealing or hiding information inside the object of interest to be protected. The hardware steganography of an IP core is employed to implant concealed information into the design in the form of stego-constraints. Thus hardware steganography provides forensic detective control to resolve conflict of ownership and to detect piracy, counterfeiting and cloning (Sengupta and Rathor, 2019). A thematic representation of protecting IP cores using hardware steganography is shown in Figure 2.1. Three different ways of using hardware steganography for forensic detective control of IP cores are depicted in Figure 2.1(a)–(c):

1. *From the end user or system integrator's perspective:* Hardware steganography enables the end user or system integrator to probe the authenticity of the IC/IP obtained from an IC vendor or IP designer. As shown in Figure 2.1(a), the end user or the system integrator inspects the steganography in the ICs/IPs and authenticates them as original if steganography is found. In this way counterfeiting of ICs/IPs can be tested (Colombier and Bossuet, 2015).
2. *From the designer or owner's perspective:* Hardware steganography enables designers to detect piracy and cloning of their IP core. As shown in Figure 2.1(b),

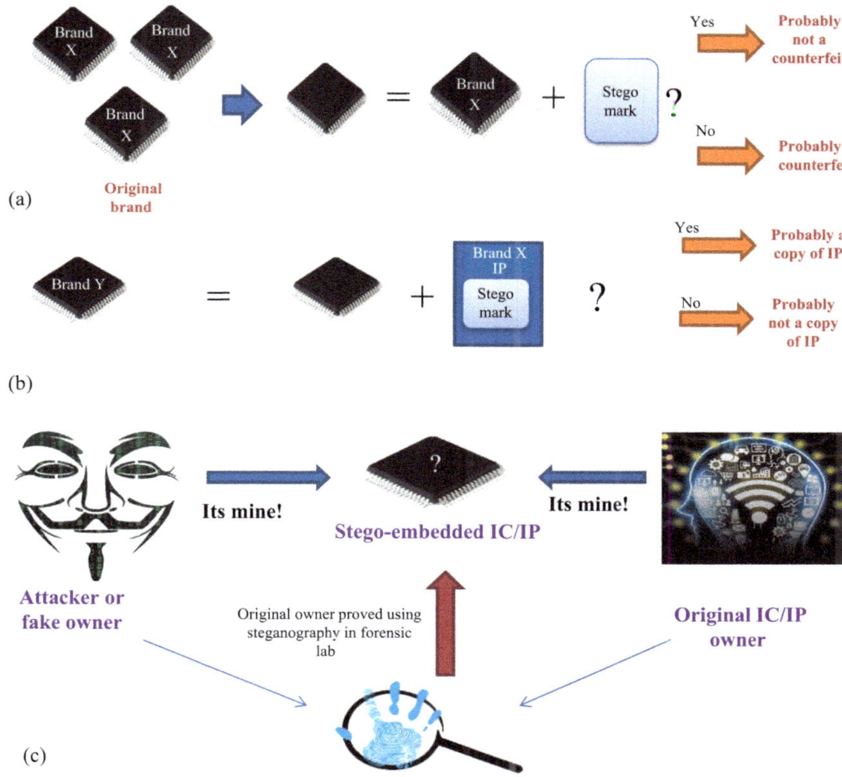

Figure 2.1 Thematic representation of protecting IP cores using hardware steganography. (a) Detection of IC counterfeiting (Colombier and Bossuet, 2015); (b) detection of IC or IP cloning (Colombier and Bossuet, 2015); and (c) proving a fraudulent claim of IC/IP ownership

designers compare their own steganography with the one which is on the IC to infer whether their own IP has been copied (and used illegally) by an adversary present in the manufacturing cycle. If their steganography is found in the IC of a different brand, then piracy or cloning of their IP core can be detected (Colombier and Bossuet, 2015).

3. *IP/IC ownership conflict:* Hardware steganography enables IP owners to prove their ownership in case of a fraudulent claim by an adversary. As shown in Figure 2.1(c), IP owners can extract the steganography from the IP/IC under a test of ownership based on some specific secret design attributes (stego-constraints) that are only known to them. Thus fraudulent claims of ownership can be nullified and ownership is awarded to the rightful IP owner (Sengupta and Rathor, 2019).

2.2 Threat model

The threat model recognizes potential vulnerabilities and the most relevant attacks to an IP core design from an attacker's point of view. Threat modelling is crucial in order to analyse a probable attacker's profile carefully and systematically, and to develop a robust defence or detection mechanism accordingly. Potential threats against IP core designs are:

1. IP/IC piracy
2. IP/IC counterfeiting and cloning
3. Fraudulent claim of ownership

Hardware steganography and watermarking targets these threats by providing robust forensic detective control to detect piracy and resolve the conflict of false claims of ownership.

2.3 Selected contemporary approaches

Hardware watermarking and cryptography are some of the contemporary approaches of hardware steganography. A comparison of contemporary approaches with hardware steganography followed by their limitations is discussed in this section.

2.3.1 Steganography versus watermarking versus cryptography

Contemporary approaches such as watermarking and crypto-driven functional obfuscation or logic encryption are also employed to protect the hardware of an IP core. Figure 2.2 compares hardware steganography with its contemporary approaches based on the features (or attributes) outlined in the illustration. As highlighted in the figure, hardware steganography, hardware watermarking and hardware cryptography are compared in terms of: abstraction level of deployment (i.e., at which very large scale integration abstraction layer it is most commonly applied), encoding style (i.e., how is the algorithm encoded), security protocol (i.e., the factor responsible for providing robustness), designer's control (i.e., how much control the designer has over the expected design overheads and security), detection process (i.e., what critical parameter is important for detection), complexity (i.e., how much effort it takes to implement the process and design overheads of the methodology).

2.3.2 Limitations of hardware watermarking

There are some limitations to hardware watermarking, as depicted in Figure 2.3. Hardware watermarking essentially depends on the signature combination and its encoding rule. Further, the optimal signature to be embedded, in terms of its length and combination as a watermark, is crucial because it dictates the design overheads and number of security constraints added. However, innumerable signature digit

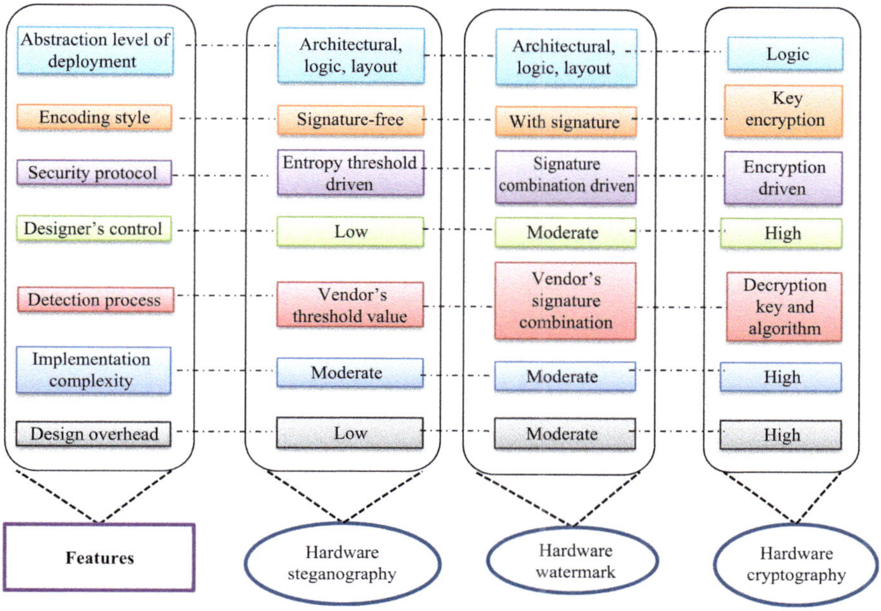

Figure 2.2 Comparison of hardware steganography with its contemporary approaches

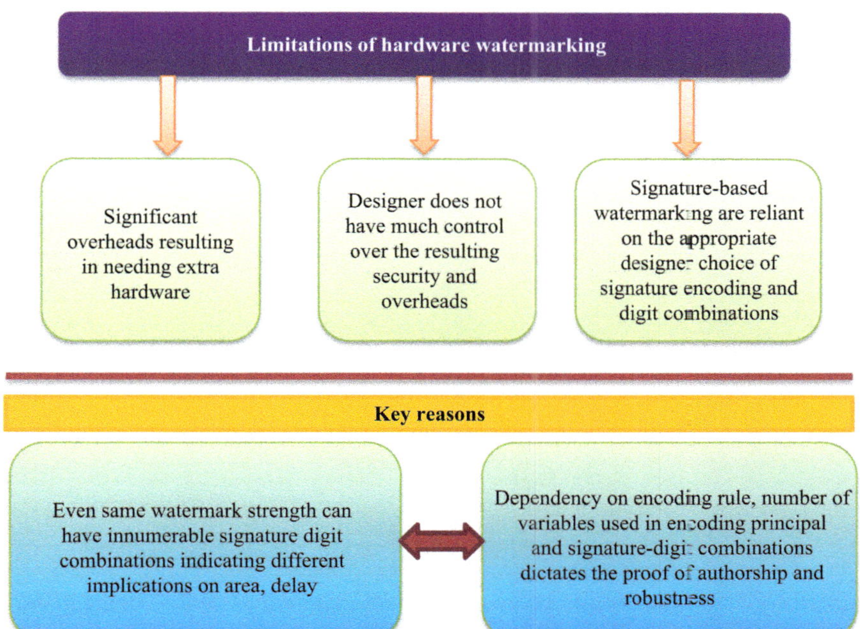

Figure 2.3 Limitations of hardware watermarking

combinations and different possible encoding styles of digits may yield the same watermark-strength causing different impacts on security and design overheads. Therefore a designer (IP/IC vendor) does not possess full control over the security strength of the watermark and design overheads. Additionally, for a chosen signature combination, security constraints to be added may already exist in the design by default. In this case, even a large size signature would result into a feebler watermark because of less security constraints added.

2.4 IP core steganography model

The IP core steganography model can be partitioned into two processes, as shown in Figure 2.4.

1. *Hardware steganography encoding process:* This process consists of a stego-system encoder that generates a stego-IP core. The stego-system encoder processes the security constraints to be added as hardware steganography and then embeds them into an RTL data path through a high-level synthesis (HLS) framework.
2. *Hardware steganography decoding process:* This process consists of a stego-system decoder that extracts the stego-constraints from the IP core and performs verification to detect steganography. Thus this process provides proof of ownership and detection of IC/IP counterfeiting and cloning.

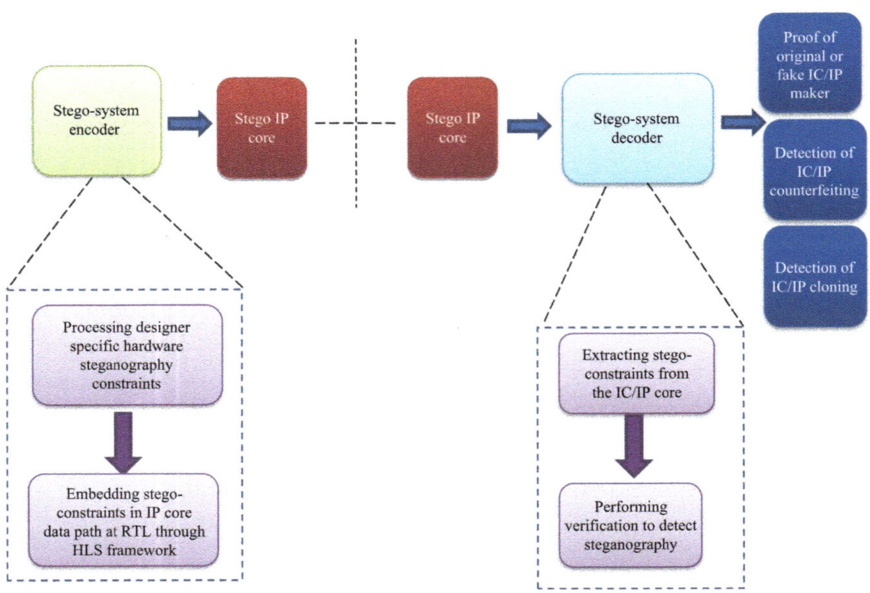

Figure 2.4 Hardware steganography encoding–decoding process (Sengupta and Rathor, 2019; Sengupta et al., 2019)

2.5 Forensic detective control using hardware (IP core) steganography

The approach for performing forensic detective control using hardware steganography is described comprehensively in Sections 2.5.1–2.5.4 followed by a demonstration.

2.5.1 Problem solved by hardware steganography

By using a vendor-specified resource configuration $N_i(R_j)$, a hardware steganography-based IP core design can be generated. A data flow graph (DFG) of the IP core design is fed as a primary input along with the resource configuration. The resource configuration $N_i(R_j)$ represents the number of resources of each resource type, R_j (e.g., if m types of resources are represented as R_j: $R_1, R_2, \ldots, R_m$ then $N_i(R_1)$, $N_i(R_2)$ and $N_i(R_n)$ represent the number of resources of type R_1, R_2 and R_m, respectively, where, i denotes a positive integer value that may be same, or distinct, for different resource types.

2.5.2 Steganography framework

The overview of the steganography framework is shown in Figure 2.5. As illustrated, the process of implanting steganography is accomplished using a generic HLS framework (Sengupta *et al.*, 2010). Input to the HLS framework is the DFG of the DSP design, and output is the steganography-embedded IP core (stego-IP core). Further, the process of implanting steganography using an HLS framework is divided into six modules where each accomplishes a certain process, as shown in Figure 2.5. Module 1 accepts the DFG of the design as input and performs the storage assignment process after scheduling. Module 2 creates a coloured interval graph (CIG) using storage variables as nodes and performs an edge-set determination process by listing all edges between two nodes of the same colour in the CIG. This edge set is input to module 3, where swapping the pair-determination processes for each edge in the set is performed. Module 4 calculates the maximum entropy for each edge in the set and module 5 performs the shortlisted edge-set determination process based on the entropy-threshold value and maximum entropies of each edge. Finally, module 6 implants the shortlisted edges into the IP core design, thereby generating a stego-IP core. The details of these six modules during the process of embedding steganography are highlighted in Figure 2.6. The embedding process of hardware steganography via the six modules is explained as

1. Module 1: Storage assignment phase in scheduling
   Input to module 1 is the DFG of the DSP IP-core design to be secured with steganography. First, this module performs scheduling of the DFG using as soon as possible (ASAP) scheduling algorithm or list scheduling and generates a scheduled DFG (SDFG). The thus-obtained SDFG allots each operation of the design to a particular control step. Next, module 1 performs storage assignment in the SDFG of the IP core design. Storage assignment is a process

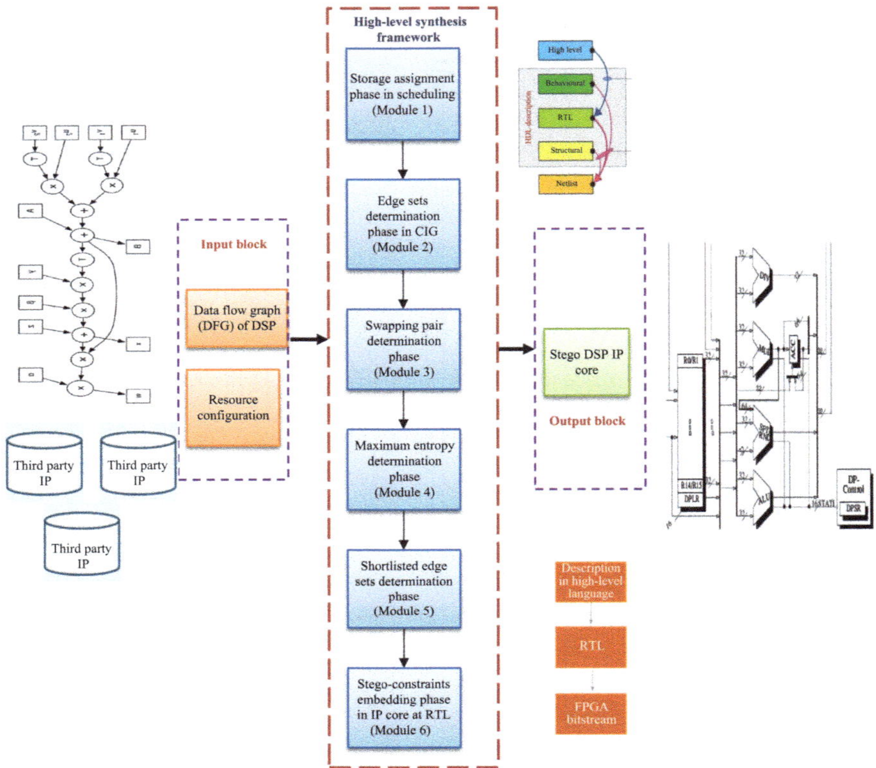

Figure 2.5 Overview of the implanting hardware steganography in IP cores (Sengupta and Rathor, 2019; Sengupta et al., 2019)

of assigning storage variables to the input and output of each operation. A storage variable in the physical data path is equivalent to a dedicated register. However, the total number of registers are either less than or equal to the total number of storage variables. Subsequently, to identify the minimum number of registers required, a CIG is created in module 2.

2. Module 2: Edge-set determination phase in the CIG
   A CIG is an undirected graph of interconnected coloured nodes, where each node represents a storage variable. Thus the total number of nodes in the CIG is equal to the total number of storage variables. The colours of the nodes represent registers and, therefore, the total number of colours is equal to the minimum number of registers required to accommodate all storage variables. An edge between two nodes indicates the overlapping of life time of two storage variables accommodated through distinct registers. Therefore edges between the same colour nodes in the CIG never exist because a single register cannot be used to accommodate two storage variables executing at the same time (in the

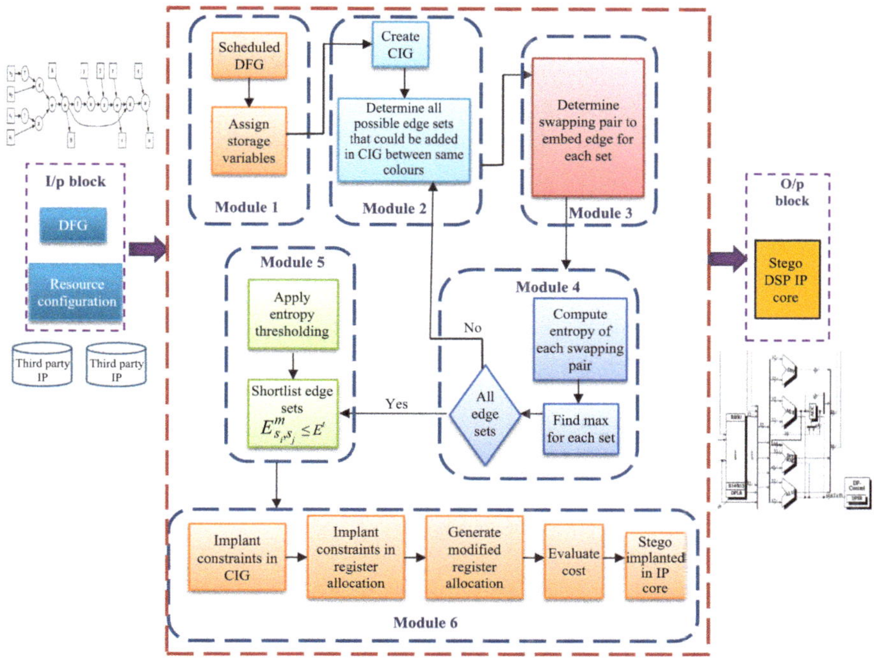

Figure 2.6 Embedding process of hardware steganography for an IP core (Sengupta and Rathor, 2019; Sengupta et al., 2019)

same control step). After determining the minimum number of registers (colours) by creating a CIG, this module identifies all the potential edges that could be embedded between the nodes of the same colour. The collection of such edges constitutes the edge set. These edges can be defined as the elements of the edge set, S, as

$$S = \{(Si, Sj) | (Si, Sj) \text{is an edge between a pair of nodes of the same colour in the CIG where } i \neq j\}$$

3. Module 3: Swapping pair-determination phases
All edges between nodes of the same colour listed in S by the previous module are the input for module 3. This module implements the swapping of pair-determination processes. A swapping pair is a potential pair of nodes in the CIG that can be swapped to embed an edge between two nodes of the same colour. Since an edge between two nodes of the same colour cannot exist, possible swapping pairs are determined to resolve the conflict while embedding an edge between two nodes of the same colour. Practically, there are multiple such potential swapping pairs available to resolve the conflict during adding an edge between a node pair (Si, Sj). This module determines all possible swapping pairs for each edge in S.

4. Module 4: Maximum entropy determination phase
This module calculates the hardware entropy corresponding to each swapping pair of each edge between node pairs (*Si*, *Sj*) in *S*. For a swapping pair of an edge between a node pair (*Si*, *Sj*), the entropy (E_{s_i,s_j}) is defined as the number of colour (register) transformations required to embed that particular edge in the CIG. Further, the maximum entropy ($E^m_{s_i,s_j}$) for an edge between a node pair (*Si*, *Sj*) is defined as

$$E^m_{s_i,s_j} = \max\left(E_{s_i,s_j}\right), \forall \text{ swapping pairs} \in (Si, Sj) \tag{2.1}$$

Thus $E^m_{s_i,s_j}$ is calculated for all the possible edges in *S* using (2.1).
5. Module 5: Shortlisted edge-set determination phase
The shortlisted edge set (*S**) is a subset of the set *S* that contains only those edges that qualify for the insertion, as secret information, in the CIG. This edge set (*S**) is obtained based on a vendor-specified threshold entropy value, E^t, according to the formula:

$$S^* = \left\{(Si, Sj)|(Si, Sj) \in S \;\; and \;\; E^m_{s_i,s_j} \leq E^t\right\} \tag{2.2}$$

According to the definition of *S**, only those edges in the set *S* qualify for insertion in the CIG that have corresponding maximum entropy $E^m_{s_i,s_j}$ less than, or equal to, E^t.
6. Module 6: Stego-constraints embedding phase in an IP core at RTL
First, this module sorts all elements in the set *S** based on the values of *i* and *j* in the corresponding node pair (Si, Sj). Thereafter all edges in *S** are implanted into the CIG according to the sorted sequence. During implanting edges, if conflict occurs then it is resolved by randomly selecting the colour swapping pair corresponding to each (Si, Sj). These edge constraints are implanted into the CIG during the register-allocation phase of HLS, therefore, implanting constraints results in a modified register allocation. After embedding edges, the design cost is evaluated to assess the effect of the constraints–implantation. If the expected design cost is not achieved, then the shortlisted edge set is modified by changing the entropy threshold. Further, a higher vendor-selected entropy threshold can be used to embed more stego-constraints (stronger robustness and security) into the IP design. Thus hardware steganography of the IP-core design is accomplished by implanting all edges of the set *S** into the CIG, while achieving the expected design cost and security. The effect of the steganography is reflected in the form of a modified register allocation, thereby transformation in the register multiplexer inputs of the RTL data path of the IP-core design. Thus the amount of hardware steganography can be controlled by the threshold entropy value E^t. This is because E^t determines the number of secret edges (artificial constraints) to be inserted as hidden information into the CIG of the IP-core design.

2.5.3 Detection (extraction) of steganography

Detection of steganography is performed to prove ownership and detect piracy. Based on the entropy threshold value (E^t), which is only known to IP owner or vendor, concealed information can be extracted from the RTL data path of the design for verification. Figure 2.7 illustrates the hardware steganography detection process in IP cores. In the case of conflict of ownership, true IP owners perform five steps, as shown on the left-hand side of Figure 2.7, to prove their ownership: (1) generate the scheduling of input DFG based on resource configuration; (2) perform register allocation; (3) create CIG; (4) determine the elements (edges between nodes of the same colour) in the set S; and (5) obtain the subset S^* (based on the designer-specified E^t) that contains the edges representing designer-specified, stego-constraints and feeds them into a verification block. On the right-hand side of Figure 2.7, the IP core under ownership testing is examined to extract the stego-constraints from the RTL data path of the design. First, multiplexers' inputs of all registers of the RTL data path are inspected and storage variables mapped to these registers are extracted. Next, the extracted stego-constraints are fed to the verification block, as shown in Figure 2.7. If the designer-specified,

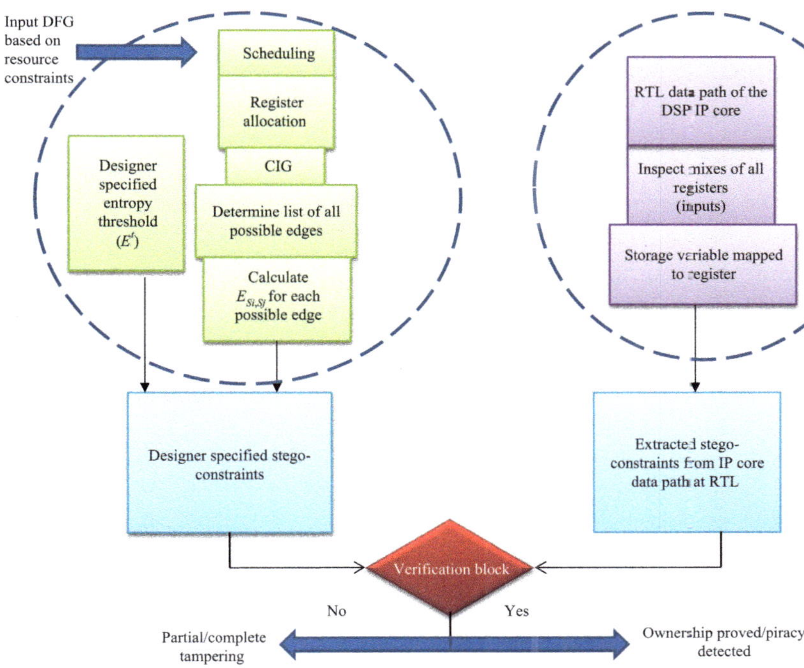

Figure 2.7 Hardware steganography detection process in IP cores (Sengupta and Rathor, 2019; Sengupta et al., 2019)

stego-constraints match the extracted ones from the RTL data path of the IP core, then true ownership is proved, and ownership is awarded to the rightful IP owner. In contrast, mismatching of some stego-constraints ascertains that the IP core design has been tampered with partially or completely.

2.5.4 Demonstration

The process of implanting hardware steganography is demonstrated here with a sample application. The CIG of the sample application and corresponding register allocation is shown in Figure 2.8 and Table 2.1, respectively. As shown in Figure 2.8, the eight storage variables ($S0, S1, \ldots, S7$) of the sample application can be accommodated using only three registers, which are represented by yellow (Y), violet (V) and red (R). An edge set, S, containing all possible (potential) edges between nodes of the same colour can be determined by observing the CIG. Thus the edge set, S, constitutes the variables (Sengupta and Rathor, 2019):

$$S = \{(S0, S3), (S0, S6), (S0, S7), (S1, S4), (S2, S5), (S3, S6), (S3, S7), (S6, S7)\}$$

Since embedding an edge between two nodes of the same colour is not possible because the same register cannot store two storage variables while overlapping in

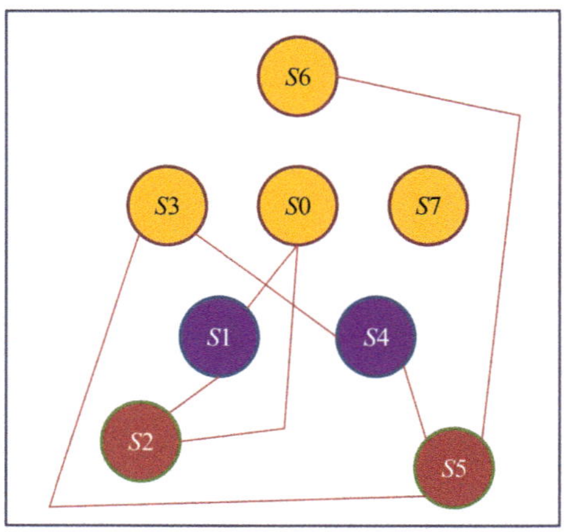

Figure 2.8 CIG of a sample application

Table 2.1 Register allocation of sample application

Clock cycle	Yellow (Y)	Violet (V)	Red (R)
T0	S0	S1	S2
T1	S3	S4	S5
T2	S6	–	S5
T3	S7	–	–

their life time. Therefore embedding an edge between a node pair (Si, Sj) invites conflict that can be resolved in several ways. For instance, embedding an edge between a node pair $(S0, S3)$ is possible through the solutions presented in Table 2.2. As shown in the table, there are four options to resolve the conflict for embedding the edge between node pair $(S0, S3)$. In the control step, T0, conflict can be resolved in two ways, namely (1) by swapping the node pair *S*0 (yellow) and *S*1 (violet) and (2) by swapping the node pair *S*0 (yellow) and *S*2 (red). Since, during the swapping of *S*0 and *S*1, two colour transformations (*S*0: Y $\Rightarrow$ V, *S*1: V $\Rightarrow$ Y) are required, therefore, the entropy of this swapping pair is 2. Similarly, the entropy of two other swapping pairs, *S*0 $\Leftrightarrow$ *S*2 and *S*3 $\Leftrightarrow$ *S*4, are 2 for each. On the other hand, the swapping pair *S*3 $\Leftrightarrow$ *S*5 alone (in control step T1) is not capable to resolve the conflict. Therefore one more swapping pair, *S*6 $\Leftrightarrow$ *S*5, needs to be performed in T2 along with *S*3 $\Leftrightarrow$ *S*5 to resolve the conflict of adding the edge between the node pair $S0, S3$. This results in three colour transformations (*S*3: Y $\Rightarrow$ R, *S*5: R $\Rightarrow$ Y, *S*6: Y $\Rightarrow$ R) resulting in the entropy of 3. The maximum entropies among all the swapping pairs of the edge between the node pair $S0, S3$ is 3.

Similarly, all possible solutions for other edges in the set *S* are presented in Tables 2.3–2.9. These tables list all the possible swapping pairs and their corresponding entropies for each edge between node pairs (Si, Sj). For designer/vendor–specified, entropy-threshold values, $E^t = 2$, the edges which qualify for insertion

Table 2.2 Swapping pairs and corresponding entropies for the edge S0, S3

Edge	Control step	Swapping pairs	Swapping colours (registers)	Hardware entropies (E_{s_i,s_j})
(*S*0, *S*3)	T0	(*S*0 $\Leftrightarrow$ *S*1)	(Y $\Leftrightarrow$ V)	2
	T0	(*S*0 $\Leftrightarrow$ *S*2)	(Y $\Leftrightarrow$ R)	2
	T1	(*S*3 $\Leftrightarrow$ *S*4)	(Y $\Leftrightarrow$ V)	2
	T1	(*S*3 $\Leftrightarrow$ *S*5)	(Y $\Leftrightarrow$ R) + (Y $\Leftrightarrow$ R)	3
	T2	(*S*6 $\Leftrightarrow$ *S*5)		

Table 2.3 Swapping pairs and corresponding entropies for the edge S0, S6

Edge	Control step	Swapping pairs	Swapping colours (registers)	Hardware entropies (E_{s_i,s_j})
(*S*0, *S*6)	T0	(*S*0 $\Leftrightarrow$ *S*1)	(Y $\Leftrightarrow$ V)	2
	T0	(*S*0 $\Leftrightarrow$ *S*2)	(Y $\Leftrightarrow$ R)	2
	T2	(*S*6 $\Rightarrow$ –)	(Y $\Rightarrow$ V)	1
	T2	(*S*6 $\Leftrightarrow$ *S*5) + (*S*3 $\Leftrightarrow$ *S*5)	(Y $\Leftrightarrow$ R)	3

Table 2.4 Swapping pairs and corresponding entropies for the edge S0, S7

Edge	Control step	Swapping pairs	Swapping colours (registers)	Hardware entropies (E_{s_i,s_j})
($S0$, $S7$)	T0	($S0 \Leftrightarrow S1$)	(Y ⇔ V)	2
	T0	($S0 \Leftrightarrow S2$)	(Y ⇔ R)	2
	T3	($S7 \Rightarrow -$)	(Y ⇒ V)	1
	T3	($S7 \Rightarrow -$)	(Y ⇒ R)	1

Table 2.5 Swapping pairs and corresponding entropies for the edge S1, S4

Edge	Control step	Swapping pairs	Swapping colours (registers)	Hardware entropies (E_{s_i,s_j})
($S1$, $S4$)	T0	($S1 \Leftrightarrow S0$)	(V ⇔ Y)	2
	T0	($S1 \Leftrightarrow S2$)	(V ⇔ R)	2
	T1	($S4 \Leftrightarrow S3$)	(V ⇔ Y)	2
	T1	($S4 \Leftrightarrow S5$) + ($S5 \Rightarrow -$)	(V ⇔ R) + (R ⇒ V)	2
	T2			

Table 2.6 Swapping pairs and corresponding entropies for the edge S2, S5

Edge	Control step	Swapping pairs	Swapping colours (registers)	Hardware entropies (E_{s_i,s_j})
($S2$, $S5$)	T0	($S2 \Leftrightarrow S0$)	(Y ⇔ R)	2
	T0	($S2 \Leftrightarrow S1$)	(R ⇔ V)	2
	T1	($S5 \Leftrightarrow S4$)	(R ⇔ V)	2
	T1	($S5 \Leftrightarrow S3$) + ($S5 \Leftrightarrow S6$)	(R ⇔ Y) + (R ⇔ Y)	3
	T2			

Table 2.7 Swapping pairs and corresponding entropies for the edge S3, S6

Edge	Control step	Swapping pairs	Swapping colours (registers)	Hardware entropies (E_{s_i,s_j})
($S3$, $S6$)	T1	($S3 \Leftrightarrow S4$)	(Y ⇔ V)	2
	T2	($S6 \Rightarrow -$)	(Y ⇒ R)	1

Table 2.8 Swapping pairs and corresponding entropies for the edge S3, S7

Edge	Control step	Swapping pairs	Swapping colours (registers)	Hardware entropies (E_{s_i,s_j})
(*S*3, *S*7)	T1	(*S*3 ⇔ *S*4)	(Y ⇔ V)	2
	T1	(*S*3 ⇔ *S*5) + (*S*5 ⇔ *S*6)	(Y ⇔ R)	3
	T3	(*S*7 ⇒ –)	(Y ⇒ V)	1
	T3	(*S*7 ⇒ –)	(Y ⇒ R)	1

Table 2.9 Swapping pairs and corresponding entropies for the edge S6, S7

Edge	Control step	Swapping pairs	Swapping colours (registers)	Hardware entropies (E_{s_i,s_j})
(*S*6, *S*7)	T2	(*S*6 ⇒ –)	(Y ⇒ V)	1
	T2	(*S*6 ⇔ *S*5) + (*S*5 ⇔ *S*3)	(Y ⇔ R)	3
	T3	(*S*7 ⇒ –)	(Y ⇒ V)	1
	T3	(*S*7 ⇒ –)	(Y ⇒ R)	1

into the CIG (of the IP core) are represented using a set, S^*, where $S^* \subset S$. The generated S^* constitutes as:

$$S^* = \{(S0, S7), (S1, S4), (S3, S6)\}$$

Only the edges in the set S^* are eligible to be embedded into the CIG. This is because the maximum entropy ($E^m_{s_i,s_j}$) of each of these edges is less than, or equal to, two, whereas the maximum entropy of the remaining edges in the set *S* is greater than two, hence, only three edges of the edge set *S* qualifies according to (2.2). These edges are embedded considering the solutions obtained from Tables 2.4, 2.5 and 2.7, respectively:

1. (*S*0, *S*7): (*S*7 ⇒ –) (Y ⇒ R) [control step T3]
2. (*S*1, *S*4): (*S*4 ⇔ *S*5) (V ⇔ R) [control step T1] + (*S*5 ⇒ –) (R ⇒ V) [control step T2]
3. (*S*3, *S*6): (*S*6 ⇒ –) (Y ⇒ R) [control step T2]

The modified-register allocation after embedding the first edge between the node pair *S*0, *S*7 is shown in Table 2.10. Further modification in register allocation after embedding the other two edges between node pairs *S*1, *S*4 and *S*3, *S*6 is shown in Tables 2.11 and 2.12, respectively. Embedding these edges in the form of concealed-sego information results in a modified CIG of the IP core, as shown in Figure 2.9 and the corresponding final, modified-register allocation is shown in Table 2.12.

Table 2.10 Register allocation after implanting S0, S7

Clock cycle	Yellow	Violet	Red
T0	*S*0	*S*1	*S*2
T1	*S*3	*S*4	*S*5
T2	*S*6	–	*S*5
T3	–	–	***S*7**

Note: The bold indicates the respective storage variables that have been moved due to implanting constraints.

Table 2.11 Register allocation after implanting S1, S4

Clock cycle	Yellow	Violet	Red
T0	*S*0	*S*1	*S*2
T1	*S*3	***S*5**	***S*4**
T2	*S*6	***S*5**	–
T3	–	–	*S*7

Note: The bold indicates the respective storage variables that have been moved due to implanting constraints.

Table 2.12 Register allocation after implanting S3, S6

Clock cycle	Yellow	Violet	Red
T0	*S*0	*S*1	*S*2
T1	*S*3	*S*5	*S*4
T2	–	*S*5	***S*6**
T3	–	–	*S*7

Note: The bold indicates the respective storage variables that have been moved due to implanting constraints.

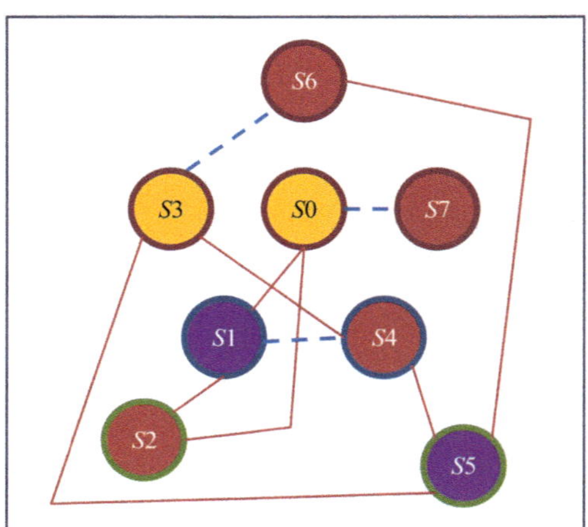

Figure 2.9 Final CIG after implanting all the edges of the hardware steganography

2.6 Design process of a steganography-embedded IP core: a case study on 8-point DCT

Highly complex designs such as DSP kernels are easier to describe at a higher abstraction level, such as architecture level or ESL. Such complex designs are transformed from their architecture description to RTL descriptions using an HLS framework. Therefore during HLS, a protection mechanism can easily be deployed in complex designs to secure them against various threats.

As discussed in Section 2.5.4, (Sengupta and Rathor, 2019; Sengupta *et al.*, 2019) have implanted hardware steganography in the register-allocation phase during HLS. Therefore more complex designs such as DSP kernels, which are initially described at architecture level, can be protected using hardware steganography during their HLS.

DSP IP cores, as a part of SoC designs, facilitate several applications such as de-noising, compression/decompression of images, audio, videos, etc. To facilitate image compression/decompression in consumer electronic devices, such as camera systems, a JPEG CODEC (a DSP IP core) is used. Further, a JPEG CODEC comprises discrete cosine transform (DCT) as an integral part (Schneiderman, 2010; Obukhov and Kharlamov, 2008; Sengupta and Mohanty, 2019c). The design process of embedding steganography into a real-life, 8-point DCT is demonstrated in this section.

The design process is accomplished by two phases, namely the (1) pre-processing phase before steganography and (2) stego-constraints implantation phase. These two phases will be discussed as follows.

2.6.1 Pre-processing phase before steganography

This phase translates the 8-point DCT design into a suitable form to employ the steganography approach. The DFG representing the 8-point DCT design along with the vendor-specified resource configuration is fed into the HLS framework. Based on the resource configuration, four multipliers (*M*1, *M*2, *M*3 and *M*4) and one adder (*A*1), the DFG is scheduled as shown in Figure 2.10. A total of eight multiplication operations and seven addition operations of 8-point DCT are scheduled in the total eight control steps **(T)** complying with the resource constraints. Further, inputs and outputs of each operation are assigned to storage variables. To store all the primary and intermediate inputs and outputs of the design, 23 storage variables (*S*0–*S*22) were opted. To determine the minimum number of registers required to accommodate all storage variables, a CIG is generated where each node represents a storage variable. As shown in Figure 2.11, eight colours are used to construct the CIG of the 8-point DCT design; hence, a minimum of eight registers is required to accommodate all the storage variables. These eight registers represented by the different colours (i.e., violet, indigo, blue, green, yellow, orange, red and black) initially accommodate eight primary inputs of the 8-point DCT design. As shown in

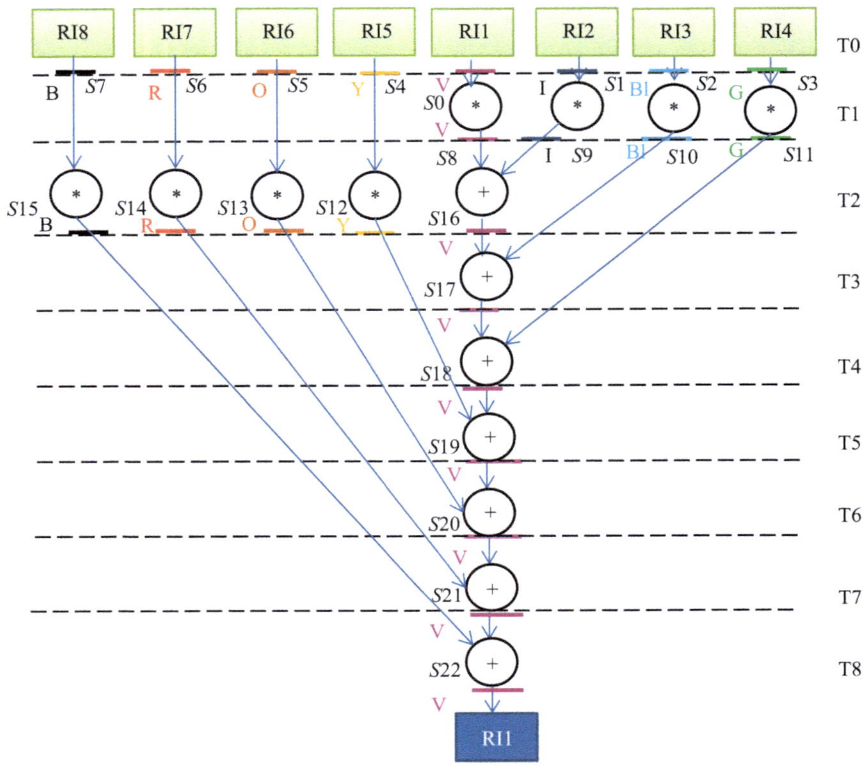

Figure 2.10 Scheduled and hardware allocated 8-point DCT using 1 (+) and 4 (*) before *implanting steganography*

Figure 2.10, these eight primary inputs are assigned to storage variables *S*0–*S*7 and in the control step T0. In further control steps, these registers are assigned to other storage variables and the final output of the design is stored into *S*22. The allocation of all storage variables to their corresponding registers is shown in Table 2.13. The RTL data path of the 8-point DCT design before implanting the steganography is shown in Figure 2.12. The multiplexing schemes of all the registers and resources of the design are presented in Tables 2.14A–E. The CIG representation of the 8-point DCT design is a suitable representation to employ the hardware steganography embedding process. Different steps of implanting steganography are described in the next phase.

2.6.2 Stego-constraints implantation phase

The steps of implanting stego-constraints to perform hardware steganography are:

Step 1: The CIG shown in Figure 2.11 is inspected to obtain all the possible edges between nodes of the same colour. Hence the set *S* containing all

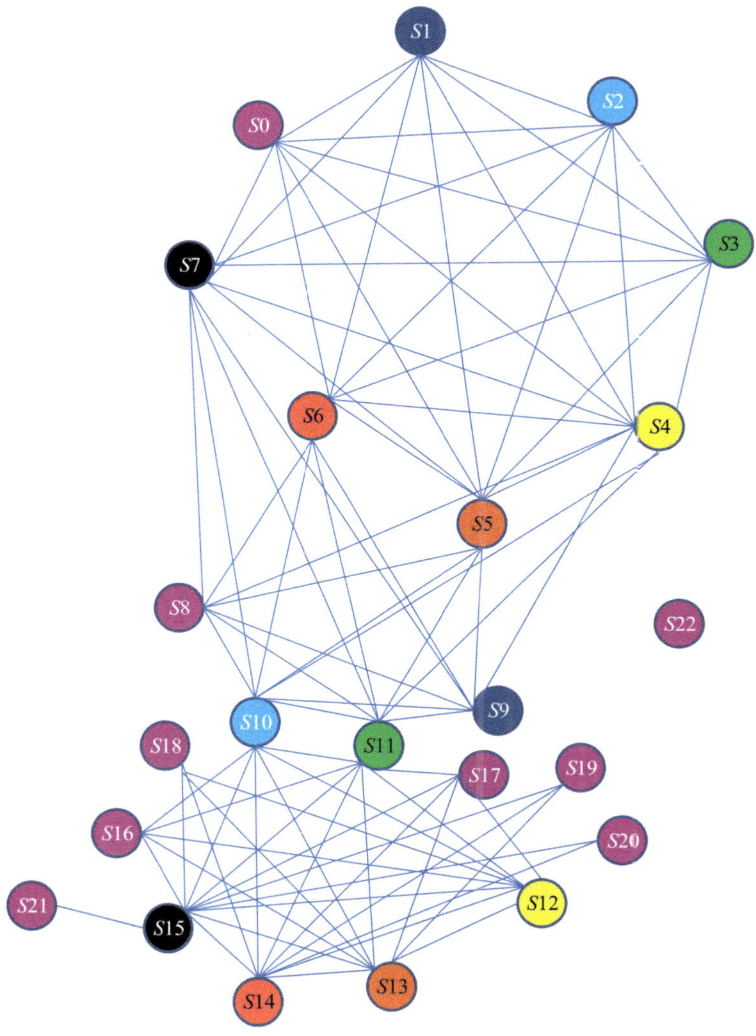

Figure 2.11 A CIG of the 8-point DCT before steganography

possible edges comprises of:

$$S = \{(S0,S8),(S0,S16),(S0,S17),(S0,S18),(S0,S19),(S0,S20),(S0,S21),(S0,S22),(S1,S9),(S2,S10),(S3,S11),(S4,S12),(S5,S13),(S6,S14),(S7,S15),(S8,S16),(S8,S17),(S8,S18),(S8,S19),(S8,S20),(S8,S21),(S8,S22),(S16,S17),(S16,S18),(S16,S19),(S16,S20),(S16,S21),(S16,S22),(S17,S18),(S17,S19),(S17,S20),(S17,S21),(S17,S22),(S18,S19),(S18,S20),(S18,S21),(S18,S22),(S19,S20),(S19,S21),(S19,S22),(S20,S21),(S20,S22),(S21,S22)\}$$

Table 2.13 Register allocation of 8-point DCT before implanting steganography

CS	Violet	Indigo	Blue	Green	Yellow	Orange	Red	Black
0	*S*0	*S*1	*S*2	*S*3	*S*4	*S*5	*S*6	*S*7
1	*S*8	*S*9	*S*10	*S*11	*S*4	*S*5	*S*6	*S*7
2	*S*16	–	*S*10	*S*11	*S*12	*S*13	*S*14	*S*15
3	*S*17	–	–	*S*11	*S*12	*S*13	*S*14	*S*15
4	*S*18	–	–	–	*S*12	*S*13	*S*14	*S*15
5	*S*19	–	–	–	–	*S*13	*S*14	*S*15
6	*S*20	–	–	–	–	–	*S*14	*S*15
7	*S*21	–	–	–	–	–	–	*S*15
8	*S*22	–	–	–	–	–	–	–

Step 2: All possible swapping pairs corresponding to each edge in the edge set *S* are determined to resolve the conflict of embedding that edge.

Step 3: Entropy corresponding to each swapping pair of each edge is determined.

Step 4: The maximum entropy value corresponding to each edge in the set is determined using (2.1) and is listed in Table 2.15. *Note: the pairs in bold indicate entropy less than threshold entropy.*

Step 5: An appropriate value of the entropy threshold is chosen by the IP vendor or designer.

Step 6: The set of shortlisted edges *S** to be embedded into the CIG as stego-constraints is obtained based on the entropy threshold value (E^t). For a vendor-selected E^t equal to five, only the following edges in the set *S** qualify for insertion into the CIG (according to (2.2)):

$$S^* = \{(S0,S8),(S0,S22),(S1,S9),(S2,S10),(S3,S11),(S4,S12),(S5,S13),\\ (S8,S22),(S16,S17),(S16,S18),(S16,S19),(S16,S20),(S17,S18),\\ (S17,S19),(S17,S20),(S18,S19),(S18,S20),(S19,S20)\}$$

These edges in the set *S** have been sorted according to their node indices (i, j).

Step 7: All eligible edges in the set *S** are embedded into the CIG as stego-constraints and the design cost is evaluated. If the design cost and robustness (number of stego constraints) is not obtained as per the expected value, then return to Step 5 and choose another E^t value. Further, Steps 5 to 7 must be repeated until the expected value of the design cost is achieved.

Step 8: For a designer-selected E^t, secret information in the form of artificial constraint edges is concealed into the IP-core design while achieving the expected design cost and security required and, hence, the process of implanting steganography into the 8-point DCT design is accomplished.

After implanting stego-constraints for E^t equal to five, the effect of steganography is reflected (at ESL) in the form of a modified CIG and register allocation, as shown in Figure 2.13 and Table 2.16, respectively. Constraint edges embedded into the CIG are shown using dashed red lines in Figure 2.13. The modified

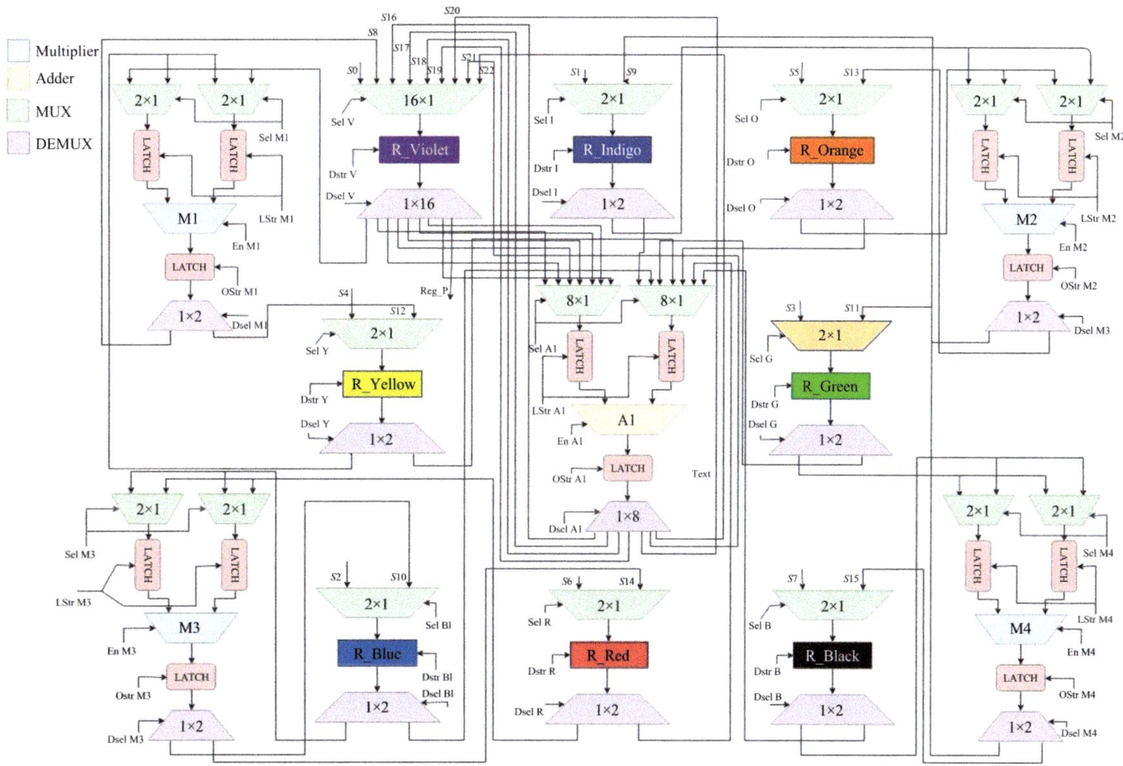

Figure 2.12   RTL data path of 8-point DCT before steganography

Table 2.14A Multiplexing tables for violet, indigo, blue and green registers before steganography

Control steps	Violet		Indigo		Blue		Green	
	Input	Output	Input	Output	Input	Output	Input	Output
T0	RI1	–	RI2	–	RI3	–	RI4	–
T1	M1_out1	M1_in1	M2_out1	M2_in1	M3_out1	M3_in1	M4_out1	M4_in1
T2	A1_out2	A1_in2	–	A1_in2	–	–	–	–
T3	A1_out3	A1_in3	–	–	–	A1_in3	–	–
T4	A1_out4	A1_in4	–	–	–	–	–	A1_in4
T5	A1_out5	A1_in5	–	–	–	–	–	–
T6	A1_out6	A1_in6	–	–	–	–	–	–
T7	A1_out7	A1_in7	–	–	–	–	–	–
T8	A1_out8	A1_in8	–	–	–	–	–	–

Table 2.14B Multiplexing tables for yellow, orange, red and black registers before steganography

Control steps	Yellow		Orange		Red		Black	
	Input	Output	Input	Output	Input	Output	Input	Output
T0	RI5	–	RI5	–	RI7	–	RI8	–
T1	–	–	–	–	–	–	–	–
T2	M1_out2	M1_in2	M2_out2	M2_in2	M3_out2	M3_in2	M4_out2	M4_in2
T3	–	–	–	–	–	–	–	–
T4	–	–	–	–	–	–	–	–
T5	–	A1_in5	–	–	–	–	–	–
T6	–	–	–	A1_in6	–	–	–	–
T7	–	–	–	–	–	A1_in7	–	–
T8	–	–	–	–	–	–	–	A1_in8

Table 2.14C Multiplexing tables multipliers M1 and M2 before steganography

Control steps	M1			M2		
	Input1	Input2	Output	Input1	Input2	Output
T0	Violet_out0	Violet_out0	–	Indigo_out0	Indigo_out0	–
T1	Yellow_out1	Yellow_out1	Violet_in1	Orange_out1	Orange_out1	Indigo_in1
T2	–	–	Yellow_in2	–	–	Orange_in2

Table 2.14D Multiplexing tables multipliers M3 and M4 before steganography

Control steps	*M3*			*M4*		
	Input1	**Input2**	**Output**	**Input1**	**Input2**	**Output**
T0	Blue_out0	Blue_out10	–	Green_out0	Green_out0	–
T1	Red_out1	Red_out1	Blue_in1	Black_out1	Black_out1	Green_in1
T2	–	–	Red_in2	–	–	Black_in2

Table 2.14E Multiplexing tables for adder A1 before steganography

Control steps	*A1*		
	Input1	**Input2**	**Output**
T0	–	–	–
T1	Violet_out1	Indigo_out1	–
T2	Violet_out2	Blue_out2	Violet_in2
T3	Violet_out3	Green_out3	Violet_in3
T4	Violet_out4	Yellow_out4	Violet_in4
T5	Violet_out5	Orange_out5	Violet_in5
T6	Violet_out6	Red_out6	Violet_in6
T7	Violet_out7	Black_out7	Violet_in7
T8	–	–	Violet_in8

Table 2.15 Sorted list of all possible edges in the set S and their respective maximum entropies

Possible edge	Maximum entropy to insert an edge	Possible edge	Maximum entropy to insert an edge
(*S*0, *S*8)	**4**	**(*S*8, *S*22)**	**4**
(*S*0, *S*16)	7	**(*S*16, *S*17)**	**3**
(*S*0, *S*17)	7	**(*S*16, *S*18)**	**4**
(*S*0, *S*18)	7	**(*S*16, *S*19)**	**4**
(*S*0, *S*19)	7	**(*S*16, *S*20)**	**5**
(*S*0, *S*20)	7	(*S*16, *S*21)	6
(*S*0, *S*21)	7	(*S*16, *S*22)	7
(*S*0, *S*22)	**3**	**(*S*17, *S*18)**	**4**
(*S*1, *S*9)	**2**	**(*S*17, *S*19)**	**4**
(*S*2, *S*10)	**3**	**(*S*17, *S*20)**	**5**
(*S*3, *S*11)	**4**	(*S*17, *S*21)	6
(*S*4, *S*12)	**4**	(*S*17, *S*22)	7
(*S*5, *S*13)	**5**	**(*S*18, *S*19)**	**4**
(*S*6, *S*14)	6	**(*S*18, *S*20)**	**5**
(*S*7, *S*15)	7	(*S*18, *S*21)	6
(*S*8, *S*16)	7	(*S*18, *S*22)	7
(*S*8, *S*17)	7	**(*S*19, *S*20)**	**5**
(*S*8, *S*18)	7	(*S*19, *S*21)	6
(*S*8, *S*19)	7	(*S*19, *S*22)	7
(*S*8, *S*20)	7	(*S*20, *S*21)	6
(*S*8, *S*21)	7	(*S*20, *S*22)	7
–	–	(*S*21, *S*22)	7

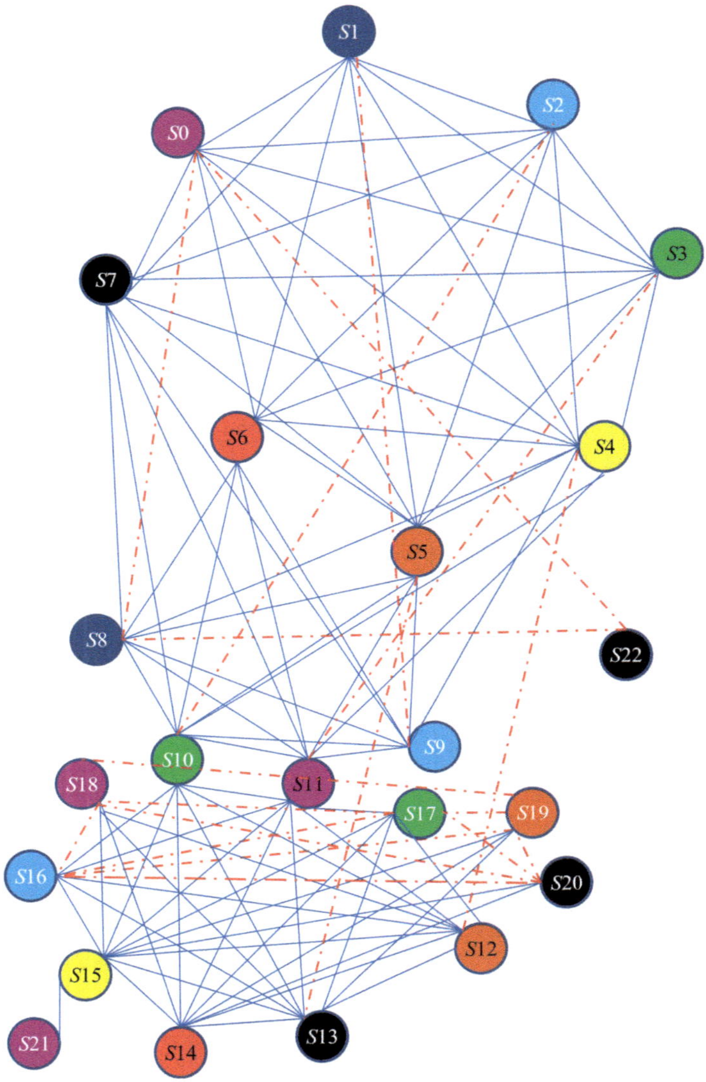

Figure 2.13 A CIG of 8-point DCT after steganography

scheduling of the 8-point DCT post-steganography implant is shown in Figure 2.14. Further, the effect of hardware steganography at RTL after HLS appears in the form of modified multiplexing schemes of different multiplexer units associated with distinct registers and resources. Tables 2.17A–E present the multiplexing schemes, showing the changes in the inputs and outputs of multiplexer units, after steganography. The effect of hardware steganography at the RTL data path of the

Table 2.16 Register allocation of 8-point DCT after *implanting steganography*

CS	Violet	Indigo	Blue	Green	Yellow	Orange	Red	Black
0	*S*0	*S*1	*S*2	*S*3	*S*4	*S*5	*S*6	*S*7
1	*S*11	*S*8	*S*9	*S*10	*S*4	*S*5	*S*6	*S*7
2	*S*11	–	*S*16	*S*10	*S*15	*S*12	*S*14	*S*13
3	*S*11	–	–	*S*17	*S*15	*S*12	*S*14	*S*13
4	*S*18	–	–	–	*S*15	*S*12	*S*14	*S*13
5	–	–	–	–	*S*15	*S*19	*S*14	*S*13
6	–	–	–	–	*S*15	–	*S*14	*S*20
7	*S*21	–	–	–	*S*15	–	–	–
8	–	–	–	–	–	–	–	*S*22

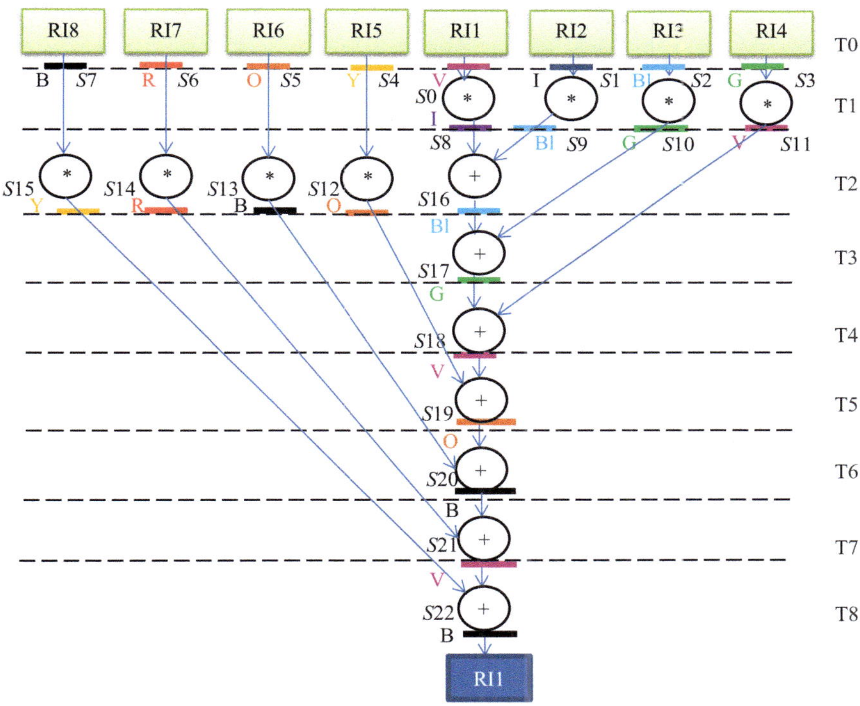

Figure 2.14 Scheduled and hardware-allocated 8-point DCT using 1 (+) and 4 (*) after *implanting steganography*

IP-core design is shown in Figure 2.15 where the corresponding changes showing the stego-constraints are enclosed using red ovals.

Observations: It can be observed from Figures 2.13 and 2.14 that the number of registers and control steps are the same as before steganography. Thus implanting constraint edges for entropy threshold value E^t of five results into zero overhead of

Table 2.17A Multiplexing tables for violet, indigo, blue and green registers after steganography

Control steps	Violet		Indigo		Blue		Green	
	Input	**Output**	**Input**	**Output**	**Input**	**Output**	**Input**	**Output**
T0	RI1	–	RI2	–	RI3	–	RI4	–
T1	*M*4_out1	*M*1_in1	*M*1_out1	*M*2_in1	*M*2_out1	*M*3_in1	*M*3_out1	*M*4_in1
T2	–	–	–	*A*1_in2	*A*1_out2	*A*1_in2	–	–
T3	–	–	–	–	–	*A*1_in3	–	*A*1_in3
T4	*A*1_out4	*A*1_in4	–	–	–	–	–	–
T5	–	*A*1_in5	–	–	–	–	–	–
T6	–	–	–	–	–	–	–	–
T7	*A*1_out7	–	–	–	–	–	–	–
T8	–	*A*1_in8	–	–	–	–	–	–

Table 2.17B Multiplexing tables for yellow, orange, red and black registers after steganography

Control steps	Yellow		Orange		Red		Black	
	Input	**Output**	**Input**	**Output**	**Input**	**Output**	**Input**	**Output**
T0	RI5	–	RI6	–	RI7	–	RI8	–
T1	–	–	–	–	–	–	–	–
T2	*M*4_out2	*M*1_in2	*M*1_out2	*M*2_in2	*M*3_out2	*M*3_in2	*M*2_out2	*M*4_in2
T3	–	–	–	–	–	–	–	–
T4	–	–	–	–	–	–	–	–
T5	–	–	*A*1_out5	*A*1_in5	–	–	–	–
T6	–	–	–	*A*1_in6	–	–	*A*1_out6	*A*1_in6
T7	–	–	–	–	–	*A*1_in7	–	*A*1_in7
T8	–	*A*1_in8	–	–	–	–	*A*1_out8	–

Table 2.17C Multiplexing tables multipliers M1 and M2 after steganography

Control steps	*M*1			*M*2		
	Input1	**Input2**	**Output**	**Input1**	**Input2**	**Output**
T0	Violet_out0	Violet_out0	–	Indigo_out0	Indigo_out0	–
T1	Yellow_out1	Yellow_out1	Indigo_in1	Orange_out1	Orange_out1	Blue_in1
T2	–	–	Orange_in2	–	–	Black_in2

Table 2.17D Multiplexing tables multipliers M3 and M4 after steganography

Control steps	*M3*			*M4*		
	Input1	**Input2**	**Output**	**Input1**	**Input2**	**Output**
T0	Blue_out0	Blue_out0	–	Green_out0	Green_out0	–
T1	Red_out1	Red_out1	Green_in1	Black_out1	Black_out1	Violet_in1
T2	–	–	Red_in2	–	–	Yellow_in2

Table 2.17E Multiplexing tables for adder A1 after steganography

Control steps	*A1*		
	Input1	**Input2**	**Output**
T0	–	–	–
T1	Indigo_out1	Blue_out1	–
T2	Blue_out2	Green_out2	Blue_in2
T3	Green_out3	Violet_out3	Green_in3
T4	Violet_out4	Orange_out4	Violet_in4
T5	Orange_out5	Black_out5	Orange_in5
T6	Black_out6	Red_out6	Black_in6
T7	Violet_out7	Yellow_out7	Violet_in7
T8	–	–	Black_in8

register/area and delay. Hence, the design cost after steganography remains the same as before, since the entropy threshold value decides the number of eligible edges to be embedded as stego constraints. Thus a much higher value of E^t causes a larger number of constraint edges to be added and, hence, may result in marginally higher design cost, but would insert a large number of secret stego constraints. Further, this observation shows that hardware steganography is a fully designer (IP vendor) controlled approach for forensic detection of IP piracy, counterfeiting and false claims of ownership.

2.7 Security features of hardware steganography

Security features of hardware steganography are discussed in terms of security properties and how threats and attacks are handled.

2.7.1 Security properties

The security of hardware steganography is assessed in terms three properties:

1. *Security and reliability*

   In the hardware steganography approach, the IP owner/designer possesses direct monitoring over the amount of security information (steganography)

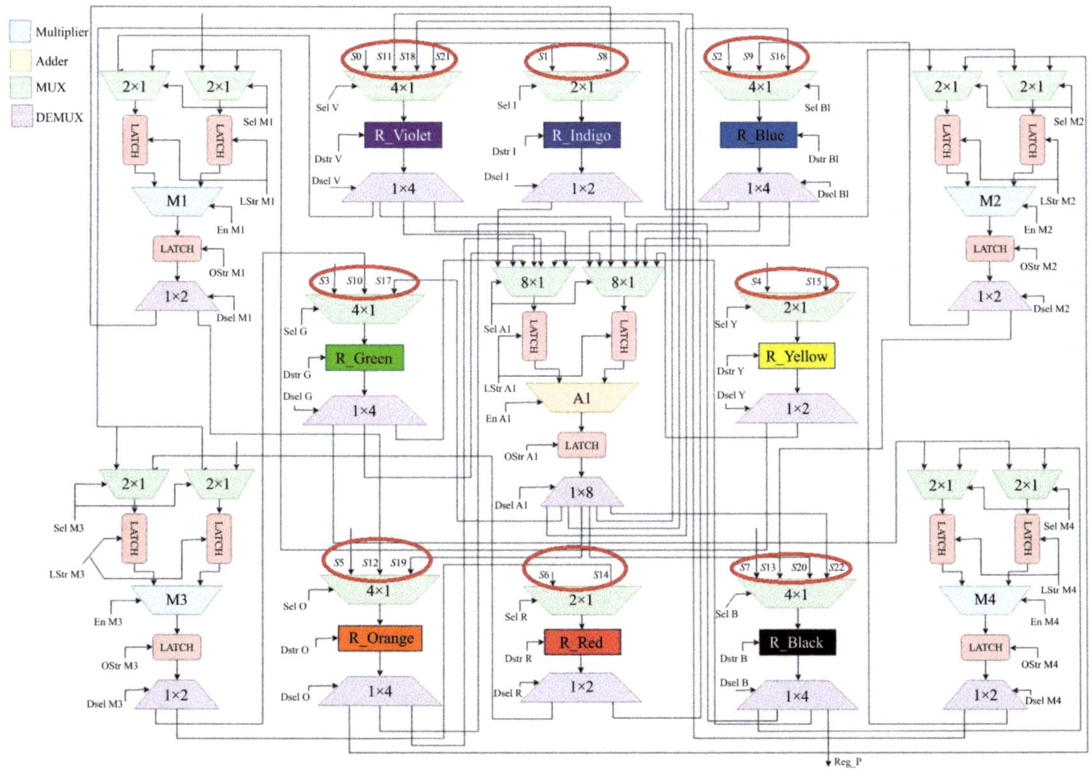

Figure 2.15 RTL data path of 8-point DCT after *steganography*

implanted into the design. This is because stego-constraints are added into the design via the entropy-threshold parameter. Further, the security of the design does not have to rely on signature digit combinations and encoding of digits, unlike the watermarking approach. Thus the uncertainty in security is eliminated and more reliable forensic detective control of the IP-core design becomes feasible. This is because during detection of steganography, the IP designer can easily produce added security constraints through the entropy threshold value to match them with the extracted stego-constraints. On the contrary, during detection of watermarking, the same security constraints extracted can be inferred as various signature combinations and, hence, renders the forensic detection less reliable.

2. *Robustness*

   All the constraint edges are definitely embedded in the steganography approach once they are shortlisted as concealed information. This is because all the shortlisted stego-constraints are the potential edges between nodes of the same colour and, hence, none of the edges can exist by default in the CIG of the design. On the contrary, in watermarking approaches some of the edges may already exist by default and, hence, a smaller number of potential constraints are added. Thus steganography results in more robust forensic detective control as it embeds more information, which renders the detection of IP core threats easier.

3. *Tamper tolerance*

   Since the stego-constraints are embedded during the HLS process, the steganography information diffuses throughout the design post-synthesis. Further, for a designer-selected threshold parameter, all the shortlisted stego-constraints are embedded and, therefore, the spread of steganography information across the RTL data path of the IP-core design is high. This characteristic of the steganography approach renders it highly tolerant against tampering by an adversary.

2.7.2 How does steganography handle threats and attacks?

Consider a scenario in which two different entities *X* and *Y* are involved where *X* is the owner/designer of the steganography-embedded IP core and *Y* is a user (SOC integrator, IC manufacturer or end user). Here, the IP core is vulnerable to several attacks and threats by *Y*. Therefore a discussion of the handling of different attacks through hardware steganography is crucial. The hardware steganography handles different attacks as follows:

1. *Unauthorized stego-constraints insertion*

   In this attack scenario, *Y* inserts unauthorized stego-constraints into a steganography-embedded IP core. Since the IP core already contains steganography embedded by *X*, entity *Y* is only able to insert unauthorized stego constraints on top of the existing steganography. Hence, it results into an IP-core design containing the steganography of *X* and *Y*. In this situation, if *Y* claims ownership of the IP core, he/she can easily be caught as the design contains the stego-constraints of *X* and *Y*, whereas the original steganography-embedded IP core only contains the stego-constraints of *X*.

2. *Original stego-information tampering*

   In this case, *Y* attempts to tamper with the original stego-information embedded by *X*. To do so, he/she tries to apply changes in the embedded-constraint edges with the intention of owning the design of *X* as his/her IP. However, tampering with steganography is not easy because the diffused stego-information across the entire design disrupts any attempt of modification in the stego-constraints by an adversary. Further, tampering with existing stego-information may change the area overhead and latency of the original steganography embedded-IP core of *X*.

3. *Unintended stego-constraints extraction*

   In this scenario, *Y* performs the reverse steganography procedure in the IP core of *X* with the intention of extracting his/her stego-constraints. *Y* can do so by claiming any existing information in the design of *X* as his/her steganography. However, *Y* fails to realize his/her intentions if the stenography is embedded in the IP core of *X*. This is because the claimed stego-constraints by *Y* are weaker compared to that of *X* as he/she tries to extract his/her constraints from the steganography-embedded IP core; whereas *X* can extract his/her potential stego-information easily based on a threshold parameter and, hence, prove his/her ownership. Thus ownership is awarded to the entity whose claim is proved to be more meaningful and strong.

2.8 Analysis of case studies

Different DSP applications along with the 8-point DCT design discussed in Section 2.7 are analysed here to evaluate the performance of the hardware steganography. Different magnitudes of entropy threshold (E^t) result in different numbers of hidden, embedded, edge constraints which, thus, affect the storage (registers) hardware requirement in the IP-core design. Therefore the impact of different values of E^t over the number of effective (non-default) constraints added and register overheads are analysed to assess the effectiveness of the hardware steganography. Further, this impact analysis for the different values of E^t helps to achieve the expected design cost and security of the stego-embedded IP cores. The following criteria are used to analyse the performance of the hardware steganography in terms of its security and design cost:

1. Probability of coincidence (P^c) metric to assess the strength of proof of ownership.
2. Effect of P^c for different values of the entropy threshold.
3. Effect on register overheads for different values of the entropy threshold.
4. Effect on post-stego–embedded design cost for different values of the entropy threshold.

2.8.1 Security analysis

The security achieved through hardware steganography is assessed in terms of the probability of coincidence (P^c) metric given as (Sengupta and Roy, 2017, 2018)

$$P^c = \left(1 - \frac{1}{n}\right)^k \tag{2.3}$$

where P^c denotes the probability of coincidence, n denotes the number of colours in the CIG before implanting steganography and k denotes the number of effective constraints (hidden edges) embedded into the CIG to process the steganography. Here the probability of coincidence signifies the proof of ownership which indicates that a non-steganography solution carries steganography (based on an entropy threshold) by coincidence. It thus reflects the strength of the steganography. The strength of proof of ownership increases with the decrease in probability of coincidence. This is because as more numbers of constraints-edges are added into the CIG of the IP core, there is much less chance that the same colouring solution would be obtained by coincidence. According to (2.3), the value of P^c decreases with an increase in effective stego-constraint edges (k) embedded into the CIG. The number of effective stego-constraint edges increases with the increasing entropy threshold, E^t, as shown in Tables 2.18A–D. Therefore to achieve stronger proof of ownership, a large value of entropy threshold is selected. The probability of

Table 2.18A Effect of entropy threshold value '$E^t = 4$' on probability of coincidence, P^c, of the steganography-embedded IP cores

DSP application	Number of storage variables	Number of registers before steganography	Number of constraints	P^c
FIR	30	8	20	6.9E−02
MPEG	41	14	21	2.1E−01
DCT	22	8	13	1.8E−01
EWF	35	7	12	1.6E−01
JPEG_IDCT	135	29	50	1.7E−01

Table 2.18B Effect of entropy threshold value '$E^t = 6$' on probability of coincidence, P^c, of the steganography-embedded IP cores

DSP application	Number of storage variables	Number of registers before steganography	Number of constraints	P^c
FIR	30	8	57	4.9E−04
MPEG	41	14	52	2.1E−02
DCT	22	8	24	4.1E−02
EWF	35	7	34	5.3E−03
JPEG_IDCT	135	29	203	8.1E−04

Table 2.18C Effect of entropy threshold value '$E^t = 8$' on probability of coincidence, P^c, of the steganography-embedded IP cores

DSP application	Number of storage variables	Number of registers before steganography	Number of constraints	P^c
FIR	30	8	57	4.9E−04
MPEG	41	14	59	1.3E−02
DCT	22	8	43	3.2E−03
EWF	35	7	57	1.5E−04
JPEG_IDCT	135	29	317	1.5E−05

Table 2.18D Effect of entropy threshold value '$E^t = 10$' on probability of coincidence, P^c, of the steganography-embedded IP cores

DSP application	Number of storage variables	Number of registers before steganography	Number of constraints	P^c
FIR	30	8	57	4.9E−04
MPEG	41	14	59	1.3E−02
DCT	22	8	43	3.2E−03
EWF	35	7	86	1.8E−06
JPEG_IDCT	135	29	355	3.9E−06

coincidence for different values of entropy thresholds (E^t = 4, 6, 8 and 10) are presented in Tables 2.18A–D. These tables show that the effective number of constraints (secret edges) embedded increases with the increase in entropy threshold. Hence, proof of ownership becomes stronger due to the decrease in probability of coincidence with an increase in entropy threshold.

2.8.1.1 Comparison with the contemporary approach (hardware watermarking) for DSP cores

The hardware steganography approach provides stronger proof of ownership compared to hardware watermarking approaches (Sengupta and Roy, 2017, 2018) in the context of DSP cores. This is because if the same number of constraints qualify to be added into the CIG for both steganography and watermarking, then all the qualified constraints are embedded in the case of steganography resulting in more effective secret constraints, while a smaller number of constraints are added in the case of watermarking. For different values of entropy thresholds, the number of effective constraints implanted for steganography and its equivalent constraints in the watermarking approach is shown in Tables 2.19A–D. As shown in the tables, the number of constraints added in the watermarking approaches are less because some of the constraints (edges) may already exist in the CIG by default. On the contrary, in the case of the steganography approach, none of the stego-constraints to be embedded exist by default in the design and are, thus, effectively added.

Table 2.19A Comparison of hardware steganography (Sengupta and Rathor, 2019) with hardware watermarking (Sengupta and Roy, 2017) in terms of the number of effective constraints imposed on the DSP core design for '$E^t = 3$'

DSP application	Number of storage variables	Number of effective constraints	Number of effective constraints imposed	
			Hardware steganography (Sengupta and Rathor, 2019)	Hardware watermarking (Sengupta and Roy, 2017)
FIR	30	17	17	1
MPEG	41	5	5	0
DCT	22	4	4	0
EWF	35	4	4	1
JPEG_IDCT	135	18	18	0

Table 2.19B Comparison of hardware steganography (Sengupta and Rathor, 2019) with hardware watermarking (Sengupta and Roy, 2017) in terms of the number of effective constraints imposed on the DSP core design for '$E^t = 4$'

DSP application	Number of storage variables	Number of effective constraints	Number of effective constraints imposed	
			Hardware steganography (Sengupta and Rathor, 2019)	Hardware watermarking (Sengupta and Roy, 2017)
FIR	30	20	20	1
MPEG	41	21	21	1
DCT	22	13	13	1
EWF	35	12	12	2
JPEG_IDCT	135	50	50	2

Table 2.19C Comparison of hardware steganography (Sengupta and Rathor, 2019) with hardware watermarking (Sengupta and Roy, 2017) in terms of the number of effective constraints imposed on the DSP core design for '$E^t = 5$'

DSP application	Number of storage variables	Number of effective constraints	Number of effective constraints imposed	
			Hardware steganography (Sengupta and Rathor, 2019)	Hardware watermarking (Sengupta and Roy, 2017)
FIR	30	57	57	9
MPEG	41	46	46	6
DCT	22	18	18	2
EWF	35	30	30	6
JPEG_IDCT	135	124	124	5

Table 2.19D Comparison of hardware steganography (Sengupta and Rathor, 2019) with hardware watermarking (Sengupta and Roy, 2017) in terms of the number of effective constraints imposed on the DSP core design for '$E^t = 6$'

DSP application	Number of storage variables	Number of effective constraints	Number of effective constraints imposed	
			Hardware steganography (Sengupta and Rathor, 2019)	Hardware watermarking (Sengupta and Roy, 2017)
FIR	30	57	57	9
MPEG	41	52	52	6
DCT	22	24	24	2
EWF	35	34	34	7
JPEG_IDCT	135	203	203	11

Therefore the hardware steganography approach provides stronger designer control over the secret constraint edges added into the design compared to hardware watermarking approaches. Further, due to achieving less probability of coincidence by embedding a higher number of constraint edges than used in watermarking approaches, the steganography approach provides stronger proof of ownership.

2.8.2 Design cost analysis

The design cost of a steganography-embedded IP core is evaluated using the following cost function:

$$C_d(R_i) = w_1 \frac{L_d}{L_{\max}} + w_2 \frac{A_d}{A_{\max}} \tag{2.4}$$

where $C_d(R_i)$ denotes the stego-embedded design cost with vendor-specified resource configuration, R_i. Here the design cost is evaluated in terms of the design latency and hardware area denoted by L_d and A_d, respectively. Further, $L_{\max}$ and $A_{\max}$ denote the maximum execution latency and hardware area, respectively, and w_1, w_2 represent the use-specified weights (both fixed at 0.5 to assign equal preference). The area and latency of an IP-core design are estimated using the NanGate library based on a 15 nm technology scale (http://www.nangate.com/?pageid=2328). The design cost of different DSP applications for a vendor-specified resource configuration is shown in Table 2.20. Further, the effect on register overheads for an increase in magnitude of entropy threshold, E^t, is shown in Table 2.21. For different values of E^t, the number of extra registers (R^e) required to embed stego-constraints is listed in the table. As shown in the table, for each DSP application, E^t is increased until further insertion of the stego-constraints is infeasible. Further, it is evident from the table that the extra register requirement for different threshold values is trivial. Hence the area/register overheads after implanting the hardware steganography is insignificant. The effect on the design

Table 2.20 *Design cost of hardware steganography for various resource configurations for (Sengupta and Rathor, 2019)*

DSP core	Resource configurations	L_d (ps)	A_d (μm)	Design cost
FIR	4A 4M	994	383.78	0.44437
MPEG	3A 7M	795	596.11	0.36169
DCT	1A 4M	928	327.156	0.47212
EWF	2A 1M	2,716	118.75	0.66231
JPEG_IDCT	12A 12M	1,988	1,155.27	0.33271

Table 2.21 *Effect of hardware steganography on register overheads with an increase in threshold entropy for (Sengupta and Rathor, 2019)*

DSP application	Number of storage variables	Number of registers before steganography	Value of threshold				
			Entropy threshold resulting into R^e more registers				
			$R^e = 0$	$R^e = 1$	$R^e = 2$	$R^e = 3$	$R^e = 4$
FIR	30	8	2–5	NA	NA	NA	NA
MPEG	41	14	2–7	NA	NA	NA	NA
DCT	22	8	2–6	7	NA	NA	NA
EWF	35	7	2–7	8	NA	9,10	11
JPEG_IDCT	135	29	2–9	NA	NA	NA	NA

Table 2.22A *Design cost of the IP core pre-steganography for (Sengupta and Rathor, 2019)*

DSP application	Number of registers	Cost
FIR	8	0.44437
MPEG	14	0.36169
DCT	8	0.47212
EWF	7	0.66231
JPEG_IDCT	29	0.33271

cost of IP cores before and after hardware steganography for $E^t = 4$ is presented in Tables 2.22A and B, respectively. As shown in the tables, design cost remains the same after the steganography is embedded for $E^t = 4$.

2.8.2.1 Comparison with the contemporary approach (hardware watermarking) for DSP cores

To compare the hardware area overheads of the two approaches, the number of extra registers required to add the same number of secret constraints edges is

Table 2.22B Design cost of the IP core post-steganography for $E^t = 4$ for (Sengupta and Rathor, 2019)

DSP application	Entropy threshold (E^t)	Number of registers	Cost
FIR	4	8	0.44437
MPEG	5	14	0.36169
DCT	5	8	0.47212
EWF	7	7	0.66231
JPEG_IDCT	6	29	0.33271

Table 2.23A Comparison of hardware steganography (Sengupta and Rathor, 2019) with hardware watermarking (Sengupta and Roy, 2017) in terms of the storage hardware for equivalent numbers of effective constraints for '$E^t = 3$'

DSP application	Number of registers before steganography	Number of effective constraints	Amount of storage hardware required after implanting equivalent number of constraints		Per cent reduction in registers
			Hardware steganography (Sengupta and Rathor, 2019)	Hardware watermarking (Sengupta and Roy, 2017)	
FIR	8	17	8	9	**11**
MPEG	14	5	14	14	–
DCT	8	4	8	8	–
EWF	7	4	7	7	–
JPEG_IDCT	29	18	29	29	–

compared. To do so, the number of constraint edges required in the case of hardware steganography is calculated for different entropy threshold values. Then, the same number of constraints is identified in the hardware watermarking approach (Sengupta and Roy, 2017, 2018) to make the comparison. While considering the same number of constraints to be embedded in both approaches, the number of extra registers required in the case of steganography is trivial compared to watermarking, as shown in Tables 2.23A–D. As shown in the tables, the per cent reduction in registers with respect to watermarking shows that steganography incurs nominal register overheads (that also occasionally) in comparison to watermarking. This is because even the same signature size (watermark strength) can have several signature-digit combinations (where each digit has its own encoded meaning), which results into different likelihoods of number of registers required to implement the signature in the design. This makes watermarking

Table 2.23B Comparison of hardware steganography (Sengupta and Rathor, 2019) with hardware watermarking (Sengupta and Roy, 2017) in terms of the storage hardware for equivalent numbers of effective constraints for '$E^t = 4$'

DSP application	Number of registers before steganography	Number of effective constraints	Amount of storage hardware required after implanting equivalent number of constraints		Per cent reduction in registers
			Hardware steganography (Sengupta and Rathor, 2019)	Hardware watermarking (Sengupta and Roy, 2017)	
FIR	8	20	8	9	**11**
MPEG	14	21	14	15	**6.6**
DCT	8	13	8	9	**11**
EWF	7	12	7	8	**12.5**
JPEG_IDCT	29	50	29	29	–

Table 2.23C Comparison of hardware steganography (Sengupta and Rathor, 2019) with hardware watermarking (Sengupta and Roy, 2017) in terms of the storage hardware for equivalent numbers of effective constraints for '$E^t = 5$'

DSP application	Number of registers before steganography	Number of effective constraints	Amount of storage hardware required after implanting equivalent number of constraints		Per cent reduction in registers
			Hardware steganography (Sengupta and Rathor, 2019)	Hardware watermarking (Sengupta and Roy, 2017)	
FIR	8	57	8	10	**20**
MPEG	14	46	14	15	**6.6**
DCT	8	18	8	10	**20**
EWF	7	30	7	8	**12.5**
JPEG_IDCT	29	124	29	30	**3.3**

uncertain (and variable) in terms of register overheads, unlike hardware steganography which is independent of signature encoding and combinations. Tables 2.23A–D report the per cent reduction in hardware register obtained through steganography after embedding the same number of edge constraints as watermarking.

Table 2.23D Comparison of hardware steganography (Sengupta and Rathor, 2019) with hardware watermarking (Sengupta and Roy, 2017) in terms of the storage hardware for equivalent numbers of effective constraints for '$E^t = 6$'

DSP application	Number of registers before steganography	Number of effective constraints	Amount of storage hardware required after implanting equivalent number of constraints		Per cent reduction in registers
			Hardware steganography (Sengupta and Rathor, 2019)	Hardware watermarking (Sengupta and Roy, 2017)	
FIR	8	57	8	10	**20**
MPEG	14	52	14	15	**6.6**
DCT	8	24	8	10	**20**
EWF	7	34	7	8	**12.5**
JPEG_IDCT	29	203	29	30	**3.3**

2.9 Conclusion

This chapter highlights the necessity for IP-core protection against various threats and presents a solution to achieve this by using forensic detective control through hardware steganography. Thus hardware steganography is a promising alternative to hardware watermarking for forensic detective control. Like watermarking, it is also employed during the hardware register allocation phase of ELS (or HLS). However, unlike watermarking, it is signature-free and provides more designer control. All the constraint edges eligible to be embedded in the steganography approach are determined using an entropy-threshold parameter. This is why the steganography approach provides full control to the designer over the amount of secret stego-constraints embedded into the design. Further, the steganography approach results in comprehensive forensic detective control while incurring nominal design overheads.

This chapter provides deep insight into hardware steganography and presents its advantages with respect to other contemporary approaches. At the end of this chapter, the reader is able to understand the following concepts:

- Utility of the hardware steganography.
- Application of the hardware steganography to protect IP-core designs against various hardware threats, such as IP piracy, counterfeiting, false claims of ownership, etc.
- Comparative study between hardware watermarking, hardware steganography and hardware cryptography based on various features.
- Hardware steganography encoder and decoder systems.

- Forensic detective control using hardware steganography to resolve the conflict of fraudulent claims of ownership and detect piracy, counterfeiting or cloning.
- Method for implanting hardware steganography is demonstrated using a sample DSP application.
- Design of hardware steganography embedded IP core is demonstrated using a real-life DSP core application used in glossy image compression.
- Performance and security features of hardware steganography.
- Security and design-cost analysis on case studies (different DSP applications).

2.10 Questions and Exercises

1. Compare hardware steganography with hardware watermarking in terms of embedding process.
2. What is hardware entropy?
3. What is threshold entropy?
4. Compare hardware steganography with hardware cryptography in terms of designer control.
5. What role does a multiplexing scheme play in designing a stego-embedded IP core?
6. What is the role of an edge set in steganography?
7. How is a qualified edge set determined?
8. What is the role of swapping pairs during the stego-constraint implanting process?
9. How does steganography provide tamper tolerance?
10. Explain the steganography detection process.
11. How is IP/IC cloning detected through steganography?
12. How is false claim of ownership detected through steganography?
13. Which approach incurs more implementation complexity during detection process, hardware steganography or hardware watermarking?
14. Explain how hardware steganography provides stronger robustness and designer control than hardware watermarking.
15. Perform scheduling of 8-point DCT using ASAP with four adders and four multipliers.
16. How would the performance of an 8-point DCT vary when the resources are reduced to one adder and four multipliers?
17. How would the performance vary when the resources are increased to four adders and five multipliers? What would be the effect on the overall design cost?
18. How is IP counterfeiting detected through hardware steganography?

References

E. Castillo, U. Meyer-Baese, A. Garcia, L. Parilla, and A. Lloris, "IPP@HDL: Efficient intellectual property protection scheme for IP cores," IEEE

Transactions on Very Large Scale Integration (VLSI) Systems, vol. 15, no. 5, Pages: 578–590, 2007.

R. S. Chakraborty and S. Bhunia, "HARPOON: An obfuscation-based SoC design methodology for hardware protection," IEEE Transactions on Computer-Aided Design of Integrated Circuits and Systems, vol. 28, no. 10, Pages: 1493–1502, 2009.

B. Colombier and L. Bossuet, "Survey of hardware protection of design data for integrated circuits and intellectual properties," IET Computers & Digital Techniques, vol. 8, no. 6, Pages: 274–287, 2015.

B. Le Gal and L. Bossuet, "Automatic low-cost IP watermarking technique based on output mark insertions," Design Automation for Embedded Systems, vol. 16, no. 2, Pages: 71–92, 2012.

M. C. McFarland, A. C. Parker, and R. Camposano, "Tutorial on high-level synthesis," in DAC '88 Proceedings of the 25th ACM/IEEE Design Automation, vol. 27, no. 1, Pages: 330–336, 1988.

A. Obukhov and A. Kharlamov, "Discrete Cosine Transform for 8×8 Blocks with CUDA," Nvidia whitepaper document, 2008.

S. M. Plaza and I. L. Markov, "Solving the third-shift problem in IC piracy with test-aware logic locking," IEEE Transactions on Computer-Aided Design of Integrated Circuits and Systems, vol. 34, no. 6, Pages: 961–971, 2015.

D. Roy and A. Sengupta, "Multilevel watermark for protecting DSP kernel in CE systems [hardware matters]," IEEE Consumer Electronics Magazine, vol. 8, no. 2, Pages: 100–102, 2019.

R. Schneiderman, "DSPs evolving in consumer electronics applications," IEEE Signal Processing Magazine, vol. 27, no. 3, Pages: 6–10, 2010.

A. Sengupta, "Intellectual Property Cores: Protection designs for CE products," IEEE Consumer Electronics Magazine, vol. 5, no. 1, Pages: 83–88, 2016.

A. Sengupta, "Hardware security of CE devices [hardware matters]," IEEE Consumer Electronics Magazine, vol. 6, no. 1, Pages: 130–133, 2017.

A. Sengupta and S. Bhadauria, "Exploring low cost optimal watermark for reusable IP cores during high level synthesis," IEEE Access, vol. 4, Pages: 2198–2215, 2016.

A. Sengupta and S. P. Mohanty, "Advanced Encryption Standard (AES) and Its Hardware Watermarking for Ownership Protection," IP Core Protection and Hardware-Assisted Security for Consumer Electronics, Pages: 317–335, 2019a, e-ISBN: 9781785618000.

A. Sengupta and S. P. Mohanty, "IP Core and Integrated Circuit Protection Using Robust Watermarking," IP Core Protection and Hardware-Assisted Security for Consumer Electronics, Pages: 123–170, 2019b, e-ISBN: 9781785618000.

A. Sengupta and S. P. Mohanty, "Trojan Security Aware DSP IP Core and Integrated Circuits," IP Core Protection and Hardware-Assisted Security for Consumer Electronics, Pages: 79–121, 2019c, e-ISBN: 9781785618000.

A. Sengupta and M. Rathor, "IP Core Steganography for Protecting DSP Kernels used in CE Systems," IEEE Transactions on Consumer Electronics, Accepted, Nov 2019.

A. Sengupta, G. Gupta and H. Jalan, "Hardware Steganography for IP Core Protection of Fault Secured DSP Cores," in Proceedings of 9th IEEE International Conference on Consumer Electronics (ICCE), Berlin, Sep 2019, IEEE. doi: 978-1-7281-2745-3/19/, available online at: https://edas.info/p25749.

A. Sengupta and D. Roy, "Antipiracy-aware IP chipset design for CE devices: A robust watermarking approach [hardware matters]," IEEE Consumer Electronics Magazine, vol. 6, no. 2, Pages: 118–124, 2017.

A. Sengupta and D. Roy, "Multi-Phase Watermark for IP Core Protection," in Proc. 36th IEEE International Conference on Consumer Electronics (ICCE), Pages: 1–3, 2018.

A. Sengupta, R. Sedaghat, and Z. Zeng, "A high level synthesis design flow with a novel approach for efficient design space exploration in case of multi-parametric optimization objective," Microelectronics Reliability, vol. 50, no. 3, Pages: 424–437, 2010.

A. Sengupta, D. Roy, and S. P. Mohanty, "Triple-phase watermarking for reusable IP core protection during architecture synthesis," IEEE Transactions on Computer-Aided Design of Integrated Circuits and Systems, vol. 37, no. 4, Pages: 742–755, 2018.

D. Ziener and J. Teich, "Power signature watermarking of IP cores for FPGAs," Journal of Signal Processing Systems, vol. 51, no. 1, Pages: 123–136, 2008.

Chapter 3

Forensic detective control using a digital signature–based watermark for IP core protection

Anirban Sengupta[1]

This chapter discusses forensic detective control using a digital signature–based watermark for intellectual property (IP) cores or integrated circuits (ICs). The chapter provides detailed insight on digital-signature generation and the embedding process that includes a discussion on signature-generation rules and algorithms, IP core protection using forensic detective control, methodology of implanting digital signature–based watermark constraints covertly into an IP core design and its respective design process.

The chapter is organized as follows: Section 3.1 introduces the need of digital signature–based watermarks; Section 3.2 discusses the threat model for which digital signature is applicable; Section 3.3 presents a comparative study on contemporary approaches; Section 3.4 explains security modules employed in the digital signature–generation process; Section 3.5 discusses forensic detective control using digital signature–based watermarks for digital signal processing (DSP) cores; Section 3.6 presents a case study of the design process of a digital signature–embedded DSP core; Section 3.7 discusses desirable security properties of digital signature–based watermarks; Section 3.8 discusses countermeasures using a digital signature against various threat scenarios; Section 3.9 presents an analysis and comparison of digital signature–based watermarking with existing watermarking for different DSP–IP cores; finally, Section 3.10 concludes the chapter.

3.1 Introduction

In the current design era, a single design house alone is not responsible for executing all the stages of an IC design flow, i.e., various stages of an IC design cycle from its system-level specification and description to an end-functional IC, are implemented at different design houses and/or foundries. The primary reason behind the globalization of the IC design cycle is the exponential rise in the cost of

[1]Computer Science and Engineering, Indian Institute of Technology Indore, Indore, India

manufacturing facilities. Additionally, the pressure of 'time to marketing' has forced designers or manufacturers to adopt this design paradigm. The design house that is responsible to produce the design of an IP core is called a fabless design house. However, because of time and economy constraints, it becomes impractical to realize their IP core into an end-fabricated IC all by themselves. Thus they transfer it to a production or manufacturing house, which are referred as foundries. Further, system-on-chip (SoC) integrators also buy the IP cores from different fabless design houses or third-party IP vendor to generate a complex electronic system. Thereby, in both the scenarios, the true IP owner or designer essentially sends the design to an SoC integrator or IC manufacture (Colombier and Bossuet, 2015).

However, the SoC/manufacturing design houses (remote foundries) may flout the genuine terms of using an IP core. Therefore if IP cores designs are unsecured, then an adversary present in the remote foundries may exploit their vulnerabilities and can generate pirated, counterfeited or cloned copies of the design illegally and may also claim ownership fraudulently. Hence robust protection during the design process is crucial in order to keep an IP core secured in the untrusted regimes (such as a foundry) (Plaza and Markov, 2015; Sengupta, 2016, 2017).

Hardware watermarking is a well-known approach used to protect the designs against hardware threats such as piracy, counterfeiting, etc. In this approach, the owners implant their signature as secret marks in the design to fortify the IP core against forgery and infringements (Sengupta and Bhadauria, 2016; Sengupta and Mohanty, 2019a, 2019b). The embedded, secret, watermark plays a crucial role in case of conflict of ownership to render the false claim void as well as to detect cloned or counterfeited ICs. The watermark can be embedded at various abstraction levels of the design flow. At a lower abstraction level, such as the gate level, Ziener and Teich (2008) embedded a signature using netlist and bitstream of the design. Further, Le Gal and Bossuet (2012) embedded a watermark during the in-synthesis process. At a higher abstraction level, such as the electronic system level (ESL), Koushanfar *et al.* (2005) implanted a dynamic watermark based on artificial design and timing constraints. Sengupta and Roy (2017) implanted a multivariable signature, encoded with pre-defined rules, during the register allocation phase of ESL. Further, Roy and Sengupta (2017) embedded a vendor watermark and a buyer fingerprint simultaneously in the IP core design to provide symmetrical protection. Furthermore, Sengupta and Roy (2018) and Sengupta *et al.* (2018) implanted watermarks simultaneously in three phases, viz. scheduling, hardware allocation and register allocation of ESL design. Besides these approaches that implanted the watermark at a specific abstraction level, Roy and Sengupta (2019) implanted a watermark during register transfer level (RTL) and ESL synthesis process (Sengupta *et al.*, 2010) of the design flow. This type of watermark is referred as a multi-level watermark.

However, in the hardware watermarking approaches, to choose an appropriate signature is still a challenge to the designer. Further, the security and design overheads that result through the watermark is not under the full control of the designer. Additionally, if the signature or watermark is leaked to adversaries, then they could claim ownership of the design fraudulently. In such cases of ownership conflict, the real owner may have to compromise with their ownership right of the IP core. These limitations encourage the designer/owner to produce a more definite proof-of-signature through some sophisticated algorithm. Sengupta *et al.* (2019) proposed a digital signature–based watermark, driven through cryptographic hashing and encryption. Cryptographic hashing is employed through secure hash algorithms (SHA-512) and private-key encryption is performed through RSA. Watermarking based on a digital signature provides more meaningful proof of ownership. This is because here the signature is not chosen arbitrarily, but is generated in more scientific way through some encoding rules, hashing and encryption. Hence in case of ownership conflict, the owner is not only capable to extract their digital signature embedded into the design, but is also efficiently able to prove the digital signature as their own in a more meaningful way. Thus a signature leaked (compromised) to an adversary does not help as he/she is not able to meaningfully or scientifically prove his/her rights to the digital signature as he/she is unaware of the encoding rules, hash digest and encryption that are only known to the true vendor or designer.

The strength of a digital signature is also assessed based on the probability of coincidence (P^c) metric. A digital signature is stronger if the probability of coincidence is lower. Digital signature–based watermarking proposed by Sengupta *et al.* (2019) achieves lower P^c with reduced area overheads compared to other contemporary watermarking approaches. Further, the length of the digital signature is scalable with respect to the size of the IP core design or DSP application.

Sengupta *et al.* (2019) employed digital signature–based watermarking to the highly complex DSP kernels designs during the ESL synthesis process. To employ the digital signature essentially at ESL is very significant because complex DSP kernels are not easy to describe directly at lower abstraction levels such as RTL, gate level, etc. Additionally, the role of DSP kernels–based IP cores cannot be undervalued in the current design era of highly compact and complex consumer electronics (CE) systems. In CE products, such as digital cameras, televisions, tablets and laptops, DSP kernels are extensively used. A diagrammatic representation of secured IP cores used in CE products is shown in Figure 3.1. As shown in the figure, attackers present in the IP/IC design chain may attack the design and generate illegal copies without authorization. These illegal copies of original IP core designs may be integrated into SoCs used in CE products. However, vendors' digital signature embedded into the IP core is capable to thwart illegal uses of the original design through forensic detective control.

3.2 Threat models of an IP core

An IP core may be misused while it is sent to other design houses for integration into an SoC or for the fabrication of a standalone IC. This is because the untrustworthy design houses may violate the original terms of use to generate illegal

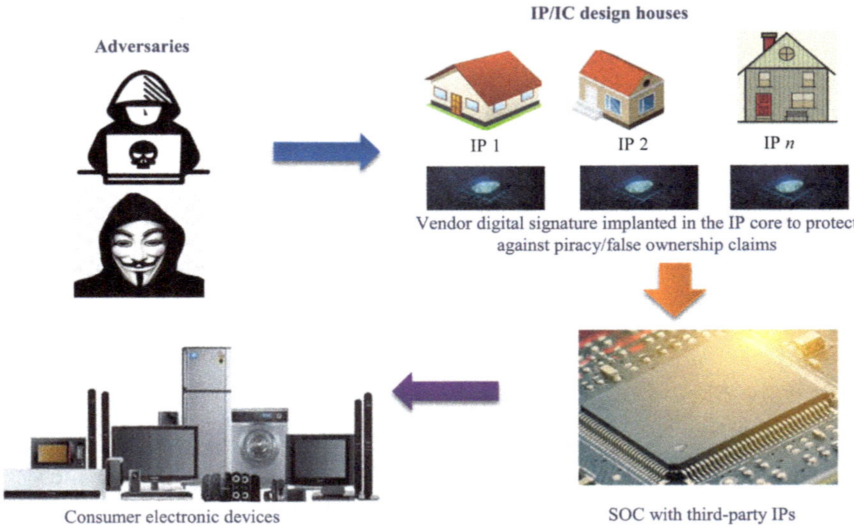

Figure 3.1 IP core protection using a digital signature

revenue. Therefore it becomes essential for the IP owner to estimate all the possible threats in advance. To do so, he/she needs to analyse a potential attacker's profile in a very sophisticated way in order to identify all the potential threats to an IP design. The IP core designs are susceptible to a number of potential threats, which can be thwarted by digital signature–based watermarking:

1. IP/IC piracy
2. IP/IC counterfeiting and cloning
3. Fraudulent claim of ownership

3.3 Selected contemporary approaches

This section discusses prior works on IP core protection using watermarking and the limitations of some of these approaches in the light of the digital signature–based approach.

3.3.1 Prior work on IP core protection using watermarking

This section discusses contemporary, hardware-watermarking approaches for DSP. Koushanfar *et al.* (2005) presented a behavioural synthesis technique for IP core protection using dynamic watermarking based on the concept of embedding artificial design and timing constraints in the design to encode the author's signature. The strength of the approach is low-design overheads, high resiliency and low-tamper probability. The authors implanted artificial design constraints into the structural properties of the IP core architecture. It also leverages a key design property called 'design space exploration' that selects a solution among numerous others competitive design variances. The approach chooses a solution to encode the author's signature at low-design cost without compromising correct functionality. The approach is based on the binary encoding of the authors signature where bit '1' is added in the interval graph of the design, such that the terminal node represents the bit value. To embed 1, the terminal node of the added edge should have an odd number. To embed 0, the terminal node of the added edge should have an even number. The approach uses the ASCII-based encoding for selecting the author's signature. The watermark presented by the author successfully satisfies the following properties: (1) low overheads, (2) strong tamper-tolerance, (3) high robustness and (4) low signature–detection time.

Sengupta *et al.* (2018) presented a triple-phase watermarking methodology that leverages a high-level synthesis (HLS) framework for implanting the vendor's signature into DSP cores. The approach utilizes seven variable encoding schemes based on (γ, α, β, i, I, T, $!$). Variable γ is responsible for encoding vendor information during the scheduling phase of HLS, variables α and β are responsible for encoding vendor information in the hardware allocation phase of HLS, while variables i, I, T and $!$ are responsible for encoding vendor information in the register allocation phase of HLS. Variable γ embeds the information in the scheduling phase by manoeuvring the operations of the non-critical path based on the mobility factor. Variables α and β embed the information in the hardware-allocation phase by re-assigning functional units based on an encoding rule and variables i, I, T and $!$ embed vendor information in the register-allocation phase by reassignment of registers in the coloured interval graph (CIG). An overview of the approach is shown in Figure 3.2. The approach is more tamper-tolerant and robust than other similar approaches, as well as satisfying major desirable properties of watermarks, such as low-embedding cost, low-probability of coincidence and low-watermark creation and detection time.

Le Gal and Bossuet (2012) present a watermarking methodology that uses the concept of 'temporally free slots' of an IP core design execution cycle. The approach generates an IP watermark that is a set of mathematical relations between the IP input data, the initial values of the internal computation and the IP output. The watermark generated in the form of sub-marks by the approach is invisible to an IP buyer, the IP integrator and an IP user. The sub-marks appear like transient output data and incurs low area and timing overheads. A sub-mark is represented as: $\{(M_1(n), C_1), \ldots, (M_p(n), C_p)\}$, where $M_i(n)$ indicates ith sub-mark or mathematical function and C_i indicates the clock-cycle number during which the sub-mark is sent to a 'temporally free' IP output port. The approach presented two types of watermarking: low-cost watermark and a costless watermark based on the level of protection required and the watermark cost permitted. An overview of this approach is shown in Figure 3.3.

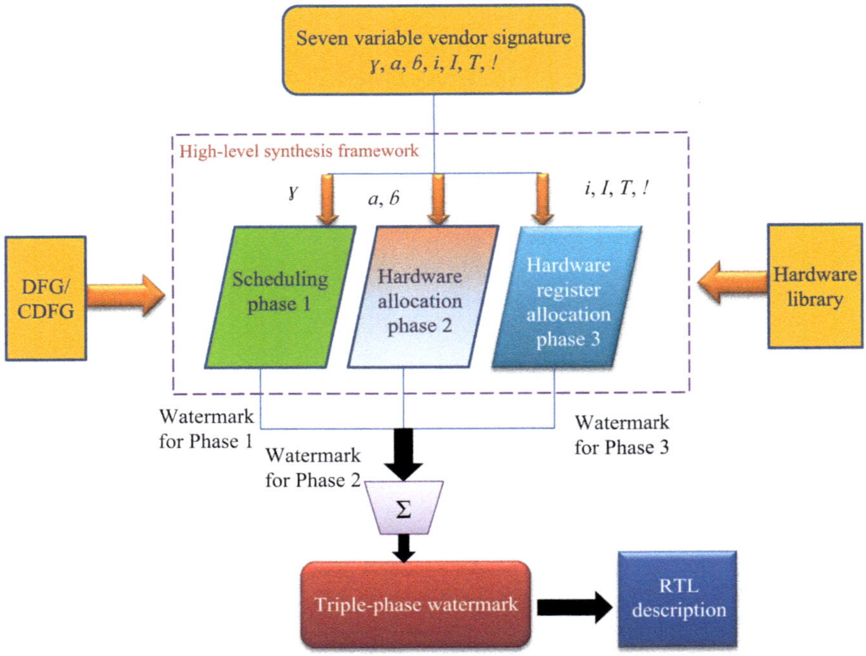

Figure 3.2 High level overview of triple-phase watermarking (Sengupta et al., 2018)

3.3.2 Limitations of contemporary hardware watermarking approaches and advantages of digital signature–based watermarking

Digital signature–based watermarking overcomes the limitations of contemporary watermarking approaches. The abstract level difference between digital signature–based watermarking and the contemporary watermarking approach is depicted in Figure 3.4. Further, the advantages of digital signature–based watermarking are discussed as follows (Sengupta *et al.*, 2019a):

1. In contemporary watermarking approaches, choosing an appropriate signature is very hard and time-consuming. This is because a large number of combinations of the same-sized signature is possible. Further, each digit of the signature can be encoded in several ways. Additionally, the impact of various combinations on the security and design cost is diverse and unpredictable. Therefore it is not easy to determine an optimal signature resulting into higher security and lower-design cost. The digital signature–based watermarking eliminates these limitations by producing a unique and definite signature through a set of scientific algorithms. The uniqueness of the digital signature is ensured by producing it through cryptographic hashing and encryption. Thus the possibility of different signature

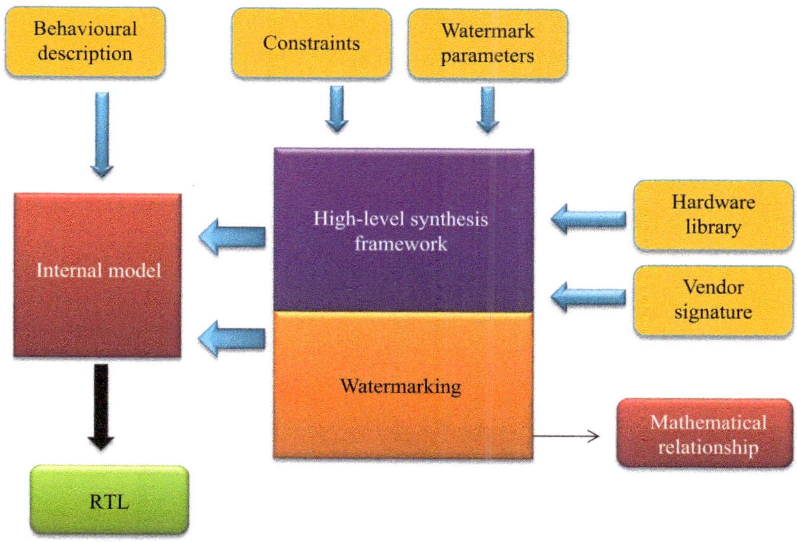

Figure 3.3 High-level overview of watermarking process of Le Gal and Bossuet (2012)

combinations, like in the case of contemporary watermarking, is eliminated. More explicitly, there is always a unique digital signature corresponding to an IP core design. This is because the IP core design to be embedded with a digital signature is first encoded into a bitstream which is converted to a hash digest of fixed length through a cryptographic hash algorithm (SHA-512). Since the probability of two distinct bitstreams having the same hash digest is almost nil, the same hash digest of an embedded digital signature cannot be produced with any other bitstream. Thus, the bitstream corresponding to the IP core design generates a unique hash digest and, hence, a unique digital signature.

2. The strength of contemporary watermarking approaches abides with the secrecy of signature encoding. If the signature is leaked or known to an adversary, then the ownership may be claimed fraudulently and proving ownership by the genuine owner can become challenging. In this case, ownership conflict is difficult to resolve because the original designer and adversary can extract the signature from the design. However, in case of digital signature–based watermarking, the real owner is comprehensively capable to prove his/her rightful ownership even if the signature is compromised. This is because the digital signature is generated through an encoding rule, followed by digest generation using a hashing algorithm and RSA, private, key-based encryption which are only known to the vendor. Hence, the genuine designer (vendor) can prove his/her right over the digital signature in the design in a more meaningful, scientific and definite way. On the contrary, the fraudulent owner, being unaware of the digital signature generation algorithm, fails to attest his/her right over the digital signature.

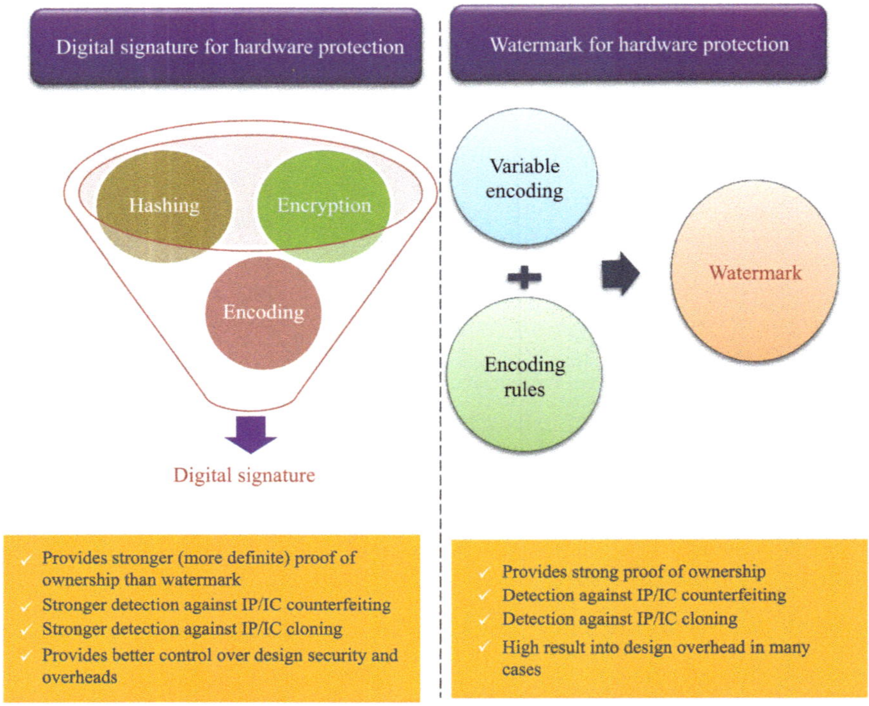

Figure 3.4 Hardware-based digital signature vs. hardware watermarking for IP core protection

3. In the case of contemporary watermarking approaches, the designer/owner signature needs to be remembered or stored safely for future reference. Unfortunately, if the designer loses or mislays his/her signature, then he/she becomes unable to authenticate his/her IP core designs. On the contrary, a digital signature is not required to be remembered. During forensic detection, the same embedded digital signature can easily be generated through pre-defined encoding rules, cryptographic hashing and RSA encryption.
4. Contemporary watermarking approaches may result into significant design overheads and lower strength of security depending on the nature of the signature. On the other hand, digital signature–based watermarking achieves low-design overheads and provides stronger proof of ownership. Further, due to the nature of the encoding rule, the size of the digital signature scales well in accordance with the size of the IP core design (number of operations) and, hence, security strength is not compromised. Further, the security features of SHA-512, such as uniformity and collision resistance, render the digital signature more robust.

3.4 Security modules employed in digital signature–based watermarks

In the process of creating a digital signature–based watermark, two primary security modules are employed (Sengupta *et al.*, 2019), viz. (1) crypto-hash function – SHA-512 (2) and crypto-encryption function – RSA. The high-level process of creating a digital signature for IP cores using these two modules is shown in Figure 3.5. As shown in the figure, the information of the IP/IC vendor's secret mark in the form of bitstream is fed to the crypto-hash function block (SHA-512), which produces a 512-bit hash digest. Further, this hash digest is encrypted through an RSA encryption block using the IP/IC vendor's secret key. This encrypted bitstream followed by a pre-defined encoding is used to form a digital signature of the IP/IC vendor. A brief description of these two major security modules entails (Sengupta *et al.*, 2019).

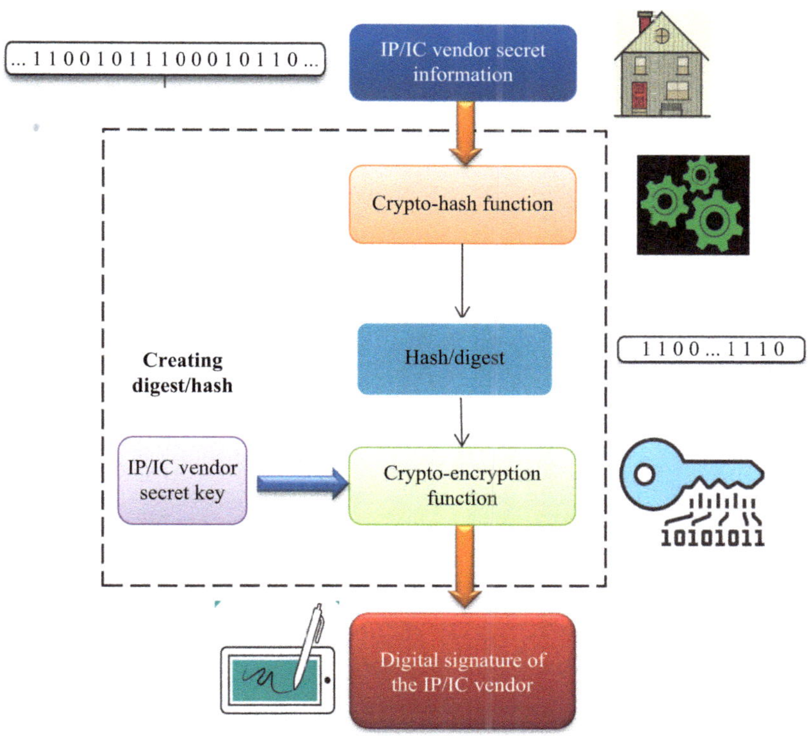

Figure 3.5 High-level process of creating a digital signature for IP cores

3.4.1 Crypto-hash function – SHA-512

SHA-512 is used as a crypto-hash function to create the hash digest of the vendor secret mark or watermark. Figure 3.6 shows the detailed process of creating a hash digest. As shown in the figure, initial information (in the form of bitstream generated through a pre-defined encoding rule) corresponding to the IP/IC vendor's watermark is fed as the primary input to the crypto-hash function. This m-bit of bitstream representing the vendor's secret mark information is first converted into 1,024 bits, which is further transformed into a 512-bit hash digest. The reason behind employing this crypto-hash function is as follows: the generated hash digest is always unique corresponding to the initial vendor secret mark, thus producing a unique digital signature. The collision resistance feature (that indicates it is very difficult to find two distinct inputs that produce same hash value through the same hash function) of SHA-512 ensures the uniqueness property of the digital signature. Additionally, the generated hash digest is always deterministic, i.e., it always produces same hash value for a given input. Thus by using the initial vendor secret mark information, the same digital signature can definitely be reproduced by the vendor during forensic detection analysis.

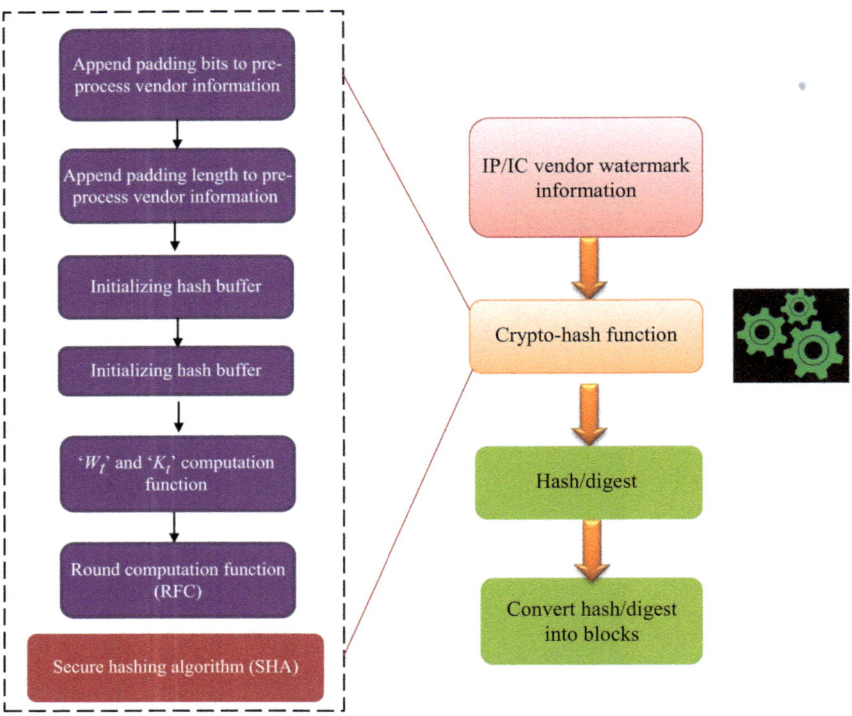

Figure 3.6 Detailed process of creating a hash-value/digest of the secret vendor mark

Further, as shown in Figure 3.6, the initial vendor secret mark information is processed through several steps to generate the final hash digest. The crypto hash-function transforms the input of arbitrarily length into a hash digest of fixed length (512 bits). However, the arbitrarily length input is first pre-processed to produce a 1,024-bit block that is transformed into the hash digest using a set of hash functions executed in 80 rounds. In the first step, an m-bit long bitstream representing the initial vendor secret mark information is padded with a fixed bit pattern ('1' followed by sequence of '000 . . . ') to form a block of length of 896-bit. In the second step, the 128-bit representation of the initial input data length is appended to the 896-bits (obtained from previous step) to form the 1,024-bit block. In the third step, a total of eight hash buffers (a, b, c, d, e, f, g and h) are initialized to be used by the crypto-hash function. The size of each hash buffer value is 64 bits. The next step is to extract/compute a 64-bit word (W_t) from the 1,024-bit block and set a predefined 64-bit constant (K_t) to be used in the further steps. The computation process of W_t generates a new W_t for each round/iteration. In the next step, round function computation is performed using the hash-buffer value, W_t and K_t. The computation process performed by the round function is depicted in Figure 3.7. This function is iterated for 80 rounds and, in each round, all hash buffers are updated. After executing all the rounds, the output of the final round (final updated hash-buffer values) and the input of the first round are added to obtain the hash digest of the bitstream. In this way, the crypto-hash function SHA-512 is employed to generate a hash value ensuring the uniqueness and definiteness of the digital signature.

3.4.2 Crypto-encryption function – RSA

RSA is an asymmetric key encryption algorithm in which two distinct keys (private key and public key) are involved in the cryptography. Sengupta *et al.* (2019a) employed an RSA encryption to generate the digital signature of initial vendor secret mark information corresponding to a DSP core design. The first employed security module (SHA-512) ensured the uniqueness of the vendor secret mark information by generating a unique hash digest. Now, RSA encryption is used to sign the hash digest of the vendor secret mark information to ensure the non-repudiation, i.e., it can be proved using a digital signature that the secret mark essentially belongs to its genuine owner and the ownership cannot be claimed by a fraudulent owner. More explicitly, signing the vendor-specific, generated-hash information using RSA encryption enables the genuine owner to verify the origin of the secret mark in a meaningful way. This produces the final digital signature–based watermark. Thus the generated digital signature followed by another encoding is used for embedding the watermark into an IP core design. Thus RSA encryption ensures that if the embedded watermark is leaked to an adversary, he/she cannot prove his/her right over the signature during ownership conflict, whereas the genuine owner can explicitly show that the digital signature belongs to him/her through scientific, mathematical and/or logical evidence.

Figure 3.8 illustrates the detailed process of creating a digital signature from the IP vendor watermark. First, a private–public key pair of the IP owner's choice is

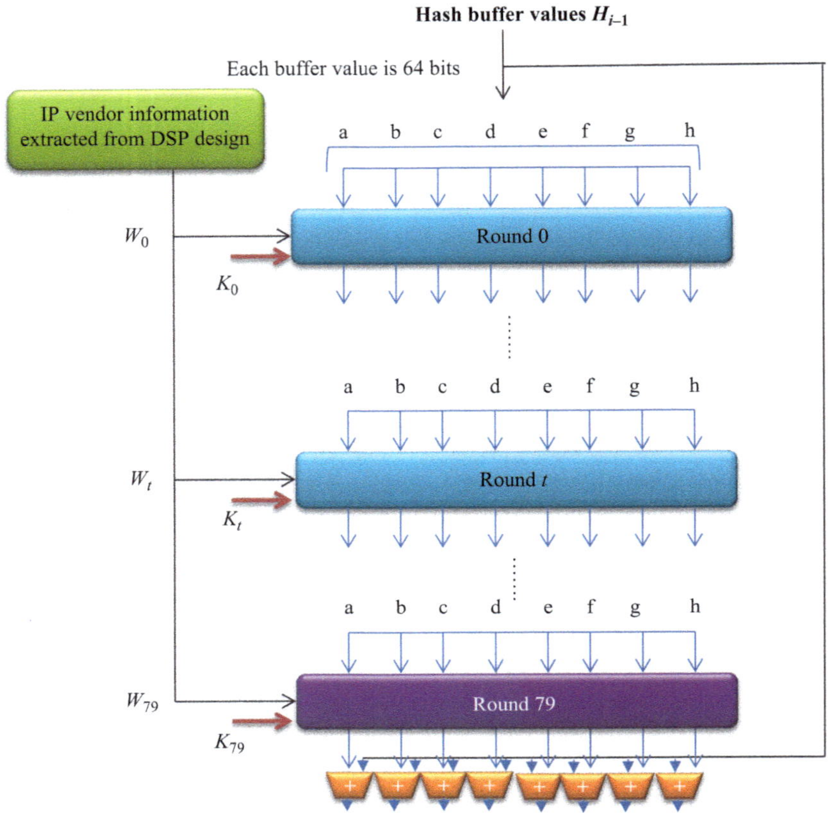

Figure 3.7 Round computation function in SHA-512

generated in an RSA module. Thereafter, the hash digest is encrypted block-wise using the private key. *It is interesting to note here that the private key is used in the RSA encryption instead of the public key for digital signature generation process. This is a standard practice employed in digital-certification applications*. The following modulo arithmetic relation is used to encrypt the hash digest of initial vendor secret mark information:

$$C = M^d \bmod n$$

where C denotes cipher-data (encrypted), M denotes the one block of hash digest, d denotes the owner private key and n represents the multiplication of two distinct prime numbers, i.e., $n = p_1 \times p_2$ and $p_1 \neq p_2$.

All the blocks of the hash digest are encrypted, subsequently, through the same private key. Thereafter, equivalent binary representation of each encrypted block is concatenated to produce the final encrypted bitstream referred to as the digital

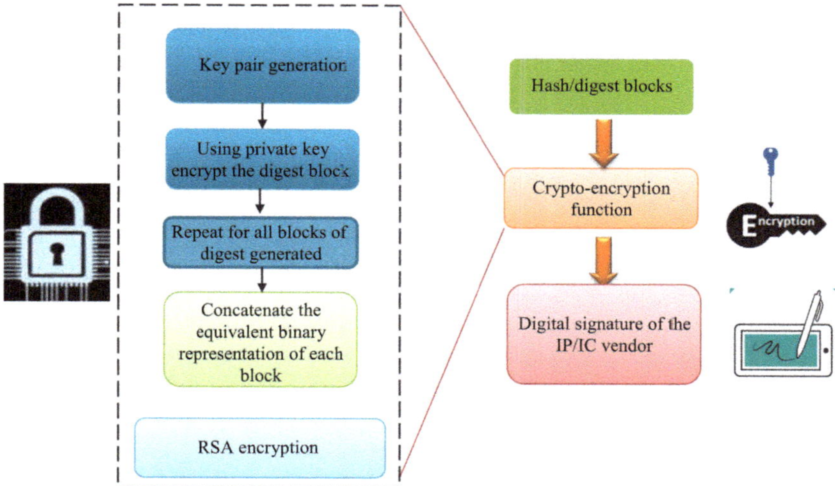

Figure 3.8 Detailed process of creating a digital signature of the IP vendor watermark

signature. In this way, initial information of the owner's watermark is signed using private key RSA encryption to preserve the rightful ownership.

3.5 Forensic detective control using a digital signature–based watermark

This section describes the process of embedding a digital signature as a watermark into the IP core design followed by the signature-detection process during forensic detection. The overview of the digital signature generation and embedding process is shown in Figure 3.9. As shown in the figure, a digital signature corresponding to the DSP–IP core is generated through some processing phases, encodings and cryptographic algorithms (hashing through SAH-512 and encryption using RSA). Consequently, it is embedded into the design through a watermark implanting process. In the process of generating a digital signature, the first step is the pre-processing of the IP core design to be fortified with the watermark. The data flow graph (DFG) representing an DSP–IP core design and user specified resource configuration are fed as input to this pre-processing phase 1, which generates the scheduled DFG (SDFG). In the next step, this SDFG is encoded into a bitstream representing the initial vendor's secret mark information. This obtained bitstream is assigned to the input of the SHA-512 algorithm, which generates the hash digest of the initial secret mark information. Thereafter, this hash digest is processed to translate the bitstream into a compatible input to RSA. The RSA encryption algorithm encrypts the decimal value of an 8-bit data block of hash digest at a time

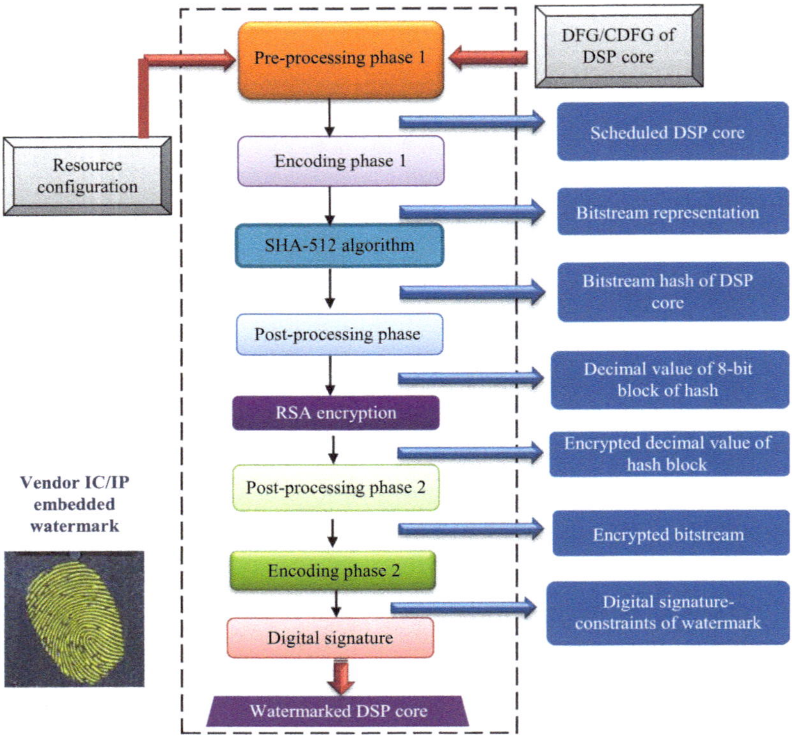

Figure 3.9 Digital signature as a watermark embedding process in a DSP-based IP core (Sengupta et al., 2019)

using the same vendor private key. Once all the decimal values representing the hash digest are encrypted one-by-one, they are further processed in the post-processing phase 2 to generate the encrypted bitstream. This encrypted bitstream is the digital signature that is encoded for embedding into the respective IP core design. This encoding represents the digital signature-constraints embedded in the form of a watermark in the IP core.

The detailed process of embedding a digital signature–based watermark is described in Section 3.5.1.

3.5.1 Digital signature framework

The complete flow of the process of generating a digital signature corresponding to the design of an IP vendor/owner is shown in Figure 3.10. As shown in the figure, the overall process is accomplished through several phases as illustrated in Figure 3.10 (Sengupta *et al.*, 2019).

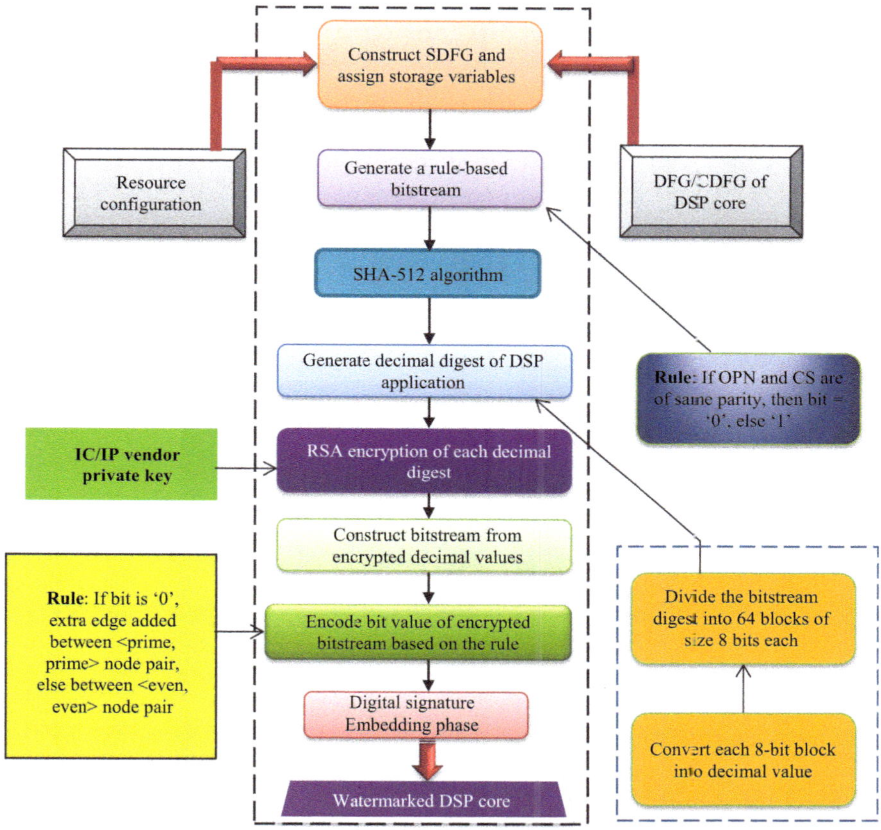

Figure 3.10 Detailed illustration of a digital signature as a watermark embedding process using encoding (Sengupta et al., 2019)

3.5.1.1 Equivalent bitstream generation of an IP core

The bitstream representing the IP core design is obtained through two steps. First, the DFG representing the IP core design is scheduled based on a LIST (or ASAP) scheduling algorithm. This obtained SDFG essentially complies with the user specified resource configuration (R_i), i.e., the assignment of a particular operation to a specific control step essentially depends on the vendor-specified resource configuration. Next, the SDFG is encoded into a bitstream using the encoding rules presented in Table 3.1. As shown in the table, if an operation falls under the control step having the same parity (i.e., the operation number and corresponding control-step number are either both odd or both even), then the operation is encoded into bit '0'. Similarly, if an operation falls under the control step bearing a different parity then the operation is encoded into bit '1'. This encoded bitstream of the IP core design represents the initial vendor's secret mark, which becomes the primary input to the hashing block.

Table 3.1 Encoding rules of a vendor's IP core

Operation number (OPN)	**Corresponding control step (CS) number**	**Encoded bit**
Even	Even	0
Odd	Even	1
Even	Odd	1
Odd	Odd	0

3.5.1.2 Hash-digest generation

In this step, the generated bitstream of an IP core design is transformed into a hash value of fixed length through a standard, secure hashing algorithm (SHA-512). The arbitrarily length bitstream is first converted into a 1,024 bit block through padding followed by appending a certain bit pattern. Further, this 1,024-bit block is transformed into the 512-bit hash digest through the execution of 80 iterations of 'round function'. The more detailed discussion on hash-digest generation process has been carried out under the description of crypto hash function in Section 3.4. Further, as discussed earlier, the collision resistance, deterministic and uniformity properties of SHA-512 ensure that the generated hash digest carrying the vendor's secret mark is unique for an IP core design and the same can always be regenerated during forensic detection analysis.

3.5.1.3 Conversion of hash digest into RSA compatible input

The RSA encryption algorithm encrypts the decimal equivalent of an 8-bit hash-block value using an IP vendor private key. Therefore a sort of post-processing of hash-bitstream is required in order to encrypt it using an RSA algorithm. To do so, the 512-bit hash-digest is divided into 64 chunks/blocks of 8-bit each. Further, each 8-bit hash block is translated into its decimal equivalent. Thus, the hash-bitstream is converted into an RSA compatible input.

3.5.1.4 RSA encryption of hash-bitstream

All the decimal values representing the hash-bitstream of the vendor's secret mark are encrypted one-by-one using the vendor's private key by the following RSA encryption function: $C = M^d \bmod n$. Here, M is the decimal value representing 8-bit block of hash-bitstream, C is the encrypted decimal value and d is the vendor's private key generated through the RSA key-generator. This RSA private key is only known to the IP vendor/owner. Once all the 64 decimal values representing the hash-bitstream are encrypted, they are passed for further processing to obtain the final encrypted bitstream of the digital signature (representing a vendor watermark).

3.5.1.5 Formation of digital signature representing a vendor watermark

Post-processing is performed over the encrypted decimal values obtained from RSA to construct an encrypted hash-bitstream or digital signature. To do so, each

Table 3.2 Encoding rules of digital signature corresponding to an IP core

A bit of digital signature	Encoding rule
0	An extra edge is embedded between a node pair (prime, prime) in the coloured interval graph (CIG)
1	An extra edge is embedded between a node pair (even, even) in the CIG

encrypted decimal value is converted into its binary equivalent. Thereafter, all binary values are concatenated to form an encrypted hash-bitstream or digital signature. This encrypted hash-bitstream is the digital signature representing the vendor's watermark corresponding to the IP core design.

3.5.1.6 Digital signature encoding

Once the digital signature is generated, it needs to be encoded into a form that represents the artificial design constraints for embedding into the corresponding IP core design. Since the digital signature is an encrypted binary bitstream, therefore each type of bit can be encoded in the form of constraints to be embedded covertly into the design. The encoding of a digital signature for conversion into its equivalent artificial constraints is performed according to the rules presented in Table 3.2. In this way, the digital signature is mapped to hidden constraints to embed the watermark.

3.5.1.7 Digital signature embedding process

As shown in Table 3.2, encoding a digital signature transforms its bits into possible extra edges (artificial secret constraints) of the CIG of an IP core design. The total number of colours used to create the CIG indicates the total number of registers required to store all primary and intermediate inputs/outputs of the IP core design. The storage variables used to accommodate all primary and intermediate inputs/outputs are represented as nodes in the CIG. Overlap in the lifetime of two storage variables is indicated by an edge in the CIG. Therefore, an edge between two nodes of the same colour can never exist in the CIG.

A digital signature encoded according to Table 3.2 is embedded into the CIG in the form of hidden constraints-edges. After embedding the generated digital signature into an IP core design, the register allocation of storage variables is modified to accommodate the presence of watermark at the RT level. Further, changes in the multiplexers input connectivity in the RTL data path of the design show the impact or presence of a digital signature–based watermark.

The process of embedding a digital signature–based watermark can be summarized in following steps:

1. The DFG of the DSP IP core is scheduled based on user-provided hardware resource configuration.
2. In the SDFG, storage variables are assigned to each primary and intermediate input/output.

3. Representing all storage variables as nodes, a CIG is created to find the minimum number of registers required to accommodate all primary and intermediate input/output.
4. A register allocation table is constructed to show the assignment of each storage variable to a particular register.
5. Constraints-edges equivalent to each digit of the digital signature (obtained through the encoding rule presented in Table 3.2) are listed.
6. All the constraints-edges representing the secret vendor digital signature are added into the CIG representing watermark.
7. The register allocation table is modified after embedding the digital signature–based watermark constraints into the CIG.

3.5.2 Detection process of digital signature–based watermarks

The detection of a digital signature is crucial to resolve ownership conflict and as well as to detect IP/IC piracy, counterfeiting and cloning. The detection process of digital signature–based watermarks is shown in Figure 3.11. As shown in the

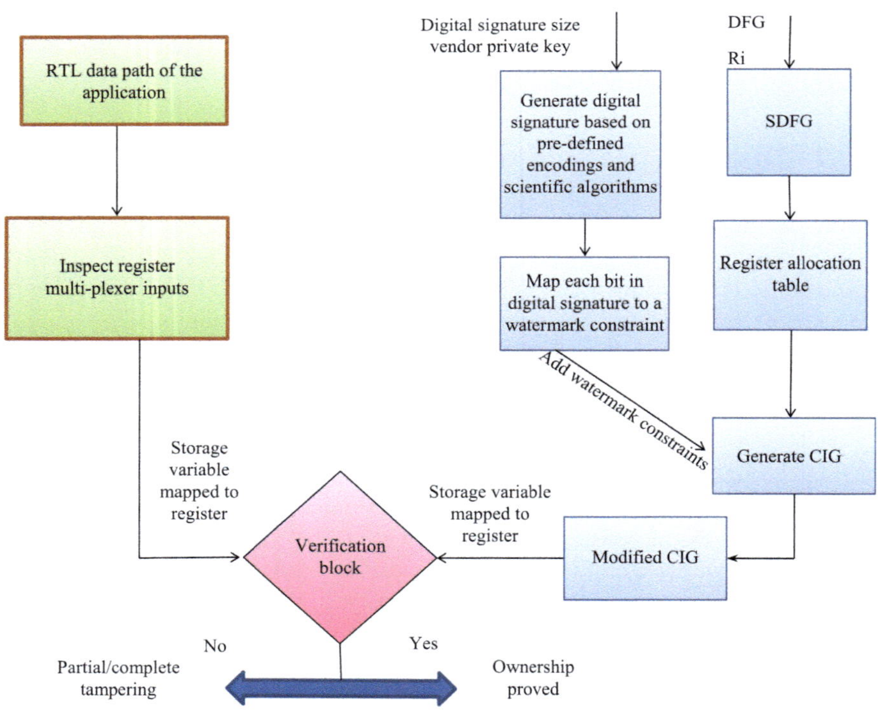

Figure 3.11 Digital signature detection process (Sengupta et al., 2019)

figure, the signature detection/verification process can be realized through following three sub-processes (Sengupta *et al.*, 2019a).

3.5.2.1 Signature-constraints' generation and mapping of storage variables to a register

As shown in the right half of Figure 3.11, the signature corresponding to the IP core design under ownership test is generated using pre-defined encodings and scientific algorithms (SHA-512, RSA encryption). The genuine IP owner can easily generate a digital signature corresponding to his/her IP core design as he/she knows his/her vendor's private key and the algorithm of generation. Thereafter, each digit of digital signature is encoded into constraints-edges. These watermark constraints are added into the CIG corresponding to the same IP core design under ownership test. This modifies the register allocation of storage variables. The modified mapping of the storage variable is recorded for verification.

3.5.2.2 Signature inspection

As shown in the left half of the figure, the RTL data path of the IP core design under test is inspected to collect the information of storage variables. To do so, inputs of all muxes associated to each register are inspected. This inspection provides the information for register allocation of all storage variables.

3.5.2.3 Signature verification

Storage variables mapping to registers obtained from the aforementioned subprocesses are fed into the verification block as shown in Figure 3.11. This block checks whether the storage variables mapping to registers obtained from signature inspection matches those generated through the implantation of digital signature constraints. If mappings of the storage variables to registers match, then the digital signature essentially belongs to the one who generated it, i.e., the true IP owner. Hence ownership is awarded to the genuine IP owner or vendor. However, mismatching of both mappings establishes that the IP core design has been tampered with by an adversary.

Additional process to resolve ownership rights conflict using digital signature decryption: In the case of the signature being compromised or leaked to an adversary, then, also, the real IP owner can prove his/her rights over the signature. To do so, he/she first decrypts the signature using a public key resulting into hash-bitstream. The real IP owner is capable to do so because only he/she has the information of point of concatenation of the bitstream. This regenerated bitstream from the digital signature is matched with the originally computed or second recomputed hash-bitstream using SHA-512. If both the hash-bitstreams (regenerated from digital signature and computed through SHA-512) are same, then the rights of the IP owner over the digital signature (and subsequently design) is verified. In this way, IP ownership conflict can also be resolved using digital signatures (Sengupta *et al.*, 2019).

3.6 Case study on 8-point DCT used in image compression

Digital signature–based watermarks are employed the during register allocation phase of ESL or HLS. Embedding a watermark during HLS is crucial in the context of DSP or multimedia IP cores because of the following reasons:

- It protects the IP core design at a higher abstraction level from IP infringements.
- The design remains secured at subsequent lower-abstraction levels, such as RTL, gate level, layout level, etc., from IP infringements.
- Highly complex electronic DSP kernels designs which are harder to describe directly at lower abstraction level can easily be secured through a watermark at a higher abstraction level (through HLS framework).
- Embedding a watermark usually always adds extra cost or overheads to the design. However, an HLS framework enables the designer to optimize the design cost or overheads through a DSE process by yielding an optimal design solution for embedding the watermark. Therefore the design constraints, such as area, delay and power, can be optimized during the watermark embedding process through an HLS framework.

Securing an 8-point discrete cosine transform (DCT) core (Schneiderman, 2010; Obukhov and Kharlamov, 2008; Sengupta and Mohanty, 2019c) through a digital signature–based watermark has been demonstrated in this section. An 8-point DCT is used as an integral part of the JPEG CODEC process for image compression and decompression. A JPEG CODEC is a DSP–IP core which is integrated in SoCs representing complex electronics systems (Schneiderman, 2010). Thereby, 8-point DCT designs are extensively used in CE systems such as digital cameras, scanners, handsets, tablets, etc. (Sengupta *et al.*, 2019). Acknowledging the wide use of 8-point DCT as DSP applications in countless electronics devices, security of the DCT design itself against hardware threats is crucial. The process of securing the 8-point DCT design using a digital signature–based watermark is illustrated in two sub-processes, discussed next.

3.6.1 Digital signature generation corresponding to an 8-point DCT core

The process of generating digital signature corresponding to an 8-point DCT design is carried out through the following steps:

1. The DFG representing an 8-point DCT is transformed into a SDFG using a LIST scheduling algorithm. The DFG is shown in Figure 3.12. The scheduling of DFG is performed based on vendor-specified resource constraints (e.g., 3 multipliers and 1 adder). The SDFG of the 8-point DCT is shown in Figure 3.13. To store the primary and intermediate inputs/outputs of the design, 23 storage variables (S0–S22) are assigned, as shown in the figure.

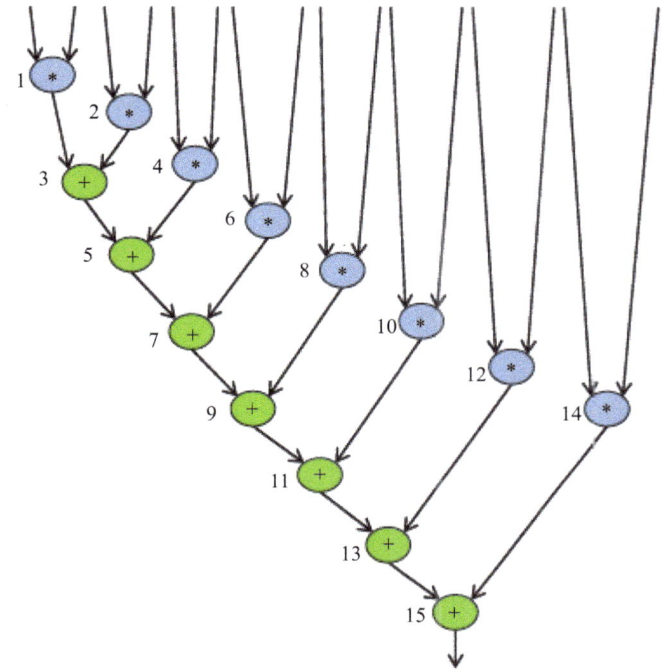

Figure 3.12 Data flow graph of 8-point DCT core used in image compression

2. In the SDFG (Figure 3.13), a total of 15 operations (multiplications and additions) have been scheduled in a total of 8 control steps. Based on the parity of operation number and its corresponding control step number, the 8-point DCT design is encoded into a bitstream (according to the rules stated in Table 3.1). The generated bitstream representing the initial vendor secret mark information is '011100100011011'.
3. The bitstream obtained from the previous step is fed to the SHA-512 algorithm, which transforms it into a 512-bit hash digest. This transformation occurs by the following sub-steps: (1) the bitstream is processed to generate an 1,024-bit block required as input for main hash algorithm – SHA-512. To do so, first the bitstream is padded with a fixed bit pattern (1 followed by sequence of 000 . . .) to construct an 896-bit block. Thereafter, 128-bit binary representation of the initial bitstream length is appended to construct a 1,024-bit block as follows: 01110010 00110111 00000000 .... 00000000 00000000 00001111 (1,024 bits); (2) all eight hash buffers are initialized with their fixed default values; (3) a 64-bit word (W_t) is extracted/computed from the 1,024-bit block for each iteration of round functions; (4) 80 iterations of round functions are executed using the W_t, hash buffer values and constant (K_t); and (5) the output of 80th round is added with the initial values of hash buffers (in the first round)

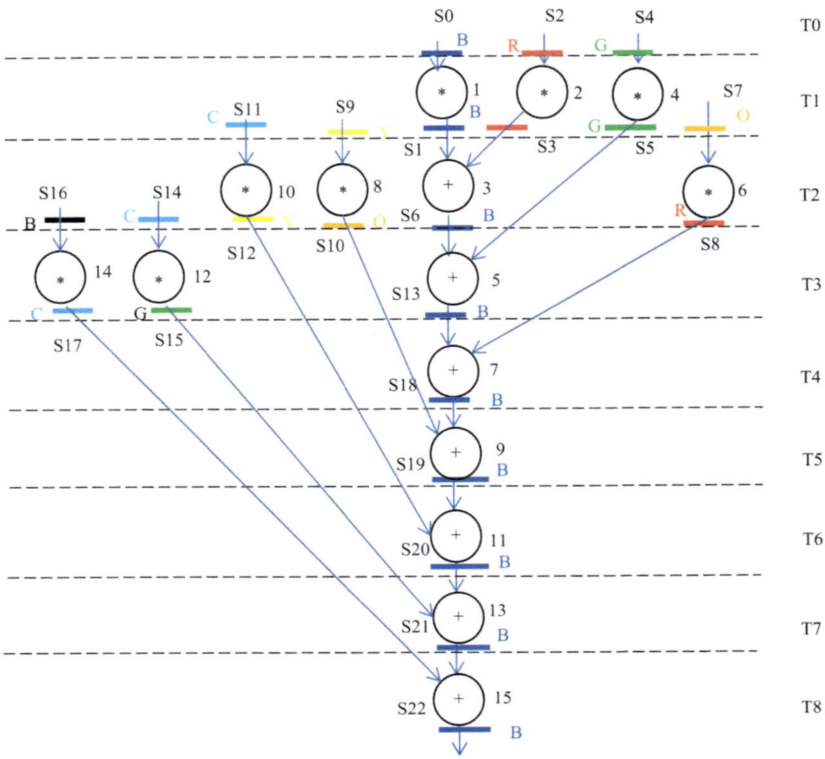

Figure 3.13 Schedule and hardware allocated to 8-point DCT using 1 (+) and 3 (*) before *implanting the digital signature*

to obtain the final hash-digest of the 512-bit. In this way, the hash-bitstream is generated.

4. The 512-bit hash-bitstream is partitioned into 64 blocks of eight bits each. Further, each block is translated into its equivalent decimal value. The first column of Table 3.3 lists the decimal equivalent of the first 16 blocks (for the sake of brevity).
5. The vendor private key (4) is computed to be 131 using RSA key generator.
6. Each decimal value of the hash-bitstream is encrypted one-by-one using the private key ($d = 131$) through RSA encryption algorithm. Once the RSA encryption is performed for all the 64 decimal values, the encrypted values are translated into their binary representations. The encrypted decimal values and their equivalent encrypted binary representations corresponding to first 16 blocks of hash-bitstream are listed in the second and third column of Table 3.3, respectively.
7. All the encrypted binary values are concatenated to construct a single, continuous, bitstream.

Table 3.3 First 16 decimal values of hash-bitstream and corresponding encrypted decimal and binary equivalent

RSA decimal input	Encrypted decimal output	Encrypted binary equivalent
247	304	100110000
59	70	1000110
175	74	1001010
18	18	10010
18	18	10010
113	56	111000
179	202	11001010
110	2	10
75	37	100101
137	188	10111100
15	230	11100110
137	188	10111100
230	32	100000
61	150	10010110
232	226	11100010
140	30	11110

8. A part of the encrypted hash-bitstream obtained from the previous step is selected as the digital signature. The respective digital signature size can be chosen by the vendor based on watermark strength required and design–cost trade-off. In this demonstration example of an 8-point DCT core, the vendor selected digital signature size (Q) is assumed to be 15 bits. The first 15 binary bits of the encrypted hash-bitstream represents the IP vendor's digital signature ($Q = 15$). Herewith, the process of generating a digital signature corresponding to the 8-point DCT core is accomplished.

3.6.2 Embedding of digital signature as watermark into the 8-point DCT core

The digital signature is processed to transform it into equivalent secret design watermark constraints. Thereafter, these digital signature-constraints are embedded as a watermark into the 8-point DCT design. The various steps of embedding the digit signature as a watermark are:

1. The CIG of the SDFG corresponding to the 8-point DCT core is created, as shown in Figure 3.14. This enables the designer to accommodate all the storage variables in a minimum number of registers. The corresponding register allocation for all storage variables (S0–S22) is shown in Table 3.4.
2. Each digit of the digital signature is encoded into watermarking constraints to be embedded into the design of the 8-point DCT core. This encoding is performed based on the rule stated in Table 3.2. According to the encoding rules, the mapping of each signature bit to an extra edge, required for embedding into the CIG, is shown in Table 3.5.

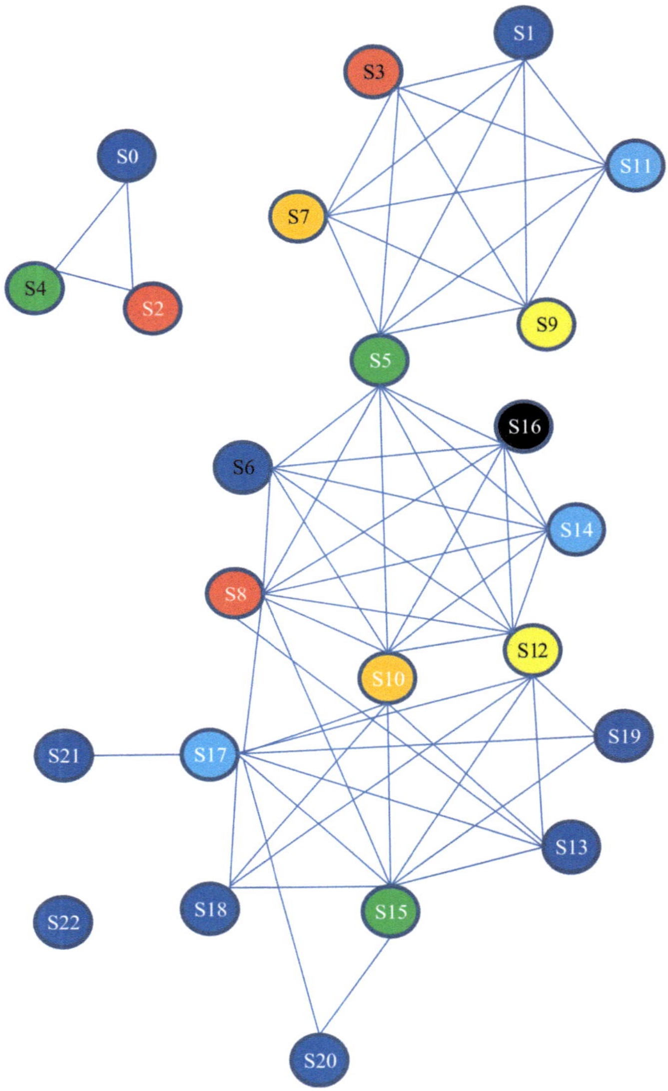

Figure 3.14 A CIG of 8-point DCT before embedding a digital signature

3. All encoded edges ⟨S*i*, S*j*⟩ listed in Table 3.5 are added into the CIG as additional secret edges. These inserted, additional, edges represent the covert watermark-constraints in the design. However, out of 15 only 12 potential edges, viz. *(S2, S3), (S2, S5), (S2, S6), (S2, S8), (S2, S7), (S2, S11), (S2, S13), (S2, S17), (S2, S10), (S2, S19), (S2, S12), (S2, S14)*, are newly added as watermark constraints and remaining three edges, ⟨*S2, S4*⟩, ⟨*S3, S5*⟩, ⟨*S3, S7*⟩,

Table 3.4 Register allocation table of 8-point DCT before *embedding a digital signature*

Control step	**B**	**R**	**G**	**O**	**Y**	**C**	**Bl**
T0	S0	S2	S4	–	–	–	–
T1	S1	S3	S5	S7	S9	S11	–
T2	S6	S8	S5	S10	S12	S14	S16
T3	S13	S8	S15	S10	S12	S17	–
T4	S18	–	S15	S10	S12	S17	–
T5	S19	–	S15	–	S12	S17	–
T6	S20	–	S15	–	–	S17	–
T7	S21	–	–	–	–	S17	–
T8	S22	–	–	–	–	–	–

Table 3.5 List of constraints-edges corresponding to the IP vendor digital signature

Bits in digital signature ($Q = 15$)	**Corresponding additional edges**
1	⟨S2, S4⟩
0	⟨S2, S3⟩
0	⟨S2, S5⟩
1	⟨S2, S6⟩
1	⟨S2, S8⟩
0	⟨S2, S7⟩
0	⟨S2, S11⟩
0	⟨S2, S13⟩
0	⟨S2, S17⟩
1	⟨S2, S10⟩
0	⟨S2, S19⟩
0	⟨S3, S5⟩
0	⟨S3, S7⟩
1	⟨S2, S12⟩
1	⟨S2, S14⟩

need not be added into the CIG as these three edges already exist in the CIG by default.

4. The register allocation of storage variables is modified after adding the aforementioned 12 secret edges into the CIG. This is because some of storage variables may need to be assigned to different registers while implanting the additional constraint-edges. For example, adding the edge ⟨S2, S3⟩ into the CIG invites conflict because both variables are initially assigned to the same register (Red) as shown in Table 3.4. However, to satisfy this edge-constraint, both storage variables (S2 and S3) must be assigned to distinct registers. To do so, the register allocation in Table 3.4 is inspected starting with control step 1 (T1). According to this inspection, swapping between S3 and S1 (Red ⇔ Blue) in T1 is possible and S3 is now assigned to the register represented by Blue.

Table 3.6 Register allocation table of 8-point DCT after *embedding the digital signature (Q = 15)*

Control step	B	R	G	O	Y	C	Bl
T0	S0	S2	S4	–	–	–	–
T1	S3	S1	S5	S7	S9	S11	–
T2	S6	S16	S5	S10	S12	S14	S8
T3	S13	–	S15	S10	S12	S17	S8
T4	S18	–	S15	S10	S12	S17	–
T5	S19	–	S15	–	S12	S17	–
T6	S20	–	S15	–	–	S17	–
T7	S21	–	–	–	–	S17	–
T8	S22	–	–	–	–	–	–

As a result of this swapping, register assignments to the storage variables S2 and S3 become distinct and, thus, the conflict of adding edge ⟨S2, S3⟩ into the CIG is resolved. Likewise, adding edge ⟨S2, S8⟩ also invites conflict. This conflict can also be resolved in similar fashion. Inspecting the register allocation in Table 3.4 shows that S8 can be swapped with S16 (Red ⇔ Black) in T2. However, swapping S8 with other storage variables, viz. S6, S5, S10, S12 and S14, in the same control step (T2) is not possible because any of these swapping actions result into a violation of other watermarking constraints. The modified register allocation after implanting all constraints-edges is shown Table 3.6. The final CIG with watermark constraints embedded is shown in Figure 3.15.

3.7 Desirable properties of a digital signature–based watermark

A watermark is embedded into designs to strengthen them against various hardware threats. A watermark is not part of the functionality of the design. Besides, embedding a watermark affects the design cost and may also violate the user-specified design constraints, such as area, delay and power requirement. In order to satisfy the security and design constraints, the digital signature–based watermark must exhibit some desirable properties (Sengupta and Mohanty, 2019b), as mentioned in Figure 3.16. A brief discussion on the desirable properties of digital signature–based watermark follows.

3.7.1 High resiliency

A highly resilient digital signature–based watermark is desirable to thwart removal attacks. More explicitly, the attacker should not be able to remove the embedded watermark constraints. To achieve high resiliency, watermark constraints require to be embedded as an integral part of the IP core design. Embedding the digital signature–based watermark during the register allocation phase of an HLS renders

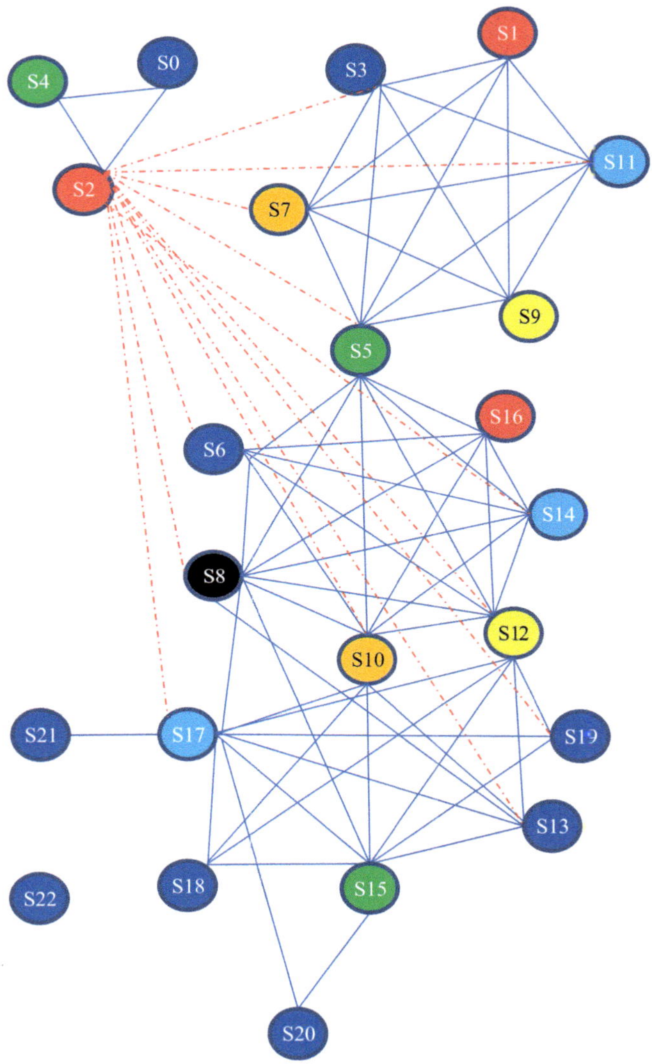

Figure 3.15 A CIG of 8-point DCT after *embedding the digital signature*

the watermark constraints of the design unobvious for the attacker and, thus, thwarts removal attacks.

3.7.2 Strong tamper tolerance

The watermarked design must be highly capable to tolerate any kind of tampering by the attacker. For example, an attacker may alter or partially remove the original watermark constraints. In this scenario, the ability of the watermark to detect

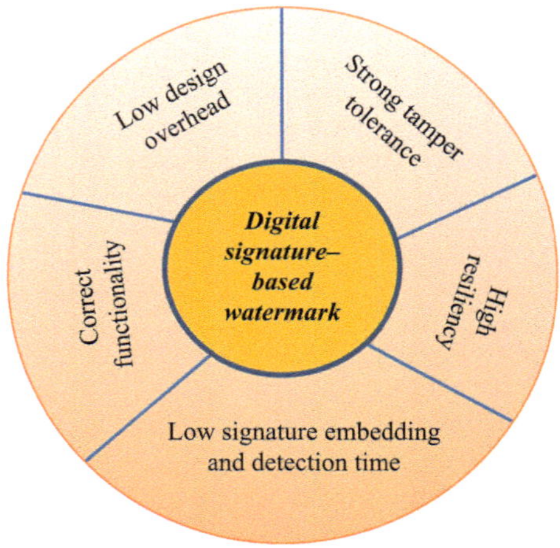

Figure 3.16 Desirable properties of a digital signature–based watermark

piracy/counterfeiting and resolve ownership conflict should not be compromised. The digital signature–based watermark is embedded during HLS, followed by the logic synthesis process which disperses the watermarking constraints throughout the entire design. Thus the digital signature–based watermark is highly tamper-tolerant.

3.7.3 Low design overheads

The watermark constraints are inserted additionally to the design (without affecting the functionality) and, hence, may result into additional design overheads. Therefore the watermark constraints should be chosen and imposed in such a way that it would incur design overheads as minimally as possible. The design overhead owing to embedding a digital signature–based watermark is significantly low because the process of implanting the watermark is essentially based on alteration of register allocation of the design's storage variables during the ESL. During imposing digital signature-constraints, zero (or nominal) extra registers are required.

3.7.4 Correct functionality

After embedding watermark constraints, the functionality of the design must be retained correctly. The embedded digital signature-constraints do not interfere in the functionality of the design; hence, the correct functionality is preserved.

3.7.5 Low-signature embedding and detection time

In order to construct a highly robust signature, an IP vendor may create complex signature encoding and decoding. Unnecessary complexity may result in higher signature embedding and detection time. However, it is crucial to create a signature with low-embedding and detection time without compromising the robustness of the design. This is because *embedding time* directly affects the ultimate production time of the design and *detection time* affects the forensic detection during ownership conflict and piracy/counterfeiting detection. The digital signature–based watermark is highly robust and can be embedded and detected seamlessly for a genuine owner in a short time.

3.8 Threat scenarios and their countermeasures using digital signature–based watermarks

Consider a scenario in which two different parties 'T' and 'F' are involved where party 'T' represents the owner/vendor of the digital signature–embedded IP core and party 'F' represents an adversary such as an SoC integrator/IC manufacturer (*Note: 'T' indicates true and 'F' indicates false*). However, F might be untrustworthy and, hence, may realize his/her malicious intents by pirating/cloning the IP core design of T and claiming ownership fraudulently. The possible attacks and their countermeasures exploiting digital signature–based watermarks are discussed next (Sengupta *et al.*, 2019b).

3.8.1 Unauthorized signature-constraints insertion

In this attack scenario, F inserts unauthorized signature-constraints into a digital signature–embedded IP core. Since the IP core already contains the digital signature of T, F is only able to insert unauthorized digital signature-constraints on the top of the existing original digital signature. Hence, it results into an IP core design containing the signature of T and F. In this situation, if F claims ownership of the IP core, his/her claim can easily be nullified as the design contains the digital signature-constraints of both parties, whereas the original digital signature–embedded IP core of T contains his/her signature-constraints only.

3.8.2 Original signature tampering

In this case, F attempts to tamper the original signature of T. To do so, he/she tries to insert changes in the embedded constraints-edges with the intention of gaining ownership of the IP core design of T. However, tampering with the original digital signature is not easy because the distributed digital signature throughout the entire design upsets the attempt of modification in all signature-constraints. Further, tampering with existing signature information may alter the area overheads and latency of the original signature–embedded IP core of T.

3.8.3 Unintended signature-constraints extraction

In this scenario, F performs back engineering of the embedding watermark in the IP core of T with the intention of extracting his/her own signature-constraints. F can do so by claiming any existing information in the design of T as his/her signature. However, F fails to realize his/her intentions in the digital signature–embedded IP core is because the claimed signature-constraints by F are weaker compared to that of T, whereas T can easily extract or prove his/her digital signature information during forensic detection analysis using the steps mentioned for signature verification (in Section 3.5.2). Thus, ownership is awarded to that party (i.e., T) whose claim is proved to be more meaningful, logical, scientific and stronger.

3.9 Analysis based on case studies

In this section, the impact of embedding a digital signature–based watermark for different DSP applications is analysed and compared with the contemporary hardware (DSP core–based) watermarking approach (Sengupta and Bhadauria, 2016). The amount of imposed signature-constraints essentially depends on the signature size chosen by the IP vendor/owner. Selecting an appropriate signature size is crucial in order to estimate the robustness of the digital signature–based watermark. Variations in signature size lead to different implications on security and design cost. The impact of different signature size on security and design cost are assessed based on the following criteria (Sengupta *et al.*, 2019a):

- Probability of coincidence (P^c).
- Extra register required after embedding digital signature as a watermark.
- Post-watermark embedded design cost.

The design cost of pre- and post-watermark embedded DSP IP cores are calculated based on 15 nm technology scale (http://www.nangate.com/?pageid=2328). An assessment of the effectiveness of the digital signature–based watermark in comparison to the watermarking approach (Sengupta and Bhadauria, 2016) is illustrated based on a security and design cost analysis in Sections 3.9.1 and 3.9.2.

3.9.1 Security analysis

Security achieved through a digital signature–based watermark is assessed in terms of the probability of coincidence (P^c) metric given as (Sengupta and Roy, 2017, 2018)

$$P^c = \left(1 - \frac{1}{£}\right)^k \tag{3.1}$$

where P^c denotes the probability of coincidence, $£$ denotes the number of colours used in the CIG before implanting digital signature and k denotes the number of digits in digital signature, i.e., the digital signature size (Q). Here, probability of coincidence signifies the proof of ownership, which indicates the probability of a non-watermark solution carrying the watermark (based on digital signature) by coincidence. Thus the strength of the digital signature–based watermark is reflected by probability of coincidence metric. As probability of coincidence decreases, the strength of proof of ownership increases. This is because as more signature-constraints are embedded into the CIG of the IP core, the likelihood of obtaining the watermarked solution by coincidence is much less. According to (3.1), the

Table 3.7 Probability of coincidence (P^c) using a digital signature–based watermark for different signature sizes

DSP applications	Number of colours used in CIG (n)	Probability of coincidence (P^c)		
		$Q = 60$	$Q = 120$	$Q = 240$
IIR Butterworth	5	1.5325×10^{-6}	2.3485×10^{-12}	5.5157×10^{-24}
BPF	6	1.7747×10^{-5}	3.1496×10^{-10}	9.9198×10^{-20}
JPEG sample	10	1.7970×10^{-3}	3.2292×10^{-6}	1.0428×10^{-11}
JPEG IDCT	29	0.1218	0.0148	2.1999×10^{-4}
MESA matrix multiplication	23	0.0695	4.8237×10^{-3}	2.3268×10^{-5}

value of P^c decreases with the increase in effective signature constraints-edges (k) embedded into the CIG. The number of effective constraints-edges essentially depends on the signature size. The number of effective signature constraints-edges increases with the increasing digital signature size, as shown in Table 3.7. The impact of three different signature sizes ($Q = 60$, 120, 240) on probability of coincidence has been shown in the table. Therefore to achieve the stronger proof of ownership, a large digital signature size needs to be selected. Hence proof of ownership becomes stronger with a decrease in probability of coincidence due to the increase in digital signature size.

Contribution of digital signature strength to the robustness of watermark: Since watermark constraints are obtained from a digital signature corresponding to an IP core design, the robustness of a watermark essentially depends on the strength of the digital signature. As discussed in Section 3.5, the digital signature is generated through a secure hash-algorithm (SHA-512) and RSA encryption. The collision resistance, uniformity and deterministic property of SHA-512 ensure that the generated digital signature is unique and robust. Besides, each bit of the hash digest is the function of every bit of input bitstream. This feature ensures that altering even a single bit of the input bitstream leads to change in a significant number of bits of the output hash digest, thus, adding strength to the digital signature. Further, RSA encryption using an IP vendor private key enhances the security strength of the digital signature.

Comparison with the contemporary watermarking approach (Sengupta and Bhadauria, 2016) for DSP cores: The digital signature–based watermarking approach provides stronger proof of ownership compared to the state-of-the-art watermarking approach (Sengupta and Bhadauria, 2016) for DSP cores. This is because between pre-watermark embedded designs of both digital signature–based and Sengupta and Bhadauria (2016) watermarking approaches, the former requires fewer colours (registers) to implement all storage variables. Thus for the digital signature-based watermark approach, probability of coincidence (computed using (3.1)) is lesser than watermarking approach (Sengupta and Bhadauria, 2016). The reason behind requiring fewer colours in the digital signature–based approach is that the DFG of DSP design (before embedding digital signature–based watermark) is scheduled based on LIST scheduling, whereas the pre-watermark embedded design in Sengupta and Bhadauria (2016) is scheduled based on ASAP scheduling. The percentage of reduction in P^c achieved through the digital signature–based watermark approach is reported in Table 3.8. The percentage of reduction in P^c in comparison to Sengupta and Bhadauria (2016) has been evaluated for the same signature sizes. It is evident from the table that percentage reduction in P^c increases with an increase in signature size ($Q = 60, 120, 240$). Hence, proof of ownership becomes stronger with the increase in percentage reduction in P^c in comparison to Sengupta and Bhadauria (2016).

3.9.2 Design cost analysis

The design cost of the digital signature–embedded IP core is evaluated using the following cost function:

$$C_d(R_i) = w_1 \frac{L_d}{L_{\max}} + w_2 \frac{A_d}{A_{\max}} \tag{3.2}$$

where $C_d(R_i)$ denotes the design cost post-embedding digital signature with vendor specified resource configuration, R_i. Here, design cost is evaluated in terms of the design latency and hardware area denoted by L_d and A_d, respectively. Further, $L_{\max}$ and $A_{\max}$ denote the maximum execution latency and hardware area respectively and w_1, w_2 represent the user-specified weights (both fixed at 0.5 to assign equal preference). After embedding the digital signature–based watermark, the design cost of a DSP–IP core depends on the impact of signature-constraints on register hardware requirements and execution latency. However, the digital signature–based watermark approach does not affect the execution latency therefore, only the potential requirement of extra registers (after embedding digital signature) may affect the overall embedding design cost. The impact on storage overhead (extra registers) due to variations in digital signature size is shown in Table 3.9. As

Table 3.8 Percentage decrement in P^c using a digital signature–based watermark in comparison to Sengupta and Bhadauria (2016)

DSP applications	**Percentage reduction in P^c (%)**		
	$S = 60$	**$S = 120$**	**$S = 240$**
IIR Butterworth	0	0	0
BPF	81.55	96.60	99.88
JPEG sample	66.74	88.94	98.78
JPEG IDCT	6.88	13.45	24.84
MESA matrix multiplication	10.9	20.31	36.5

Table 3.9 Impact of variation in digital signature size (Q) on storage overhead

DSP applications	Number of storage variables	Number of registers before applying proposed digital signature	Number of edges added resulting into R^e more registers			
			$R^e = 0$	$R^e = 1$	$R^e = 2$	$R^e = 3$
IIR Butterworth	14	5	15	NA	NA	NA
BPF	36	6	15	30, 60	NA	NA
JPEG sample	45	10	15	30, 60, 120	NA	NA
JPEG IDCT	136	29	15, 30, 60, 120, 240	–	–	–
MESA matrix multiplication	108	23	15, 30, 60, 120, 240	–	–	–

signature size is increased, the equivalent signature-constraints also increase. Therefore extra registers may be required to satisfy these signature-constraints. As shown in Table 3.9, the impact on register overheads owing to various signature sizes is trivial. The extra registers (R^e) requirement is either zero or nominal even for a large sized signature ($Q = 240$), as shown in the table. Thus the area/register overheads after embedding the digital signature–based watermark constraints is insignificant. Hence the design cost of the DSP core does not increase significantly after embedding the digital signature.

Comparison with the contemporary watermarking approach (Sengupta and Bhadauria, 2016) for DSP cores: To assess the design cost of digital signature–embedded DSP cores, the number of extra registers required to embed the secret constraints edges is compared with contemporary watermarking approach (Sengupta and Bhadauria, 2016) for same signature sizes. Table 3.10 presents comparative studies of the digital signature–based watermark approach and Sengupta and Bhadauria (2016) in terms of number of registers required. Both approaches have been compared for various signature sizes ($Q = 30, 60, 120, 240$). As shown in the tables, the digital signature–based watermarking approach always requires less registers than watermarking approach (Sengupta and Bhadauria, 2016) for a particular signature size. This is because the digital signature is generated through a unique hash digest and, hence, this uniqueness ensures that different combinations of the signature are not possible. On the contrary, the signature proposed by Sengupta and Bhadauria (2016) can have numerous possible combinations even for a same signature size. These various combinations of a signature may lead to different impressions on design overheads and security as well. Therefore it becomes harder to estimate the suitable signature combination resulting into lower overheads and higher security. Additionally, the digital signature–based watermarking approach exploits LIST scheduling in contrast to ASAP used by Sengupta and Bhadauria (2016) during the HLS process. Therefore it also results into a potential decrement in storage hardware (register) requirements when using the digital signature–based watermarking approach. Table 3.11 shows the percentage

Table 3.10 Comparative study of a digital signature–based watermark and Sengupta and Bhadauria (2016) in terms of number of registers required

DSP applications	Signature size $Q = 30$		Signature size $Q = 60$	
	Number of registers required (Sengupta *et al.*, 2019)	Number of registers required (Sengupta and Bhadauria, 2016)	Number of registers required (Sengupta *et al.*, 2019)	Number of registers required (Sengupta and Bhadauria, 2016)
IIR Butterworth	NA	NA	NA	NA
BPF	7	9	7	9
JPEG sample	11	13	11	14
JPEG IDCT	29	30	29	30
MESA matrix multiplication	23	24	23	25
DSP applications	**Signature size $Q = 120$**		**Signature size $Q = 240$**	
	Number of registers required (Sengupta *et al.*, 2019)	**Number of registers required (Sengupta and Bhadauria, 2016)**	**Number of registers required (Sengupta *et al.*, 2019)**	**Number of registers required (Sengupta and Bhadauria, 2016)**
IIR Butterworth	NA	NA	NA	NA
BPF	NA	NA	NA	NA
JPEG sample	11	14	NA	NA
JPEG IDCT	29	31	29	31
MESA matrix multiplication	23	26	23	26

decrement in number of registers required using digital signature–based watermark in contrast to Sengupta and Bhadauria (2016). Further, the fewer register requirements lead to lower design cost of DSP–IP cores (post-embedding digital signature–based watermark). A comparison of the digital signature–based watermark approach with Sengupta and Bhadauria (2016) in terms of design cost is presented in Table 3.12 for various signature sizes. As shown in the tables, the design cost of the digital signature–based watermark approach is estimated to be less than that of Sengupta and Bhadauria (2016) for a particular signature size. Hence, it can be concluded that the digital signature–based approach implants the watermark at a lower design overhead/cost (while simultaneously achieving higher security) compared to the contemporary watermarking approach (Sengupta and Bhadauria, 2016).

Table 3.11 Percentage decrement in number of registers required using a digital signature–based watermark compared to Sengupta and Bhadauria (2016)

DSP applications	Percentage reduction in number of registers required (%)		
	$Q = 60$	$Q = 120$	$Q = 240$
IIR Butterworth	0	0	0
BPF	22.22	0	0
JPEG sample	21.43	21.43	0
JPEG IDCT	3.33	6.45	6.45
MESA matrix multiplication	8	11.54	11.54

Table 3.12 Comparative study of a digital signature–based watermark and Sengupta and Bhadauria (2016) in terms of design cost

DSP applications	Signature size $Q = 30$		Signature size $Q = 60$	
	Design cost (Sengupta *et al.*, 2019)	**Design cost (Sengupta and Bhadauria, 2016)**	**Design cost (Sengupta *et al.*, 2019)**	**Design cost (Sengupta and Bhadauria, 2016)**
IIR Butterworth	NA	NA	NA	NA
BPF	0.4134	0.4145	0.4134	0.4143
JPEG sample	0.4294	0.4506	0.4294	0.4512
JPEG IDCT	0.2160	0.2333	0.2160	0.2333
MESA matrix multiplication	0.2687	0.2790	0.2687	0.2792
DSP applications	**Signature size $Q = 120$**		**Signature size $Q = 240$**	
	Design cost (Sengupta *et al.*, 2019)	**Design cost (Sengupta and Bhadauria, 2016)**	**Design cost (Sengupta *et al.*, 2019)**	**Design cost (Sengupta and Bhadauria, 2016)**
IIR Butterworth	NA	NA	NA	NA
BPF	NA	NA	NA	NA
JPEG sample	0.4294	0.4512	NA	NA
JPEG IDCT	0.2160	0.2335	0.2160	0.2335
MESA matrix multiplication	0.2687	0.2795	0.2687	0.2795

3.10 Conclusion

This chapter highlights the vulnerabilities of DSP–IP core designs against various hardware threats, such as IP piracy, counterfeiting, cloning and false claim of ownership, and presents a robust solution to protect against these threats using the digital signature–based watermarking approach. Like existing watermarking approaches, the presented approach also embeds watermarking constraints during the hardware register allocation phase of electronic-level synthesis (or HSL). However, the presented approach protects the design with higher security and lower embedding–design cost in contrast to existing watermarking approaches for DSP cores. Unlike existing watermarking approaches, the digital signature–based watermarking approach is capable of providing more a definite and meaningful proof of signature. The standard secure hashing algorithm (SHA-512), RSA encryption and pre-defined encoding rules enable the digital signature–based watermarking approach to produce a more definite, scientific and meaningful signature and, hence the randomness of a signature is eliminated completely in this approach. Forensic detective control using the digital signature–based watermarking approach has been described comprehensively in this chapter.

This chapter provides deep insight into digital signature–based watermarking and highlights its strengths with respect to existing watermarking approaches for DSP cores. At the end of this chapter, the reader is able to understand the following concepts:

- Utility of the digital signature–based watermarking.
- Application of digital signature–based watermarking to protect IP core designs against various hardware threats, such as IP piracy, counterfeiting, false claim of ownership, etc.
- Comparative study between digital signature–based watermarking and existing watermarking approaches for DSP cores of various features.
- The digital signature–generation process through pre-defined encoding rules and sophisticated algorithms.
- Method for implanting the digital signature as watermarking constraints.
- Forensic detective control using the digital signature–based watermarking approach to resolve the conflict of a fraudulent claim of ownership and detect IP/IC piracy, counterfeiting, cloning, etc.
- Design of a digital signature–embedded IP core using a real-life DSP core application used in glossy image compression.
- Performance and security features of digital signature–based watermarking.
- Security and design cost analysis on case studies (different DSP applications).

3.11 Questions and exercises

1. What is the threat model for digital signature–based IP core protection?
2. What are the variables used in triple-phase watermarking?
3. Which signature variable is assigned to which phase in triple-phase watermarking?
4. What is a sub-mark?
5. What is mobility?
6. How is a signature variable for watermarking encoded in a non-critical path?
7. What is a digital signature? How is it different from a regular signature?

8. What are the advantages of digital signature–based watermarking in comparison to conventional hardware watermarking?
9. What is the role of hashing in a digital signature?
10. What is the role of encryption on digital-signature formulation?
11. Explain the major phases of a crypto-hash function such as SHA-512.
12. How is a hash buffer initialized?
13. What is round computation function in hashing?
14. How is the input for SHA-512 prepared? What is the size of the data?
15. How many encodings are used in a digital signature–based watermark?
16. How is a DSP core data flow graph converted into its equivalent bitstream representation?
17. Explain the digital-signature detection process.
18. What are the advantages of inserting digital signature–based watermark during HLS?
19. What are the desirable properties of a digital signature–based watermark?
20. How does signature tampering happen and how does a digital signature protect against such attacks?
21. What is the role of private key bitstream concatenation in the ownership authentication process?

References

B. Colombier and L. Bossuet, "Survey of hardware protection of design data for integrated circuits and intellectual properties," IET Computers & Digital Techniques, vol. 8, no. 6, Pages: 274–287, 2015.

F. Koushanfar, I. Hong, and M. Potkonjak, "Behavioral synthesis techniques for intellectual property protection," ACM Transactions on Design Automation of Electronic Systems, vol. 10, no. 3, Pages: 523–545, 2005.

B. Le Gal and L. Bossuet, "Automatic low-cost IP watermarking technique based on output mark insertions," Design Automation for Embedded Systems, vol. 16, no. 2, Pages: 71–92, 2012.

A. Obukhov and A. Kharlamov, "Discrete Cosine Transform for 8 × 8 Blocks with CUDA," Nvidia Whitepaper Document, 2008.

S. M. Plaza and I. L. Markov, "Solving the third-shift problem in IC piracy with test-aware logic locking," IEEE Transactions on Computer-Aided Design of Integrated Circuits and Systems, vol. 34, no. 6, Pages: 961–971, 2015.

D. Roy and A. Sengupta, "Low overhead symmetrical protection of reusable IP core using robust fingerprinting and watermarking during high level synthesis," Future Generation Computing Systems, vol. 71, Pages: 89–101, 2017.

D. Roy and A. Sengupta, "Multilevel watermark for protecting DSP kernel in CE systems [hardware matters]," IEEE Consumer Electronics Magazine, vol. 8, no. 2, Pages: 100–102, 2019.

R. Schneiderman, "DSPs evolving in consumer electronics applications," IEEE Signal Processing Magazine, vol. 27, no. 3, Pages: 6–10, 2010.

A. Sengupta, "Intellectual property cores: Protection designs for CE products," IEEE Consumer Electronics Magazine, vol. 5, no. 1, Pages: 83–88, 2016.

A. Sengupta, "Hardware security of CE devices [hardware matters]," IEEE Consumer Electronics Magazine, vol. 6, no. 1, Pages: 130–133, 2017.

A. Sengupta and S. Bhadauria, "Exploring low cost optimal watermark for reusable IP cores during high level synthesis," IEEE Access, vol. 4, Pages: 2198–2215, 2016.

A. Sengupta and S. P. Mohanty, "Advanced Encryption Standard (AES) and Its Hardware Watermarking for Ownership Protection," IP Core Protection and Hardware-Assisted Security for Consumer Electronics, Pages: 317–335, 2019a, e-ISBN: 9781785618000.

A. Sengupta and S. P. Mohanty, "IP Core and Integrated Circuit Protection Using Robust Watermarking," IP Core Protection and Hardware-Assisted Security for Consumer Electronics, Pages: 123–170, 2019b, e-ISBN: 9781785618000.

A. Sengupta and S. P. Mohanty, "Trojan Security Aware DSP IP Core and Integrated Circuits," IP Core Protection and Hardware-Assisted Security for Consumer Electronics, Pages: 79–121, 2019c, e-ISBN: 9781785618000.

A. Sengupta and D. Roy, "Antipiracy-aware IP chipset design for CE devices: A robust watermarking approach [hardware matters]," IEEE Consumer Electronics Magazine, vol. 6, no. 2, Pages: 118–124, 2017.

A. Sengupta and D. Roy, "Multi-Phase Watermark for IP Core Protection," in Proc. 36th IEEE International Conference on Consumer Electronics (ICCE), Pages: 1–3, 2018.

A. Sengupta, R. Sedaghat, and Z. Zeng, "A high level synthesis design flow with a novel approach for efficient design space exploration in case of multi-parametric optimization objective," Microelectronics Reliability, vol. 50, no. 3, Pages: 424–437, 2010.

A. Sengupta, D. Roy, and S. P. Mohanty, "Triple-phase watermarking for reusable IP core protection during architecture synthesis," IEEE Transactions on Computer-Aided Design of Integrated Circuits and Systems, vol. 37, no. 4, Pages: 742–755, 2018.

A. Sengupta, E. R. Kumar, and N. P. Chandra, "Embedding digital signature using encrypted-hashing for protection of DSP cores in CE," IEEE Transactions on Consumer Electronics, vol. 65, no. 3, Pages: 398–407, 2019.

D. Ziener and J. Teich, "Power signature watermarking of IP cores for FPGAs," Journal of Signal Processing Systems, vol. 51, no. 1, Pages: 123–136, 2008.

Chapter 4

Protection of fault-secured IP cores using digital signature–based watermarks

Anirban Sengupta[1]

This chapter discusses forensic detective control using digital signature–based watermarks for fault-secured digital signal processing (DSP) cores or integrated circuits. The chapter provides detailed insight on the utility of the digital signature process for fault-secured DSP cores, which is an extension of Chapter 3. This chapter includes a discussion on transient fault security, generating loop unrolled control data flow graphs representing DSP cores, fault-secured IP core protection using forensic detective control of a digital signature, the methodology of implanting digital signature – watermarking constraints covertly into an IP core design, and the respective design process.

The chapter is organized as follows: Section 4.1 introduces the background to IP cores and their countermeasures as well as the background on fault-secured IP cores; Section 4.2 discusses the threat model for which a digital-signature approach is applicable; Section 4.3 provides a discussion on contemporary IP core-protection approaches for DSP cores; Section 4.4 explains forensic detective control using a digital signature–based watermark; Section 4.5 presents a case study of the digital signature process for securing fault-secured FIR filters; Section 4.6 presents an analysis and comparison of digital signature–based watermarks for various fault-secured DSP cores; Section 4.7 concludes the chapter.

4.1 Introduction

In today's technology-driven world, consumer electronic (CE) devices have become an integral part of life. CEs encompass a broad set of electronic devices such as televisions, smart phones, personal computers, camera systems, tablets and audio-output devices that provide multiple functionalities to an individual and assist in everyday tasks ranging from office work, entertainment, medical purposes, communications and home activities. Every CE device has a system-on-chip (SoC) as underlying hardware. Reusable IP cores – specifically DSP and multimedia cores – are part of almost every SoC in CE products. It is very difficult to find CE devices

[1]Computer Science and Engineering, Indian Institute of Technology Indore, Indore, India

without DSP kernels (Schneiderman, 2010); they can be found in various applications, viz. digital image processing, audio processing (compression/decompression), digital video broadcasting (DVB) and video compression. For example, the joint photographic experts group (JPEG) compression process used in image and video compression of camera systems, discrete cosine transform filter used in lossy compression of audio and image files, fast Fourier transform (FFT) used in digital video broadcasting, finite impulse response (FIR) filters used in audio devices such as headsets and infinite impulse response filters used in audio processors for equalization of speakers.

An IP core, by definition, is a reusable unit of design logic or functionality that is generally developed with the idea of reusing them as building blocks in different designs. Current design challenges in the domain of CEs include greater complexity and stiff time to market pressure. Reusable IP cores, often delivered by a third-party vendor, serve as a viable solution for maximizing the design productivity while minimizing design time for CE products.

Currently, the CE industry heavily depends on global SoC supply chains as using SoC designs that employ reusable IP cores has now become a de-factor standard in the industry. As a consequence, the global semiconductor supply chain for SoC design is highly vulnerable to piracy, IP forgery or fraudulent ownership abuse. Therefore owing to the value an IP core represents, it is security and protection are recognized as major challenges and are being considered as a priority design objective alongside area, delay, etc., during high-level synthesis (HLS) design process for DSP applications (Koushanfar *et al.*, 2005; Sengupta and Bhadauria, 2016; Sengupta *et al.*, 2017, 2018; Yu and Zhu, 2011; Liang *et al.*, 2015; Roy and Sengupta, 2017; Nie *et al.*, 2013; Sengupta, 2016a, 2016b, 2017).

4.1.1 Background on different IP core threats and countermeasures

In this sub-section, a background is provided on prevailing hardware threats affecting the reusable IP cores in CE devices and different countermeasures available in the literature.

- *IP piracy and counterfeiting:* A well-known threat to DSP-based IP cores and chip sets are direct piracy, where fraudulent means can result in direct theft or duplication of the IP design for illegal distribution. After obtaining the design, the adversary can even falsely claim its ownership. In such circumstances, a false claim of ownership can be invalidated by embedding a secret watermark in the design (Koushanfar *et al.*, 2005; Sengupta and Bhadauria, 2016; Sengupta *et al.*, 2017, 2018; Yu and Zhu, 2011; Liang *et al.*, 2015; Roy and Sengupta, 2017; Nie *et al.*, 2013; Le Gal and Bossuet, 2012; Ziener and Teich, 2008). For example, Koushanfar *et al.* (2005), Sengupta and Bhadauria (2016), Sengupta *et al.* (2018), Sengupta and Roy (2017, 2018), Sengupta and Mohanty (2019) embed a robust watermark in the IP core design process. Yu and Zhu (2011) present a watermarking method for soft IP core protection in which a special watermark module for copyright detection is added. A watermarking approach based on two-dimensional chaotic mapping model is proposed in Liang *et al.* (2015). The approach by Roy and Sengupta (2017) proposes a symmetrical protection method, wherein both seller watermark and buyer fingerprint are embedded in an IP design. There are few other non-watermark types of protection mechanisms against IP piracy in the literature. For instance, Alkabani and Koushanfar (2007) propose the concept of IP metering to identify the difference between an original circuit and an illegal copy.

- *Reverse engineering (RE) and Trojan insertion:* RE is a popular attack employed to extract highly relevant information such as specification and architecture of an IP core design. Once the required information is available, the adversary can duplicate the design by reconstructing the controller, or introducing malicious Trojan in the original design. Watermarking approaches cannot protect IP cores against RE attacks. Obfuscation is a frequently employed approach to prevent RE attacks. It is a mechanism that accepts a circuit design as input and hides the structure/functionality through a secret-key sequence without affecting the actual functionality of the application, thus hindering the RE process (Chakraborty and Bhunia, 2009). For example, high-level transformations (HLTs)–based structural obfuscation is presented in Lao and Parhi (2015) to protect DSP cores against RE threats. Further, IP core locking–based functional obfuscation used in Sengupta *et al.* (2019a) and a dual modular redundant (DMR) design–based HLS (Sengupta *et al.*, 2010) methodology for Trojan detection are also presented in Sengupta *et al.* (2017).

4.1.2 Brief description of fault security for IP cores

Integrated circuit (IC) packages contain trace amounts of uranium and thorium impurities that emit alpha particles. Ionizing radiation from these alpha particles causes a single-event upset (SEU), leading to a single-event transient. In simple terms, a transient fault generally occurs when a particle with moderate linear-energy transfer (LET) value strikes an electronic device. Due to the presence of α-particles in IC packages, fault-secured IP cores are becoming popular for SoC applications. An SEU can result in functional failure such as producing an erroneous output. The increase in density per unit area in a chip – due to massive technology scaling – is increasing the rate of occurrence of transient faults. Moreover, the advent of faster (high frequency) devices has resulted in multi-cycle transient faults that last more than one control step (CS). A transient fault-secured IP core implies that the design is secured against faults that have maximum transient fault strength, that is, worst case impact of a transient fault (Sengupta, 2016; Atienza *et al.*, 2008; Choudhury and Mohanram, 2009; Wu and Karri, 2001; Rusu *et al.*, 2007; Miskov-Zivanov, 2010; Lakshminarayana *et al.*, 2000; Normand, 1996; Lala, 1985; Polian *et al.*, 2005; Mukherjee *et al.*, 2005; Zhang *et al.*, 2006; Gomaa and Vijaykumar, 2005; Pignol, 2005; Mizan *et al.*, 2007; Martinez-Alvarez *et al.*, 2016; Fazeli *et al.*, 2011). A detailed description on how to transform a DSP design into a transient fault-secured one, is provided in Section 4.4.2 later.

4.2 Threat model

The threat models for digital signature–based watermarking are protection of fault-secured, DSP-based IP cores from false claim of ownership and detection of IP piracy and IP cloning. The threat model is described in detail in Chapter 3.

4.3 Selected contemporary approaches

4.3.1 Discussion on similar approaches

So far in the literature, few watermarking techniques have been proposed in the literature for protection of DSP-based IP cores (Sengupta and Bhadauria, 2016; Sengupta *et al.*, 2018; Yu and Zhu, 2011; Roy and Sengupta, 2017). Further, there exists no work that proposes protection of fault-secured DSP cores using robust watermarks. Watermarking is a passive mode of protection, wherein a hidden mark is implanted in the design to protect reusable IP cores against ownership abuse. One of the key factors considered by authors in proposing a watermark-based protection scheme is at which level of the design flow a signature has to be embedded. This is non-trivial, since the choice of design abstraction level affects the design-cost overheads, security and implementation complexity. Watermarking has been proposed at lower levels of abstraction such as gate level or register transfer level in (Nie *et al.*, 2013; Le Gal and Bossuet, 2012; Ziener and Teich, 2008). For instance, a hierarchical watermarking scheme for protection of field programmable gate array (FPGA)–IP cores is presented in (Nie *et al.*, 2013), wherein the netlist and bitstream of an IP design are used to insert a vendor watermark.

In-synthesis process of design is selected to embed the signature in Le Gal and Bossuet (2012). Moreover, Ziener and Teich (2008) proposed a powerful watermarking technique that detects the signature at the power supply pins of FPGA. However, in the context of fault-secured DSP cores, implanting the signature at lower abstraction levels increases design complexity, does not provide comprehensive security and enhances chances of design overhead. Thus these approaches are not capable of handling fault-secured DSP cores, unlike the presented approach. Further, the approach using a crypto-based digital signature process during HLS provides stronger security (more definite proof of ownership) and lower design overheads.

Besides, watermarking schemes have also been proposed at higher abstraction levels such as during high level synthesis (Koushanfar *et al.*, 2005; Sengupta and Bhadauria, 2016; Sengupta *et al.*, 2018). Explicitly, Koushanfar *et al.* (2005) and Sengupta and Bhadauria (2016) implant the watermark in the register allocation phase of HLS, whereas Sengupta *et al.* (2018) embed the vendor signature at three phases of HLS design. Further, Koushanfar *et al.* (2005) employ a binary encoding scheme to implant the watermarking constraints in the IP core design, while Sengupta and Bhadauria (2016) and Sengupta *et al.* (2018) use a multi-variable encoding scheme. However, the digital signature–based watermarking approach provides more definite proof of authorship (stronger security) using a crypto-based signature process and incurs lower overheads than (Koushanfar *et al.*, 2005;

Sengupta and Bhadauria, 2016; Sengupta *et al.*, 2018). Additionally, Koushanfar *et al.* (2005), Sengupta and Bhadauria (2016) and Sengupta *et al.* (2018) did not handle security of fault-secured DSP cores.

4.4 Forensic detective control of fault-secured IP cores using digital signature–based watermarks

4.4.1 Input/output

The list describes the different types of input to the digital signature based watermark approach for security of fault secured IP cores.

Input/output: Inputs: (1) Control data flow graph (CDFG) of a DSP core; (2) Strength of transient fault (K_c); (3) Loop unrolling factor (UF); (4) Module library; and (5) User-specified hardware resources set $(X) = N(R_1)$, $N(R_2)$, $N(R_D)$, where $N(R_D)$ is the number of hardware resources of type R_D. *Output*: Fault-secured watermarked DSP design.

4.4.2 Fault-secured DSP–IP cores protection framework

An overview of the digital signature–based protection approach (Sengupta *et al.*, 2019b, 2019c) for a fault-secured IP core is illustrated in Figure 4.1. The overall approach can be divided into four phases. Inputs/outputs and functions executed in each phase are explained in this sub-section. To begin with, the inputs to the presented algorithm are: CDFG of loop-based DSP core, strength of transient fault, loop UF, resource configuration, and module library. First, based on the UF value, given loop-based CDFG is unrolled in phase 1. The next phase comprises three steps. In the first step, DMR design of loop unrolled CDFG (output of phase 1) is constructed. Subsequently, scheduling of the DMR design based on given K_c value and resource configuration is performed using a scheduling algorithm to obtain a transient fault-secured scheduled DSP design. As a final step of phase 2, storage variables are assigned in the scheduled DMR design before it is passed as input to

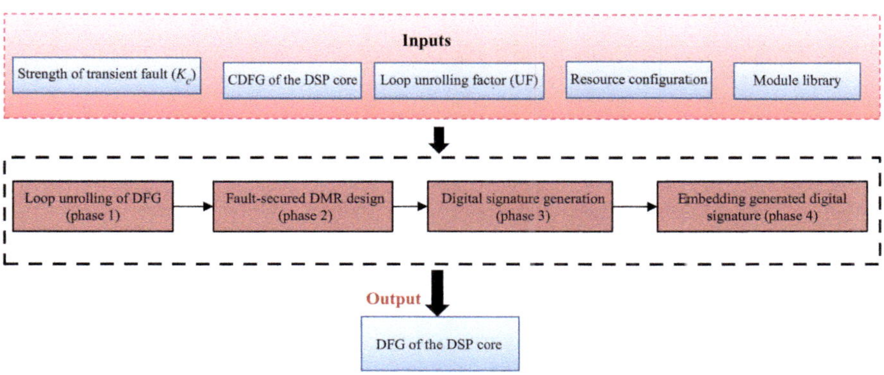

Figure 4.1 Overview of digital signature embedding approach in terms of different phases

phase 3. In phase 3, first, a bitstream is constructed from the scheduling information, which is then processed in an SHA-512 module to obtain a bitstream digest. Subsequently, this digest is divided into blocks and the decimal value equivalent of each block is encrypted using a vendor-selected Rivest, Shamir, and Adelman (RSA) private key. A bitstream of the encrypted decimal values is constructed and passed as input to phase 4. In phase 4, a coloured interval graph (CIG) is constructed representing the register allocation of all the storage variables in the scheduled fault-secured DSP design. The bitstream obtained from the preceding phase is termed the digital signature and each bit in it is encoded in order to embed it in the form of design constraints. These additional constraints are added into the register allocation step of HLS to finally obtain a transient fault-secured watermarked DSP core.

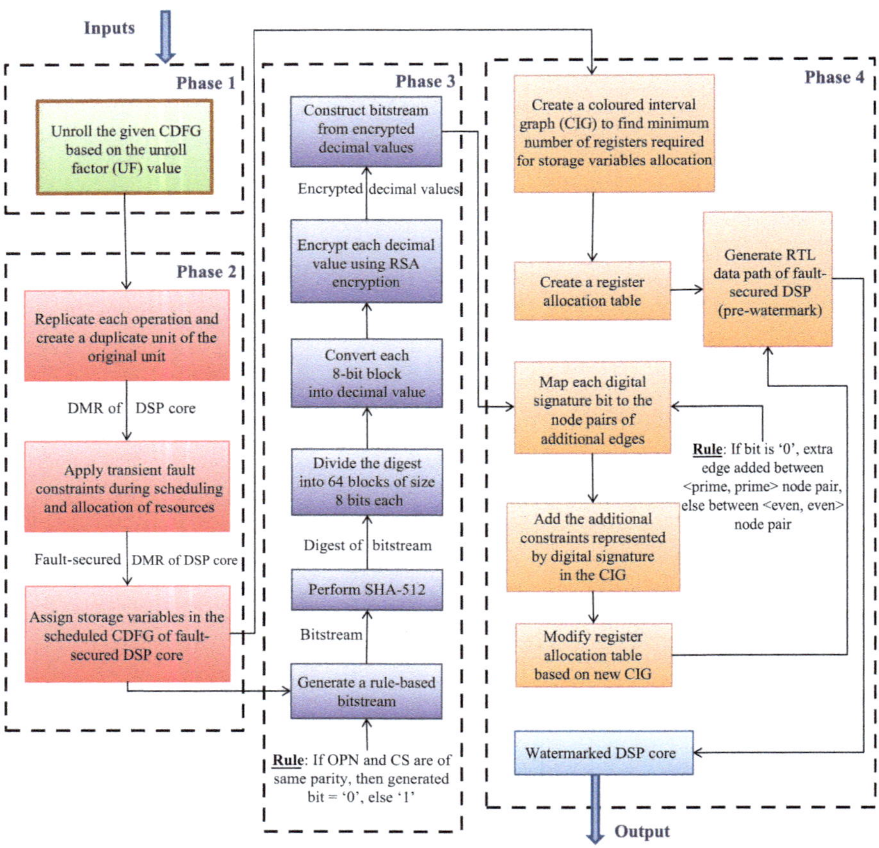

Figure 4.2 Details of the digital-signature embedding approach for a fault-secured DSP design

A schematic depiction of the details of digital signature–based watermark approach for a fault-secured IP core is shown in Figure 4.2. This sub-section elaborates on the steps of all the aforesaid phases.

- *Phase 1:* In phase 1, the primary input – loop-based CDFG – is unrolled based on the given UF value. Here, it is the responsibility of designer to carefully select an appropriate UF value before unrolling is performed. The desired performance properties of UF value are as follows: (1) maximize loop parallelization and (2) minimize sequential execution. Hence, as a preliminary step, the designer has to perform pre-processing of the UF that takes the loop iteration count as input and eliminates unsuitable UF values that result in sequential loops during unrolling. After the suitable UF value is selected, the body of CDFG is unrolled 'UF' number of times. Unrolled CDFG is then provided as input to phase 2.
- *Phase 2:* In this phase, a transient fault-secured scheduled DMR design ($\mathrm{SDFG}^{\mathrm{DMR}}$) is constructed from the unrolled CDFG provided as input. Prior to describing the steps on how to create $\mathrm{SDFG}^{\mathrm{DMR}}$, a brief description of transient fault strength is provided. As the name indicates, transient faults are faults that occur only for a specific time period. A resource affected by a transient fault temporarily produces incorrect results for K_c number of cycles. K_c indicates the fault strength. In a practical scenario, the worst case impact of transient fault lasts between 1,000 ps ($K_c = 1$) and 2,000 ps ($K_c = 2$) (Sengupta and Kachave, 2017). For example, if 1 GHz frequency is considered, then for $K_c = 2$ and the fault effect lasts up to 2 CSs. In this chapter, $\mathrm{SDFG}^{\mathrm{DMR}}$ is designed considering the worst-case impact of a transient fault, that is, $K_c = 2$.

  There are two steps for transforming a given DSP design into a transient fault-secured DSP design. First, all operations in the given CDFG (unrolled CDFG in case of loop-based DSP core) are replicated in order to construct a duplicate unit resulting in a DMR design of DSP core. Subsequently, based on the input resource configuration, operations in the original and duplicate unit are scheduled. In the second step, the scheduling and hardware allocation is modified by applying the transient fault security constraint (K_c) to obtain a transient fault-secured scheduled DMR design.

  More explicitly, the following rules are imposed to modify the scheduling and hardware allocation (Sengupta, 2016):

  - In the existing schedule of a DMR design, if the difference between the CS of sister operations of the original and duplicate units is greater than K_c, the same resource can be allocated to both sister operations.
  - In the existing schedule of DMR design, if the difference between the CS of sister operations of the original and duplicate units is less than K_c, the different resources must be allocated to both operations (if available).
  - In the case of the second rule, if distinct resources are not available, then duplicate operation has to be rescheduled by shifting it to the next CS until the fault security constraint is satisfied. Thereafter the same hardware resource can be allocated to both sister operations.

Finally, the transient fault-secured DSP design is constructed. Storage variables are assigned at the intermediate inputs and outputs in the $SDFG^{DMR}$ before it is provided as input to phase 3.

- *Phase 3:* In phase 3 the steps involved in the digital signature creation process (Sengupta *et al.*, 2019a, 2019b) are described. To begin with, the scheduling information from $SDFG^{DMR}$ is used to generate a bitstream (stream of '0' and '1' bits) using encoding rule 1 (Sengupta *et al.*, 2019b) as depicted in Figure 4.3.

  The bitstream thus generated is passed as input to the SHA-512 module. In this module, first, the input bitstream is pre-processed so that its length becomes a multiple of 1,024, followed by computing word computation values (W). Next, a buffer (512 bit), used to store intermediate hash values, is initialized with a predefined value. Subsequently, the main hash computation is performed that involves executing a function named 'round function' for 80 iterations. Finally, the output of the final iteration of round function is added to the input of the first iteration to produce the bitstream digest (Sengupta *et al.*, 2019b).

  Further, this bitstream digest is processed before performing encryption in the RSA module. First, the bitstream digest (512 bit) is divided into 64 blocks of equal size (8 bits), followed by converting each 8-bit block into an equivalent decimal value. These individual decimal values are passed as input to the RSA module.

  The final module of phase 3 involves encrypting output decimal values of the preceding step using RSA (Sengupta *et al.*, 2019b, 2019c). The first step in the encryption process is it to generate a private key as per the vendor's choice.

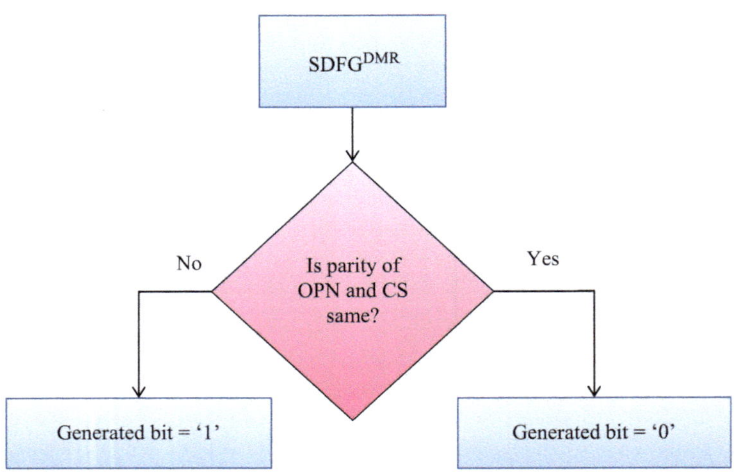

Figure 4.3 Encoding rule 1

Here the private key is used for encryption because standard digital certification applications use private keys instead of public keys in the digital signature–generation process. Once the private key is generated, the decimal values are encrypted using the following equation:

$$E = H^k \bmod n \tag{4.1}$$

where E = encrypted decimal value, H = decimal equivalent of hashed bitstream block, k = vendor private key, $n = p \times q$, and p and q are both prime, $p \neq q$. Thereafter, a bitstream is constructed by concatenating a binary representation of all the encrypted decimal values. This marks the end of the digital signature creation process and generated encrypted–hashed bitstream, also referred to as 'digital signature', is passed as input to phase 4.

- *Phase 4:* In the final phase of the presented framework, only embedding the digital signature into the $SDFG^{DMR}$ is left. To accomplish this, the digital signature obtained from the previous phase needs to be encoded in the form of hidden constraints that can be imposed in the design during any of the high-level synthesis tasks. These hidden constraints act as additional constraints as they are added besides the regular constraints of the design. The register-allocation step in HLS is selected for adding additional constraints in the design. Before describing how additional constraints are imposed in the register-allocation step, a brief background is provided on how register allocation is performed in HLS. Register allocation can be performed by creating a CIG, wherein the nodes of the CIG represent storage variables and an edge between two nodes denotes the overlapping lifetime of the two variables, represented by those nodes, in a particular CS. It implies that the same register (colour) cannot be allocated to two storage variables if there is an edge between them in the CIG (Sengupta *et al.*, 2019b).

  Additional constraints represented by the encoding of the digital signature are added in the form of extra edges between nodes in CIG. The encoding scheme (Sengupta *et al.*, 2019b, 2019c) used to encode each bit of the digital signature is shown in Figure 4.4. Storage variables connected by the newly added edges are forced to be executed through different registers, thereby affecting design data path interconnectivity and architecture. This concludes phase 4 as well as the presented approach to finally obtain the digital signature embedded in the transient fault-secured DSP design.

4.4.3 Signature detection process

To resolve the conflict of a false claim of ownership or detect IP piracy/counterfeiting, the vendor/owner signature can be extracted from the fault-secured IP core as shown in Figure 4.5. The detailed process of signature detection has been illustrated in Chapter 3.

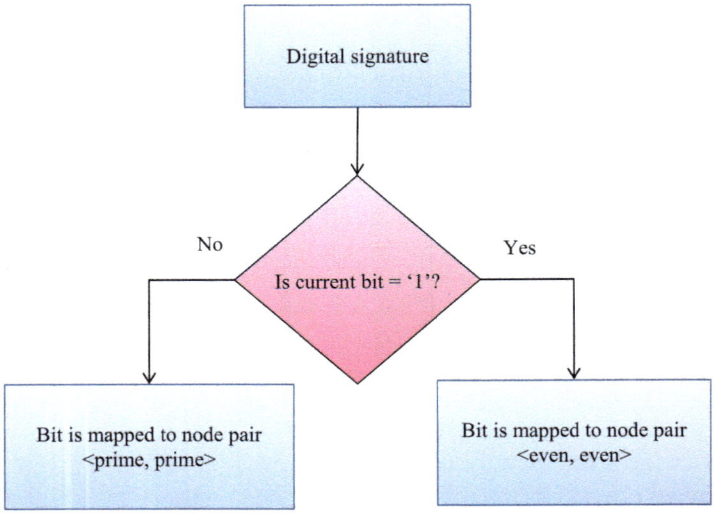

Figure 4.4 Encoding rule 2

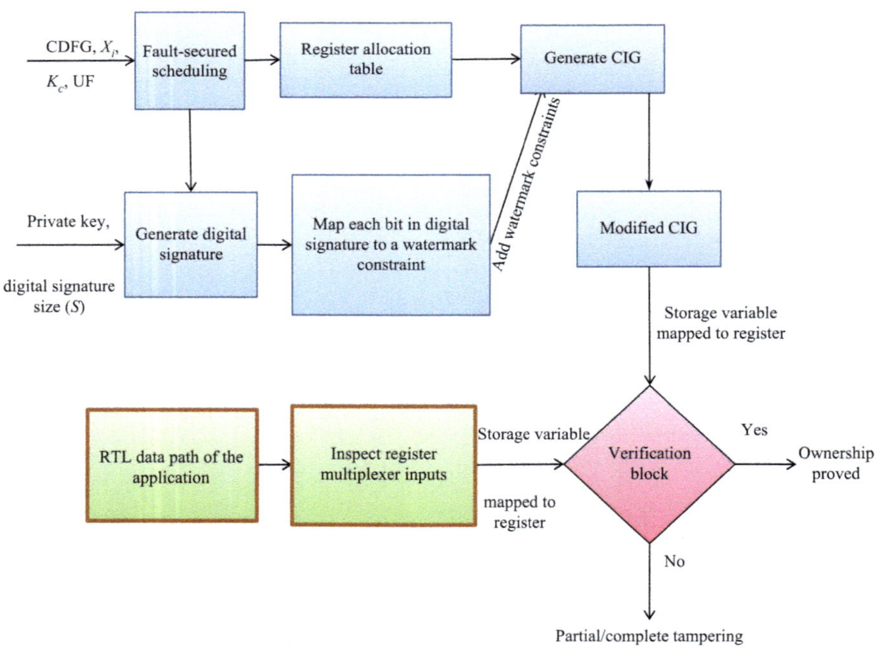

Figure 4.5 Process of detecting a digital signature for fault-secured IP cores

4.5 Case study on a fault-secured, finite impulse response (FIR) filter

In this section, a motivational example is described to demonstrate the approach for producing a digital signature implanted $SDFG^{DMR}$ of a fault-secured FIR filter core.

The equation for a loop-based FIR filter is shown as follows:

$$y[n] = \sum_{i=0}^{8} h[i] \times x[n-i] \tag{4.2}$$

where $y[n]$ is the output of FIR filter, $x[n - i]$ denotes the input impulse, $h[i]$ is constant value and variable 'i' iterates from 0 to 8.

4.5.1 Constructing a transient fault secured DSP design

The loop-based CDFG represented in (4.2) and the UF value are given as input to phase 1 of the presented approach. For this demonstration, a UF value = 3 and resource configuration: 3(×), 1(+), 1(>) are considered. The given CDFG is unrolled based on the input UF value and passed to phase 2 where the transient fault-secured FIR design is created based on fault strength (K_c = 2) and the aforementioned resource configuration using a list scheduling algorithm. The $SDFG^{DMR}$ of FIR filter with storage variables (V_0–V_{37}) assigned to inputs and outputs of intermediate operations is shown in Figure 4.6 and is passed as input to phase 3.

4.5.2 Generating the digital signature

Phase 3 begins with generating a bitstream using the scheduling information of $SDFG^{DMR}$ according to the encoding rule 2. The bitstream generated in this demonstration for the FIR filter is 0011100000001100001. Second, a bitstream digest is formed by providing the bitstream generated in the preceding step as input to SHA-512. In the SHA-512 module, the output of the initial pre-processing step (stream of 1,024 bits) is 0011100000001100001100...0010010. W values are computed from the updated bitstream obtained after the pre-processing step. Thereafter, the 512-bit buffer is initialized and the main hash computation is performed. The hashed bitstream output obtained after performing SHA-512 is 10000100010111101001110...111001010100010011 (512 bits). This 512-bit hashed bitstream is divided into 64 blocks of equal length and each block is converted to its equivalent decimal value. The first 10 blocks and their equivalent decimal values are shown in the first and second column of Table 4.1, respectively. *Note: Owing to space constraints, only the first 10 blocks are shown in Table* 4.1. Next, these decimal values are encrypted using RSA encryption. The private key considered in this demonstration is (d, n) = (139,299). The encrypted decimal values for the first 10 input blocks are shown in the third column of Table 4.1. Finally, all encrypted binary values (the last column in Table 4.1) are concatenated to form a single bitstream. This encrypted hashed bitstream is termed the digital

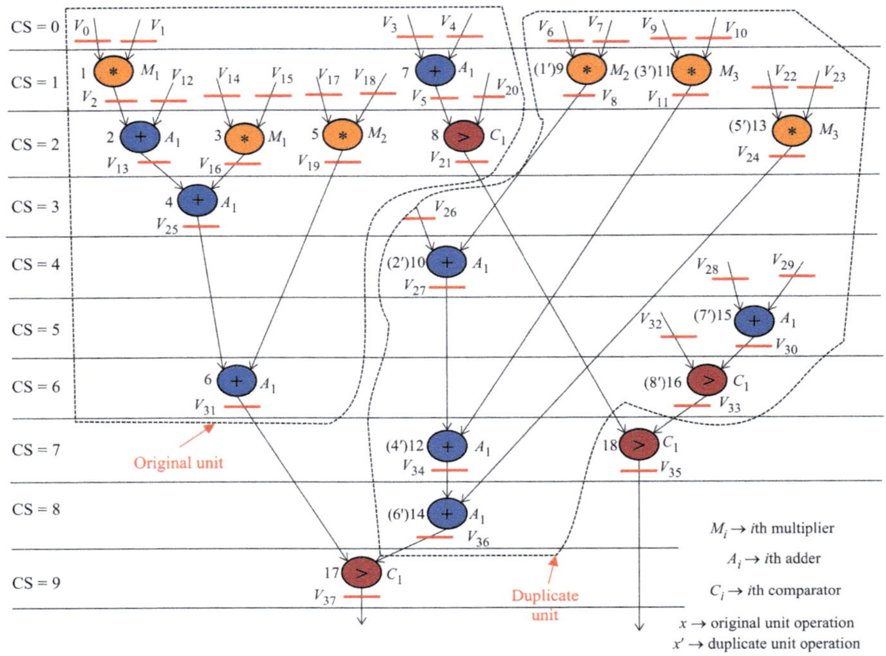

Figure 4.6 Transient fault-secured design ($SDFG^{DMR}$) of an FIR filter (UF = 3)

Table 4.1 RSA encryption module results

Hashed bitstream blocks	Decimal equivalent input	Encrypted decimal output	Binary equivalent of output
10000100	132	89	1011001
01011110	94	289	100100001
10011101	157	222	11011110
01100000	96	8	1000
11001100	204	113	1110001
00010100	20	136	10001000
10111110	190	187	10111011
01101111	111	84	1010100
10010010	146	81	1010001
01110100	116	116	1110100

signature. The first S bits of the bitstream can be used as the digital signature and the value of S is chosen by the vendor based on the robustness required. The higher the digital signature size (S), more additional edges (constraints) will be added and the greater will be the security (robustness) of the digital signature. For this demonstration, a digital signature size of $S = 15$ is considered. The digital signature 101100110010000 is embedded in the $SDFG^{DMR}$ in phase 4.

Table 4.2 Digital signature bit-to-edge constraint mapping

Bits in digital signature ($S = 15$)	Corresponding additional edges
1	$\langle V_2, V_4 \rangle$
0	$\langle V_2, V_3 \rangle$
1	$\langle V_2, V_6 \rangle$
1	$\langle V_2, V_8 \rangle$
0	$\langle V_2, V_5 \rangle$
0	$\langle V_2, V_7 \rangle$
1	$\langle V_2, V_{10} \rangle$
1	$\langle V_2, V_{12} \rangle$
0	$\langle V_2, V_{11} \rangle$
0	$\langle V_2, V_{13} \rangle$
1	$\langle V_2, V_{14} \rangle$
0	$\langle V_2, V_{17} \rangle$
0	$\langle V_2, V_{19} \rangle$
0	$\langle V_2, V_{23} \rangle$
0	$\langle V_2, V_{29} \rangle$

Table 4.3 Register allocation table of fault-secured FIR loop ($UF = 3$)

Control step	*R*1	*R*2	*R*3	*R*4	*R*5	*R*6	*R*7	*R*8	*R*9	*R*10	*R*11	*R*12
0	V_0	V_1	V_3	V_4	V_6	V_7	V_9	V_{10}	–	–	–	–
1	V_2	V_5	V_8	V_{11}	V_{12}	V_{14}	V_{15}	V_{17}	V_{18}	V_{20}	V_{22}	V_{23}
2	V_{13}	V_{16}	V_8	V_{11}	V_{19}	V_{21}	V_{24}	–	–	–	–	–
3	V_{25}	V_{26}	V_8	V_{11}	V_{19}	V_{21}	V_{24}	–	–	–	–	–
4	V_{25}	V_{27}	V_{28}	V_{11}	V_{19}	V_{21}	V_{24}	V_{29}	–	–	–	–
5	V_{25}	V_{27}	V_{30}	V_{11}	V_{19}	V_{21}	V_{24}	V_{32}	–	–	–	–
6	V_{31}	V_{27}	V_{33}	V_{11}	–	V_{21}	V_{24}	–	–	–	–	–
7	V_{31}	V_{34}	V_{35}	–	–	–	V_{24}	–	–	–	–	–
8	V_{31}	V_{36}	V_{35}	–	–	–	–	–	–	–	–	–
9	V_{37}	–	V_{35}	–	–	–	–	–	–	–	–	–

4.5.3 Embedding the digital signature

To embed the digital signature, first the CIG is created representing the register allocation information of the transient fault-secured, scheduled, FIR benchmark. The register allocation information (corresponding to the created CIG) is provided in Table 4.3. A register transfer level (RTL) data path for the transient fault-secured FIR filter using the register allocation information is generated. Next, the digital signature bits are mapped to the extra edges that are to be added in the CIG to impose the watermarking constraints. This mapping between digital signature bits and node pairs of additional edges is shown in Table 4.2.

Figure 4.7 shows the modified CIG of transient fault-secured FIR filter after adding the additional edges representing the digital signature constraints. It can be observed from Figure 4.7 that the additional edges included are of two types. The

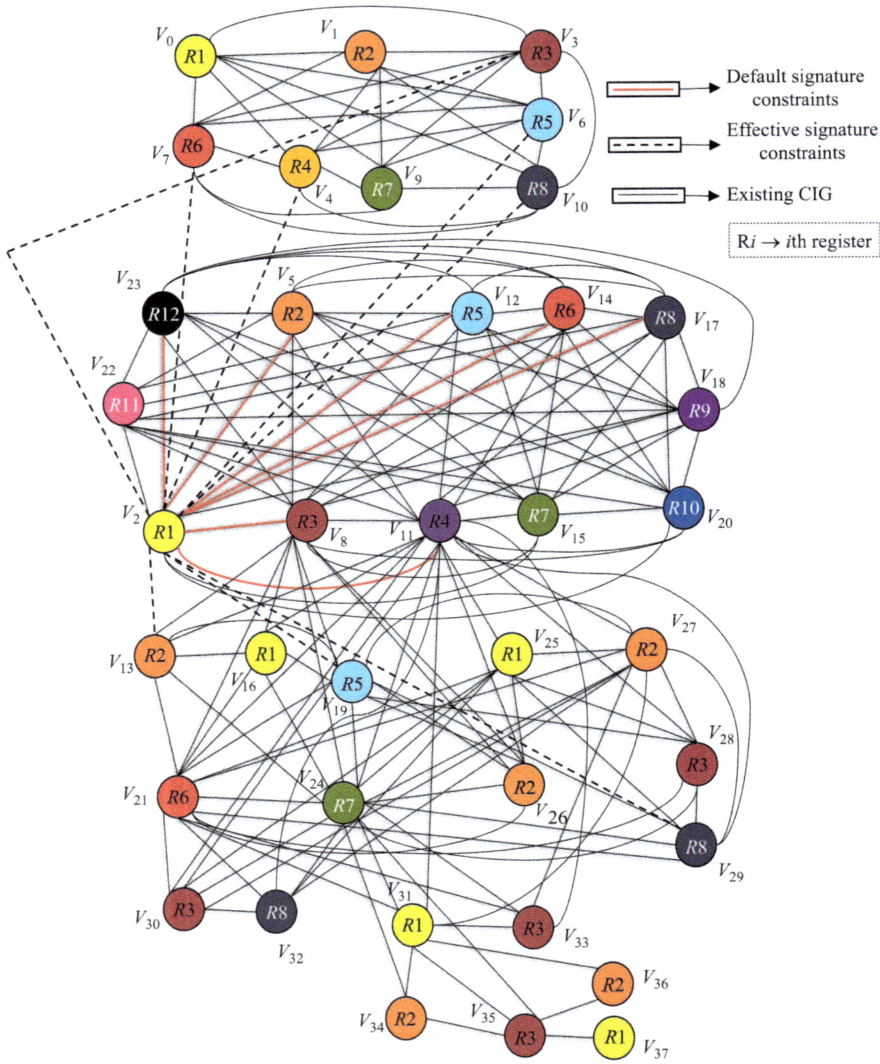

Figure 4.7 CIG of $SDFG^{DMR}$ of fault-secured FIR filter (UF = 3) after embedding the digital signature (S = 15)

edges represented by dash edges (lines) are the effective constraints and the solid edges (lines) in red represent default constraints. The effective edges are newly added edges in the CIG, whereas default edges are already present in the CIG (before embedding the digital signature). Out of 15 additional edges that are to be added in the CIG to embed the digital signature constraints, only eight edges are added (effective edges) because seven edges are already present in the CIG (default

Table 4.4 Register allocation table of fault-secured FIR loop (UF = 3) after embedding the digital signature (S = 15)

Control step	*R1*	*R2*	*R3*	*R4*	*R5*	*R6*	*R7*	*R8*	*R9*	*R10*	*R11*	*R12*
0	V_0	V_1	V_3	V_4	V_6	V_7	V_9	V_{10}	–	–	–	–
1	V_2	V_5	V_8	V_{11}	V_{12}	V_{14}	V_{15}	V_{17}	V_{18}	V_{20}	V_{22}	V_{23}
2	V_{16}	V_{13}	V_8	V_{11}	V_{19}	V_{21}	V_{24}	–	–	–	–	–
3	V_{25}	V_{26}	V_8	V_{11}	V_{19}	V_{21}	V_{24}	–	–	–	–	–
4	V_{25}	V_{27}	V_{28}	V_{11}	V_{19}	V_{21}	V_{24}	V_{29}	–	–	–	–
5	V_{25}	V_{27}	V_{30}	V_{11}	V_{19}	V_{21}	V_{24}	V_{32}	–	–	–	–
6	V_{31}	V_{27}	V_{33}	V_{11}	–	V_{21}	V_{24}	–	–	–	–	—
7	V_{31}	V_{34}	V_{35}	–	–	–	V_{24}	–	–	–	–	–
8	V_{31}	V_{36}	V_{35}	–	–	–	–	–	–	–	–	–
9	V_{37}	–	V_{35}	–	–	–	–	–	–	–	–	–

edges). The solid edges (red) between node pairs (V_2, V_5), (V_2, V_8), (V_2, V_{11}), (V_2, V_{12}), (V_2, V_{14}), (V_2, V_{17}), (V_2, V_{23}) were already present in the CIG and, therefore, the constraints imposed by them are satisfied by default. Hence the register allocation table has to be modified only to satisfy the digital signature constraints represented by the effective edges between node pairs (V_2, V_3), (V_2, V_4), (V_2, V_6), (V_2, V_7), (V_2, V_{10}), (V_2, V_{13}), (V_2, V_{19}), (V_2, V_{29}). Among the aforementioned node pairs, conflict occurs when a new edge is to be added between (V_2, V_{13}) as they both are allocated the same register – yellow. Therefore V_2 and V_{13} should be assigned to a different register in order to satisfy the digital-signature constraint. To accomplish this, it is to be determined if it is possible to swap one variable (either V_2 or V_{13}) with another variable in their respective CS without violating other digital-signature constraints, eventually necessitating them to be executed by different registers. Inspecting from left to right in the CS 2 of Table 4.3, it can be seen that V_{13} can be swapped with V_{16} (currently assigned to green) without violating other watermark constraints. Thus, as shown in modified register allocation table (Table 4.4), V_{13} is swapped with V_{16} (V_{13} re-allocated to the green register) while V_2 remains assigned to the yellow register. Once the register allocation table is modified, the RTL data path generated earlier is modified to finally obtain a watermarked, transient, fault-secured FIR filter.

4.6 Analysis of case studies – DSP cores

This section presents an analysis of case studies of the digital signature–based watermark for secured DSP–IP cores. The original transient, fault-secured design before embedding the digital signature–based watermark is called the baseline design. The 15 nm technology scale based on the NanGate library was adopted to evaluate the area and latency of a watermarked, transient, fault-secured DSP core design.

The metrics used to evaluate the digital signature–based watermark are:

1. Design cost of the presented approach of protecting fault-secured DSP IP cores using a digital signature–based watermark.

2. Robustness of the presented approach.
3. Impact on storage overheads on increasing the digital signature size.
4. Comparison of robustness and storage overheads with a contemporary approach.
5. Comparison with baseline design in terms of storage overheads.

The security and design cost analysis of the fault-secured DSP–IP core design protected using a digital signature–based watermark are:

4.6.1 Security analysis

The probability of coincidence (P_c) adopted from Koushanfar *et al.* (2005) is used as the security metric for the presented approach. It measures the probability of generating the same solution before, and after, embedding the digital signature and indicates proof of ownership. The lower the value of P_c, the stronger the proof of authorship. The equation to compute the P_c value is given as

$$P_c = \left(1 - \frac{1}{c}\right)^S \tag{4.3}$$

where c = number of registers used in the original (baseline) design, S = digital signature size. P_c values for the presented approach for five different digital signature sizes and seven fault-secured DSP applications are reported in Table 4.5. It can be observed from the table that for all fault-secured DSP cores, as digital signature size increases, the P_c value decreases. This variation of P_c with S is meaningful because an increase in digital signature size indicates more additional edges are included in the design, which can result in more storage overheads, thereby reducing the possibility of generating the same design. Further, the

Table 4.5 Variation of probability of coincidence for different digital signature sizes for the digital signature–based watermark approach

Benchmarks	**c**	**Size of digital signature (S)**				
		$S = 15$ P_c	**$S = 30$ P_c**	**$S = 60$ P_c**	**$S = 120$ P_c**	**$S = 240$ P_c**
Band pass filter (BPF)	12	0.2711	0.0735	5.4036×10^{-3}	2.9199×10^{-5}	8.526×10^{-10}
JPEG sample	22	0.4977	0.2477	0.0613	3.7635×10^{-3}	1.4164×10^{-5}
MESA feedback points	33	0.6303	0.3973	0.1578	0.0249	6.2037×10^{-4}
MESA interpolate aux	77	0.8219	0.6756	0.4564	0.2083	0.0434
Inverse discrete cosine transform (IDCT)	16	0.3798	0.1443	0.0208	4.3307×10^{-4}	1.8755×10^{-7}
FIR loop (UF = 2)	9	0.1709	0.0292	8.5280×10^{-4}	7.2727×10^{-7}	5.2893×10^{-13}
FIR loop (UF = 4)	12	0.2711	0.0735	5.4036×10^{-3}	2.9199×10^{-5}	8.526×10^{-10}

percentage reduction in the P_c value for the presented approach compared to Sengupta and Bhadauria (2016) is highlighted in Figure 4.8. This implies that the presented approach provides stronger (more definite) proof of ownership compared to Sengupta and Bhadauria (2016) for the same signature sizes. The reason for reduction in P_c values can be explained from the relation between P_c and c. From (4.3), it is clear that the P_c value is lower for small c values. The number of registers used in the baseline design is higher in case of Sengupta and Bhadauria (2016) than the presented approach, hence the reduction in P_c value.

4.6.2 Design cost analysis

The embedding cost of the presented approach is evaluated in terms of design solution area and latency. The function used to compute the embedding cost (Sengupta and Bhadauria, 2016) is as follows:

$$C_f(X_i) = \phi_1 \frac{L_T}{L_{\max}} + \phi_2 \frac{A_T}{A_{\max}} \tag{4.4}$$

where $C_f(X_i)$ is the cost of the embedding watermark in a transient, fault-secured, design solution for resource configuration input X_i; A_T and L_T denote the area and latency of a fault-secured watermarked design solution, respectively; $A_{\max}$ and $L_{\max}$ are maximum possible area and latency of a design solution and φ_1 and ϕ_2 are user-chosen weights both equal to 0.5 for equal priority of area and delay. The design-cost values computed after embedding the digital signature of size 60 for different fault secured DSP cores are reported in Table 4.6.

Sub-figures in Figures 4.9 and 4.10 plot the number of extra registers required (£) for the presented approach for different fault-secured DSP cores when doubling the digital signature size. The minimum signature size considered was 15 bits and the size was doubled till 240 bits. For more than half of the reported benchmarks, such as the JPEG sample, IDCT, MESA feedback points and FIR loop (UF = 2), there is zero increment in register hardware even for the maximum signature size applicable. For the rest of the benchmarks, such as MESA interpolate aux and FIR loop (UF = 4), there is nominal increase in k value for large digital signature sizes.

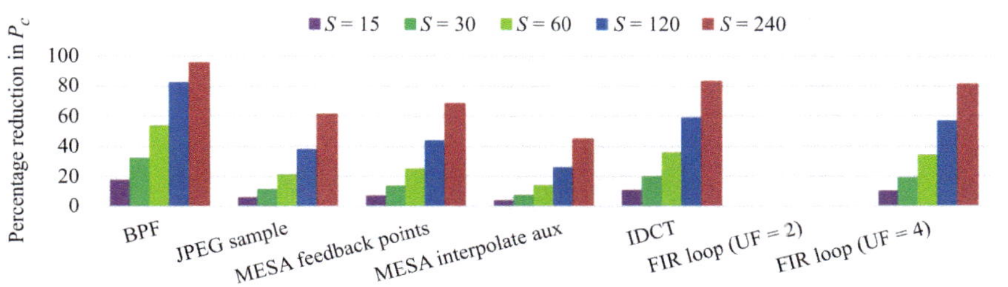

Figure 4.8 Percentage of reduction in P_c (i.e. enhancement of proof of ownership) of digital signature–based watermark approach compared to Sengupta and Bhadauria (2016)

Table 4.6 Results for a digital signature (S = 60) solution through the digital signature–based watermark approach

DSP applications	Presented solution	A_T (μm²)	L_T (ps)	Design cost
BPF	3(+), 4(×), 1(>)	372.0808	15,000	0.3139
JPEG sample	8(+), 2(×), 1(>)	322.5352	14,000	0.2849
IDCT	8(+), 4(×), 3(>)	475.3	11,000	0.1709
FIR loop (UF = 2)	1(+), 2(×), 1(>)	180.1912	8,000	0.5656
FIR loop (UF = 4)	1(+), 3(×), 1(>)	258.048	11,000	0.5165
MESA feedback points	6(+), 8(×), 4(>)	756.1544	9,000	0.2415
MESA interpolate aux	6(+), 16(×), 3(>)	1,391.4928	20,000	0.2140

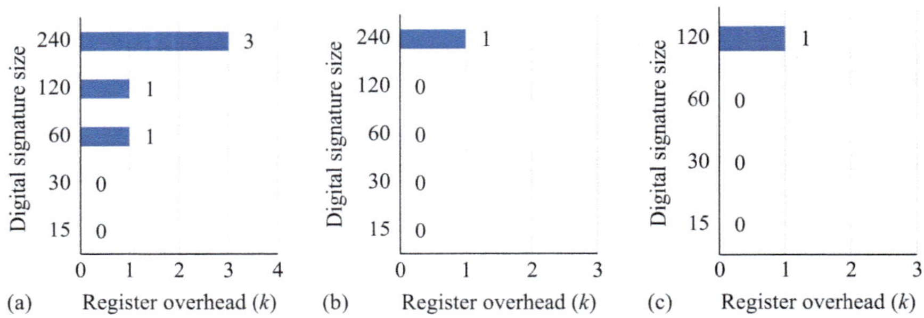

Figure 4.9 Impact on storage overheads on doubling the size of the digital signature in the digital signature–based watermark approach for (a) BPF, (b) MESA interpolate aux and (c) FIR loop (UF = 4)

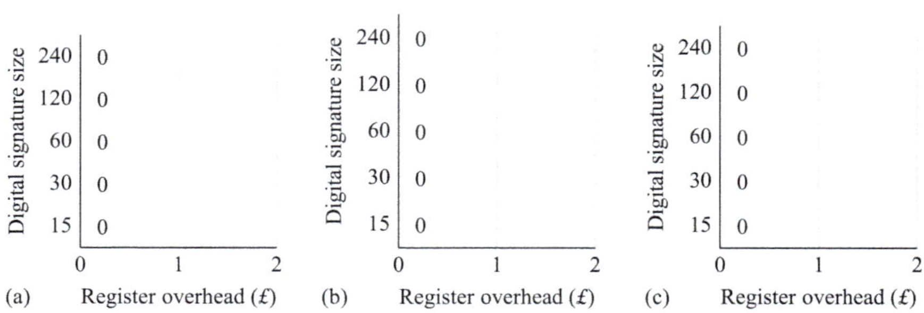

Figure 4.10 Impact on storage overheads on doubling the size of the digital signature in digital signature–based watermark approach for (a) JPEG sample, (b) IDCT and (c) MESA feedback points

Table 4.7 compares the number of registers required post-embedding the digital signature for the presented approach and Sengupta and Bhadauria (2016) for different fault-secured DSP cores and w.r.t. different digital signature sizes (watermark strengths). It is evident that less register hardware is required in the

Table 4.7 Storage hardware required in digital signature–based watermark approach and Sengupta and Bhadauria (2016) for different signature sizes

DSP applications	Number of registers required in the presented approach					Number of registers required in (Sengupta and Bhadauria, 2016)				
	$S = 15$	$S = 30$	$S = 60$	$S = 120$	$S = 240$	$S = 15$	$S = 30$	$S = 60$	$S = 120$	$S = 240$
BPF	12	12	13	13	15	14	15	16	17	17
JPEG sample	22	22	22	22	22	24	24	25	25	26
MESA feedback points	33	33	33	33	33	39	39	39	40	40
MESA interpolate aux	77	77	77	77	78	95	95	95	95	96
IDCT	16	16	16	16	16	18	18	18	19	19
FIR loop (UF = 2)	9	9	9	NA	NA	9	10	10	NA	NA
FIR loop (UF = 4)	12	12	12	13	NA	13	13	14	14	NA

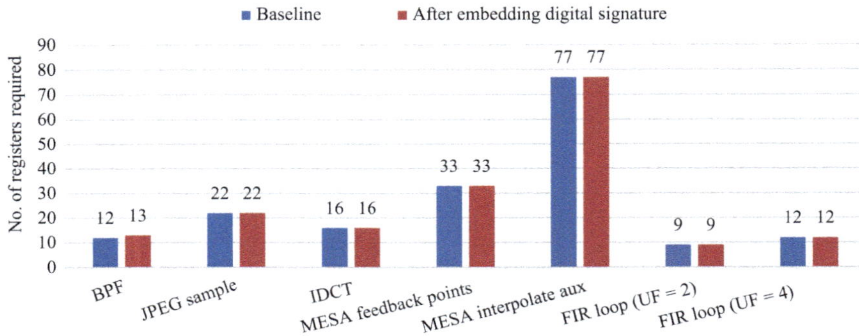

Figure 4.11 Comparison of the amount of register hardware used in the baseline design and after embedding the digital signature (S = 60)

case of the presented approach compared to Sengupta and Bhadauria (2016) for all fault-secured DSP cores reported, irrespective of the digital signature size. Additionally, from Table 4.7, the percentage reduction of register hardware for the presented approach compared to Sengupta and Bhadauria (2016) for all benchmarks at digital signature size $S = 60$ are evaluated and the average reduction obtained is 14.35%. The reason for the reduction in the number of registers required after embedding the digital signature can be logically explained. The watermark embedded in an IP core design in Sengupta and Bhadauria (2016) is a random combination of variables chosen by the vendor. It is imperative that the vendor selects a watermark that results in low overheads. But, given a watermark size, there is a very large number of combinations possible for a watermark, therefore rendering it very complex and challenging to choose a watermark at random that can guarantee high robustness and low-storage overheads. The approach presented in this chapter eliminates the aforesaid limitation by always generating a dedicated digital signature for a given IP core through an algorithm using SHA-512 and RSA encryption.

4.6.2.1 Comparison with baseline design in terms of storage hardware

Figure 4.11 plots the register hardware required in a baseline design and for the digital signature approach (i.e., after embedding the digital signature of size 60). It can be observed that only in BPF is there a nominal increment in number of registers required after imposing the additional constraints (which is only one extra register). For rest of the benchmarks, zero additional registers are required for the presented approach, even for a considerably large digital signature size of 60.

4.7 Conclusion

This chapter presented a multi-level encoding and encrypted-hash-based digital signature for protection of fault-secured DSP cores against IP piracy and false claims of ownership. The digital signature embedding approach achieved stronger proof of ownership, while incurring less register hardware overheads compared to contemporary works. The analysis on case studies revealed that the digital signature–based watermark on average obtained a reduction of Pc and storage hardware of 26.27% and 14.35%, respectively, compared to Sengupta and Bhadauria (2016). This chapter provides new insight into the utility of digital signature–based watermarks for protection of fault-secured IP cores from digital signal processing domain. A reader at the end of this chapter is able to understand the following concepts:

- Transient fault.
- Process of transient fault security.
- Encoding algorithm used for forming the digital signature of fault-secured IP cores.
- Design of colouring graph for fault-secured IP cores.
- Case study on fault-secured FIR filter.
- Application of SHA-512 and RSA in deriving digital signature for fault-secured IP cores.
- Comparison and analysis of case studies between digital signature–based watermark and traditional watermarking for fault-secured IP cores.
- Overhead analysis of digital signature–based watermark on fault-secured IP cores.

4.8 Questions and exercises

1. What is a transient fault?
2. What is the typical strength of a transient fault?
3. How is a multi-cycle transient fault different from a transient fault?
4. What is IP piracy?
5. What is the utility of IP metering?
6. What is reverse engineering?
7. What is SEU?
8. What is LET?
9. What is HLT?
10. What is loop unrolling factor?
11. How is a CDFG different from a DFG?
12. How does RA produce cipher-data? What is the encryption function?
13. What is the transfer function of an FIR filter?
14. Explain the process of detecting a digital signature for fault-secured IP cores.
15. How can you construct a fault-secured IP core schedule?
16. How is a private key in RSA generated?
17. Explain the encoding algorithm for embedding digital signature–based watermark constraints in the IP design.
18. Explain the encoding algorithm for converting a DSP core into a bitstream.
19. Explain the fault-security algorithm.

20. Explain the digital signature concatenation process.
21. How does a digital signature watermark perform for fault-secured IP cores in terms of probability of coincidence and design cost/overheads?

References

Y. M. Alkabani and F. Koushanfar, "Active hardware metering for intellectual property protection and security," in Proc. 16th USENIX Security Symposium, 2007, Art. no. 20.

D. Atienza, *et al.*, "Reliability-aware design for nanometer-scale devices," in Asia and South Pacific Design Automation Conference, Seoul, Pages: 549–554, 2008.

R. S. Chakraborty and S. Bhunia, "HARPOON: An obfuscation-based SoC design methodology for hardware protection," IEEE Transactions on Computer-Aided Design of Integrated Circuits and Systems, vol. 28, no. 10, Pages: 1493–1502, 2009.

M. R. Choudhury and K. Mohanram, "Reliability analysis of logic circuits," IEEE Transactions on Computer-Aided Design of Integrated Circuits and Systems, vol. 28, no. 3, Pages: 392–405, 2009.

M. Fazeli, S. N. Ahmadian, S. G. Miremadi, H. Asadi, and M. B. Tahoori, "Soft error rate estimation of digital circuits in the presence of Multiple Event Transients (METs)," in Design, Automation and Test in Europe Conference and Exhibition (DATE), Pages: 1–6, 2011.

M. Gomaa and T. N. Vijaykumar, "Opportunistic transient-fault detection," in Computer Architecture, ISCA'05. Proceedings. 32nd International Symposium on. IEEE, 2005, 2005.

F. Koushanfar, I. Hong, and M. Potkonjak, "Behavioral synthesis techniques for intellectual property protection," ACM Transactions on Design Automation of Electronic Systems, vol. 10, no. 3, Pages: 523–545, 2005.

G. Lakshminarayana, A. Raghunathan, and N. K. Jha, "Behavioral synthesis of fault secure controller/datapaths based on aliasing probability analysis," IEEE Transactions on Computers, vol. 49, Pages: 865–885, 2000.

P. K. Lala, Fault-Tolerant and Fault-Testable Hardware Design, Prentice-Hall, Upper Saddle River, NJ, 1985.

Y. Lao and K. K. Parhi, "Obfuscating DSP circuits via high-level transformations," IEEE Transactions on Very Large Scale Integration (VLSI) Systems, vol. 23, no. 5, Pages: 819–830, 2015.

B. Le Gal and L. Bossuet, "Automatic low-cost IP watermarking technique based on output mark insertions," Design Automation for Embedded Systems, vol. 16, no. 2, Pages: 71–92, 2012.

W. Liang, K. Wu, Y. Xie, and J. Duan, "TDCM: An IP watermarking algorithm based on two dimensional chaotic mapping," Computer Science and Information Systems, vol. 12, Pages: 823–841, 2015.

A. Martinez-Alvarez, F. Restrepo-Calle, S. Cuenca-Asensi, L. M. Reyneri, A. Lindoso, and L. Entrena, "A hardware-software approach for on-line soft error mitigation in interrupt-driven applications," IEEE Transactions Dependable and Secure Computing, vol. 13, no. 4, Pages: 502–508, 2016.

N. Miskov-Zivanov, "Multiple transient faults in combinational and sequential circuits: A systematic approach," IEEE Transactions on Computer-Aided Design of Integrated Circuits and Systems, vol. 29, no. 10, Pages: 1614–1627, 2010.

E. Mizan, T. Amimeur, and M. F. Jacome, "Self-imposed temporal redundancy: An efficient technique to enhance the reliability of pipelined functional units," in Computer Architecture and High Performance Computing, SBAC-PAD 2007. 19th International Symposium on. IEEE, 2007, 2007.

S. S. Mukherjee, J. Emer, and S. K. Reinhardt, "The soft error problem: An architectural perspective," in High-Performance Computer Architecture, HPCA-11. 11th International Symposium on, Pages: 243–247, 2005.

T. Nie, L. Zhou, and Y. Li, "Hierarchical watermarking method for FPGA IP protection," IETE Technical Review, vol. 30, no. 45, Pages: 367–374, 2013.

E. Normand, "Single-event effects in avionics," IEEE Transactions on Nuclear Science, vol. 43, Pages: 461–474, 1996.

M. Pignol, "How to cope with SEU/SET at system level?," in On-Line Testing Symposium, 2005. IOLTS. 11th IEEE International, vol. no., Pages: 315–318, 2005.

I. Polian, J. P. Hayes, S. Kundu, and B. Becker, "Transient fault characterization in dynamic noisy environments," in Test Conference, Proceedings. ITC 2005. IEEE International, pp. 10, pp. 1048, 2005.

D. Roy and A. Sengupta, "Low overhead symmetrical protection of reusable IP core using robust fingerprinting and watermarking during high level synthesis," Journal on FGCS, vol. 71, Pages: 89–101, 2017.

C. Rusu, *et al.*, "Multiple event transient induced by nuclear reactions in CMOS logic cells," in 13th IEEE International On-Line Testing Symposium, Pages: 137–145, 2007.

R. Schneiderman, "DSPs evolving in consumer electronics applications," IEEE Signal Processing Magazine, vol. 27, no. 3, Pages: 6–10, 2010.

A. Sengupta, "Soft IP core design resiliency against terrestrial transient faults for CE products [hardware matters]," IEEE Consumer Electronics Magazine, vol. 5, no. 4, Pages: 129–131, 2016a.

A. Sengupta, "Intellectual property cores: Protection designs for CE products," IEEE Consumer Electronics Magazine, vol. 5, no. 1, Pages: 83–88, 2016b.

A. Sengupta, "Hardware security of CE devices [hardware matters]," IEEE Consumer Electronics Magazine, vol. 6, no. 1, Pages: 130–133, 2017.

A. Sengupta and S. Bhadauria, "Exploring low cost optimal watermark for reusable IP cores during high level synthesis," IEEE Access, vol. 4, Pages: 2198–2215, 2016.

A. Sengupta and D. Kachave, "Low cost fault tolerance against kc-cycle and km-unit transient for loop based control data flow graphs during physically aware high level synthesis," Elsevier Journal on Microelectronics Reliability, vol. 74, Pages: 88–99, 2017.

A. Sengupta and S. P. Mohanty, "IP Core and Integrated Circuit Protection Using Robust Watermarking," IP Core Protection and Hardware-Assisted Security for Consumer Electronics, Pages: 123–170, 2019, e-ISBN: 9781785618000.

A. Sengupta and D. Roy, "Antipiracy-aware IP chipset design for CE devices: A robust watermarking approach [hardware matters]," IEEE Consumer Electronics Magazine, vol. 6, no. 2, Pages: 118–124, 2017.

A. Sengupta and D. Roy, "Multi-phase watermark for IP core protection," in Proc. 36th IEEE International Conference on Consumer Electronics (ICCE), Pages: 1–3, 2010.

A. Sengupta, R. Sedaghat, and Z. Zeng, "A high level synthesis design flow with a novel approach for efficient design space exploration in case of multi-parametric optimization objective," Microelectronics Reliability, vol. 50, no. 3, Pages: 424–437, 2010.

A. Sengupta, S. Bhadauria, and S. P. Mohanty, "TL-HLS: Methodology for low cost hardware Trojan security aware scheduling with optimal loop unrolling factor during high level synthesis," IEEE Transactions on CAD of Integrated Circuits and Systems, vol. 36, no. 4, Pages: 655–668, 2017.

A. Sengupta, D. Roy, and S. P. Mohanty, "Triple-phase watermarking for reusable IP core protection during architecture synthesis," IEEE Transactions on Computer-Aided Design of Integrated Circuits and Systems, vol. 37, no. 4, Pages: 742–755, 2018.

A. Sengupta, D. Kachave, and D. Roy, "Low cost functional obfuscation of reusable IP cores used in CE hardware through robust locking," IEEE Transactions on Computer-Aided Design of Integrated Circuits & Systems (TCAD), vol. 38, no. 4, Pages: 604–616,

A. Sengupta, P. Chandra, and R. Kumar, "Robust digital signature to protect IP core against fraudulent ownership and cloning," in Proceedings of 9th IEEE International Conference on Consumer Electronics (ICCE), Berlin, accepted, Sep 2019b.

A. Sengupta, E. R. Kumar, and N. P. Chandra, "Embedding digital signature using encrypted-hashing for protection of DSP cores in CE," IEEE Transactions on Consumer Electronics, vol. 65, no. 3, Pages: 398–407, 2019c.

K. Wu and R. Karri, "Algorithm level recomputing—a register transfer level concurrent error detection technique," in Proc. IEEE/ACM Int. Conf. Computer-Aided Design, Pages: 537–543, 2001.

T. Yu and Y. Zhu, "A new watermarking method for soft IP protection," in Proc. International Conference on Consumer Electronics, Communications and Networks (CECNet), XianNing, Pages: 3839–3842, 2011.

M. Zhang, *et al.*, "Sequential element design with built-in soft error resilience," IEEE Transactions on Very Large Scale Integration (VLSI) Systems, vol. 14, no. 12, Pages: 1368–1378, 2006.

D. Ziener and J. Teich, "Power signature watermarking of IP cores for FPGAs," Journal of Signal Processing Systems, vol. 51, no. 1, Pages: 123–136, 2008.

Chapter 5
Multi-level watermark for IP protection

Anirban Sengupta[1]

The chapter describes a multi-level watermarking process for intellectual property (IP) cores that leverages the electronic-system level (ESL) and register-transfer level (RTL). A detailed elaboration is provided on the salient features of this multi-level watermarking approach followed by its encoding technique, embedding process, detection process and, finally, implementation details using case studies.

The chapter is organized as follows: Section 5.1 discusses the abstraction levels of a digital design followed by some IP protection basics; Section 5.2 introduces some of the prior works in this domain; Section 5.3 presents the salient features and advantages of a multi-level watermarking process; Section 5.4 explains the signature-embedding process for digital signal processor (DSP) cores; Section 5.5 discusses the design process of a multi-level watermarked IP core using a finite impulse response (FIR) filter; Section 5.6 presents the signature detection details of a multi-level watermark; Section 5.7 presents an analysis based on case studies; Section 5.8 concludes the chapter.

5.1 Introduction

An integrated circuit (IC) is designed using various (abstraction) levels of design flow. As this chapter describes the deployment of a multi-level watermark using multiple-abstraction levels, a brief discussion of each level is made to build the foundation of this topic. In the very large scale integration design flow of application-specific ICs, a design (or application) is first described at system level in the form of system specifications/input–output. The system specifications represent the design at the highest level of abstraction. The next level of the design flow is the architectural or behavioural level. This level is also referred as the ESL. At this level, an application can be represented in the form of a behavioural description, while the control-data flow graph provides the intermediate representation of an application at the high-level (McFarland *et al.*, 1988). The design process at this level is integrated with the design space exploration (DSE) technique to identify the optimal resource (number of multipliers, adders/subtractors, comparators, etc.)

[1]Computer Science and Engineering, Indian Institute of Technology Indore, Indore, India

configuration (Mishra and Sengupta, 2014). Obtaining the optimal resource configuration through DSE is crucial in order to satisfy design constraints (such as delay, power and area). Better performance in a design can be achieved by performing a high-level design process (scheduling, allocation and binding) using an optimal-resource configuration obtained through DSE. Further, the high-level design process identifies the interconnection between different units such as the arithmetic and logic unit, control unit, and storage hardware (Sengupta *et al.*, 2010). Thus a high-level design process results in the RTL description of the design flow. The subsequent levels of the design flow are the logic level and physical level. The logic level represents the design in terms of optimized logic gates. At the physical level, the design is represented in the form of transistors and, hence, is also known as the transistor or layout level.

However, in the context of complex application such as DSP cores, ESL and RTL may be more significant levels of the design flow during optimization of design objectives. Moreover, these levels offer high flexibility than lower levels during employing security or protection into the design against various threats such as IP piracy, counterfeiting, cloning and false claims of ownership (Sengupta, 2016, 2017; Plaza and Markov, 2015; Castillo *et al.*, 2007). These emerging threats to IP cores for a multimedia processor (MP) and DSP have drawn attention of researchers and designers, leading to efforts for better protection of the designs. Since these IP cores are extensively used in modern electronics devices such as digital cameras, laptops and tablets, their protection against false ownership is crucial (Schneiderman, 2010). Different abstraction levels of the design flow can be leveraged to formulate a protection mechanism against threats. There are three reasons for employing protection at higher abstraction levels is more advantageous than lower levels, namely: (1) post-synthesis, security constraints get distributed throughout the design at lower levels; (2) representation of complex designs at a higher level offers lesser complexity than lower levels (e.g. complex applications such as DSP cores are easier to describe at architectural level than directly at gate level) and (3) higher abstraction levels offer the possibility of design trade-off and optimization of design objectives during the process of embedding watermarks (Sengupta and Mohanty, 2019a, 2019b).

The most common approach used for protecting IP cores against IP piracy, counterfeiting and false claims of ownership is watermarking. For example, the approach by Le Gal and Bossuet (2012) and Ziener and Teich (2008) deployed watermarking only at lower abstraction level such as gate level. On the contrary, Koushanfar *et al.* (2005), Sengupta and Bhadauria (2016), Sengupta and Roy (2017), Sengupta and Roy (2018) and Sengupta *et al.* (2018) employed watermarking only at ESL or during high-level synthesis (HLS). However, combining two consecutive abstraction levels for implanting watermarking constraints renders the watermark highly robust, thus providing enhanced design protection. Sengupta *et al.* (2019) and Roy and Sengupta (2019) employed watermarks at two consecutive abstraction levels, viz. ESL and RTL, to protect DSP–IP cores against threats. These watermarks can be referred to as multi-level watermarks. As a watermark in Sengupta *et al.* (2019) is implanted at RTL an ESL, extracting signature constraints during forensic detection becomes more convenient compared to approaches that only implant watermarks at ESL. An overview of the possible attacks on IP cores and its protection using multi-level watermarking is shown in Figure 5.1.

5.2 Discussion on selected approaches

This chapter discusses the protection of DSP–IP cores using multi-level watermarks (Sengupta *et al.*, 2019). As mentioned in the previous section, a multi-level watermark is implanted at both ESL and RTL. Besides this approach, no other approaches exist in the literature that employ watermarks at two levels concurrently. Moreover, only this approach leverages RTL to implant watermarks for DSP applications. Thus the related approaches to Sengupta *et al.* (2019) fall under the set of approaches that implant watermarks at ESL only. The approach by Roy and Sengupta (2017) proposed symmetrical protection for IP cores only at ESL using the seller's watermark and buyer's fingerprint simultaneously. Further approaches (Koushanfar *et al.*, 2005; Sengupta and Bhadauria, 2016; Sengupta and Roy, 2017) implanted watermarks during the register-allocation phase of HLS. Such watermarks are referred to as single-phase watermarks as it they are implanted in a single phase of an abstraction level. Koushanfar *et al.* (2005) implanted a dynamic watermark by adding some design and timing constraints during ESL (behavioural) synthesis. These design and timing constraints represent the vendor's signature.

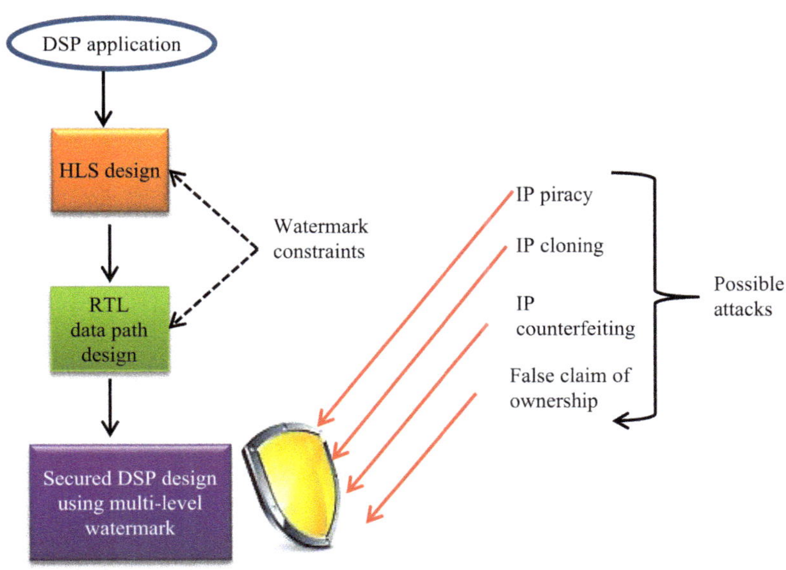

Figure 5.1 Shielding of DSP–IP cores against possible attacks (Sengupta et al., 2019)

Koushanfar *et al.* (2005) referred to the watermarking approach as dynamic because the watermarking constraints are embedded as pre-processing steps before performing the ESL synthesis and, thus, influence the design at lower-abstraction levels. Figure 5.2 shows the watermarking approach of Koushanfar *et al.* (2005). The authors' idea of watermarking (Koushanfar *et al.*, 2005) is exploited from the concept of DSE in ESL synthesis that results into several competitive solutions with identical natures. Koushanfar *et al.* (2005) leveraged the coloured interval graph (CIG) framework to embed the vendor's signature in the form of extra constraints. The CIG framework is used to solve the problem of accommodating all storage variables into a minimum number of registers. Embedding each extra edge leads to

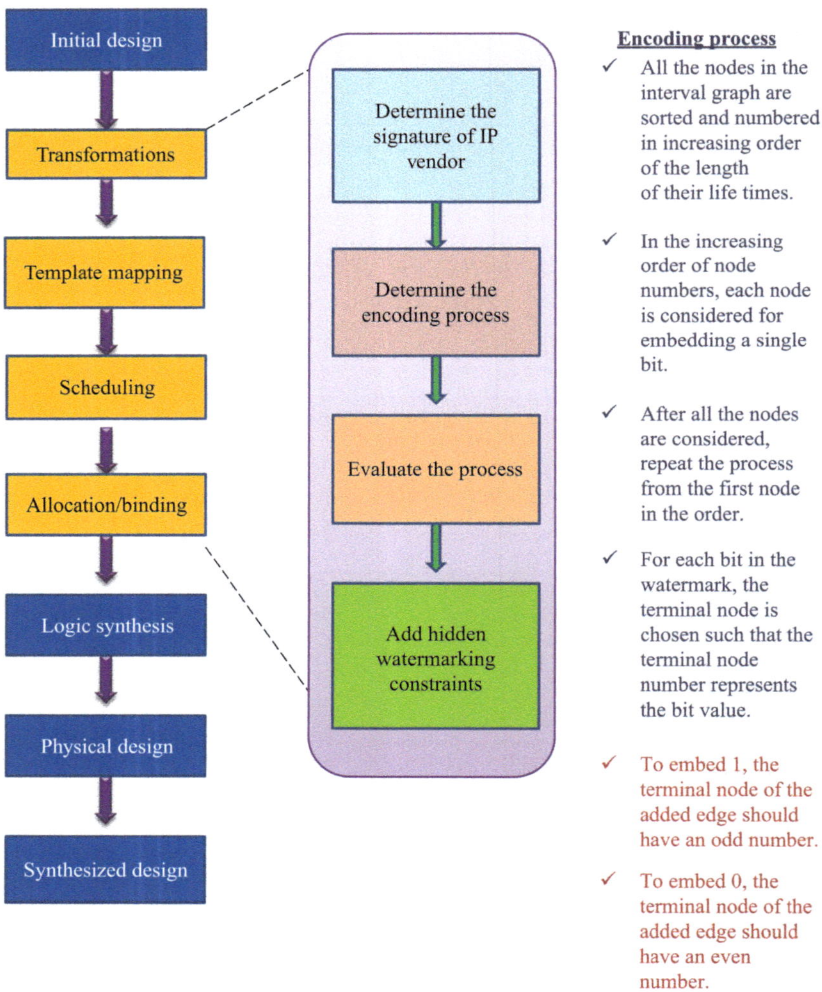

Figure 5.2 Behavioural synthesis techniques for intellectual property protection (Koushanfar et al., 2005)

forced execution of a storage variable through a different register. This results into adding an extra edge in the CIG in the form of a constraint. Koushanfar *et al.* (2005) demonstrated watermarking by using, for example, a 7-bit ASCII equivalent for embedding each character of the vendor's signature into the IP core design. In the process of embedding the watermark, the first step is to sort and assign numbers to each node in the CIG based on their lifetime. As the sorted order is traversed, the lifetime of nodes increases from the first node to the last node. Each bit of the watermark is embedded by considering each node in increasing order. Further, each bit of the watermark is represented by a terminal node of the CIG, therefore carefully choosing terminal nodes is crucial. The embedded extra–constraint edge corresponding to bit '0' should be terminated at an even-numbered node and the embedded extra–constraint edge corresponding to bit '1' should be terminated at an odd-numbered node. The watermarking approach of Koushanfar *et al.* (2005) is more suitable for larger designs. This is because the CIG of larger designs has many nodes, therefore the probability that the same colouring solution after embedding watermark is generated is very low. Koushanfar *et al.* (2005) embedded watermarks by considering eight objectives: (1) the embedded watermark should not affect correctness of the functionality, (2) it should have minimal design overheads, (3) it should be transparent in order to be used for other designs and for any computer-aided design (CAD) tool, (4) the vendor should be able to prove his/her ownership using the watermark, (5) it should be extremely hard to locate (perceptually invisible) for an attacker, (6) it should be resilient against removal through any mechanism, (7) it should be distributive in nature so that not only the entire design, but also its various portions, can be protected and (8) it should be easily detectable at low cost.

Koushanfar *et al.* (2005) assessed the effectiveness of a watermark based on further metrics, namely (1) strength of the ownership proof using probability of coincidence, (2) resilience of the watermark (i.e. how difficult it is to remove it) and (3) degradation of design metrics in terms of percentage changes of the relevant design metrics owing to embedding-watermark constraints.

Furthermore, Sengupta and Roy (2018) and Sengupta *et al.* (2018) embedded watermarks at three different phases (scheduling, hardware allocation, register allocation) of ESL (during HLS). Such watermarks are referred as multi-phase watermarks. However, these watermarks are employed only within a single-abstraction level of the design flow.

5.3 Salient features and advantages of multi-level watermarks

The multi-level watermark exploits both ESL and RTL of design flow to implant the signature constraints into the IP core design. This nature of the watermark provides a number of salient features and advantages (Roy and Sengupta, 2019; Sengupta *et al.*, 2019):

- It leverages RTL along with ESL for implanting signature constraints' distributivity throughout the design.

- The signature encoding of a multi-level watermark is distributive and exclusive by nature.
- Because of the covert signature-embedding process, higher security and robustness can be achieved.
- The implanted multi-level watermark results into a marginal area overhead even at maximum possible signature strength.

5.4 Embedding signatures as secret marks

The overall process of embedding a signature as a secret mark is described in Sections 5.4.1–5.4.2.3.

5.4.1 Problem formulation

A multi-level watermark-protected IP core is designed using an optimal resource configuration. A control data flow graph representing a DSP application is given as the input. The resource configuration used by the design can be represented using the set X as follows:

$$X = \{N_1(R_1), N_2(R_2), \ldots, N_m(R_m)\}$$

where N_1, N_2 and N_m denote the number of resources of type R_1, R_2 and R_m, respectively. Here, m denotes the total number of types of resources.

5.4.2 Details of multi-level watermarks

An overview of the multi-level watermark and its signature encoding, followed by the signature embedding process is provided next (Roy and Sengupta, 2019; Sengupta *et al.*, 2019).

5.4.2.1 Overview of multi-level watermarks

A multi-level watermark is a set of signature constraints that are implanted at two consecutive abstraction levels, viz. ESL and RTL. An abstract view of the multi-level watermark at two abstraction levels is shown in Figure 5.3. A part of the signature constraints is implanted during ESL and the remaining signature constraints are implanted during RTL. The implantation of signature constraints at ESL is performed during the *register-sharing phase*. On the contrary, for implanting signature constraints at RTL, multiplexers (muxes) and de-multiplexers (demuxes) are *fragmented into their next hierarchical size* (e.g. one 8:1 mux is fragmented into two 4:1 muxes and one 2:1 mux, while a one 4:1 mux is fragmented into three 2:1 muxes). Fragmenting the component size is often a common design practice employed in order to reduce design complexity during implementation. Additionally, there are multiple ways to employ register sharing. Thus multi-level watermarks render the detection of signature constraints non-trivial for an adversary. In the multi-level watermarking approach, three variables are used for constituting the signature. The variables are encoded with the aim of implanting the watermark at ESL and RTL. The encoding of each variable is stated next.

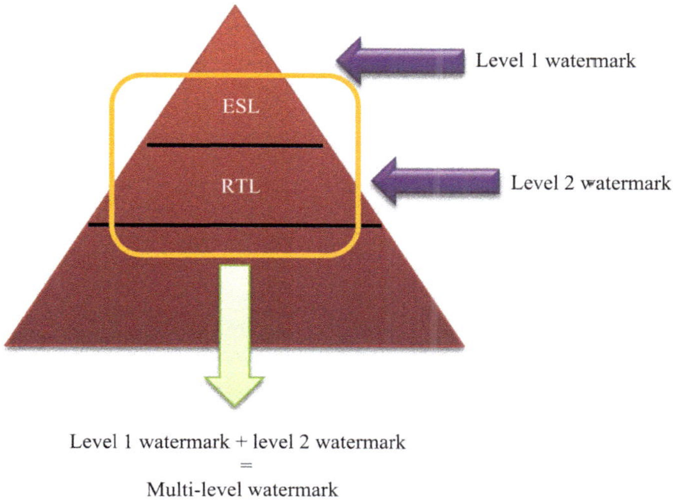

Figure 5.3 Multi-level watermarking for DSP cores

Table 5.1 Multi-level watermark signature encoding

Signature variable	Encoding
ω	Sharing two or three storage variables (executed in different control steps through different registers) to a single register (during register allocation phase of ESL)
θ	Forcing break-up of multiplexers into next hierarchical size (at RTL)
φ	Forcing break-up of de-multiplexers into next hierarchical size (at RTL)

5.4.2.2 Signature encoding

The three variables of multi-level watermark signatures are ω, θ and φ. The encoding of each signature variable is presented in Table 5.1. The signature variable ω is implanted into the design during the register sharing phase of ESL. Forced sharing of two or three storage variables (executed in different control steps through different registers) using a single register represents the security constraint corresponding to ω. Further, the signature variables θ and φ are implanted at RTL. Forcing break-up of muxes into the next lower hierarchical size represents the security constraint corresponding to θ and forcing break-up of demuxes into the next lower-hierarchical size represents the security constraint corresponding to φ.

The impact of insertion of signature variables ω, θ and φ into the IP core design is shown in Figure 5.4. This figure demonstrates the encoding of signature constraints ω, θ and φ in terms of the modifications in the RTL data path of the IP core design.

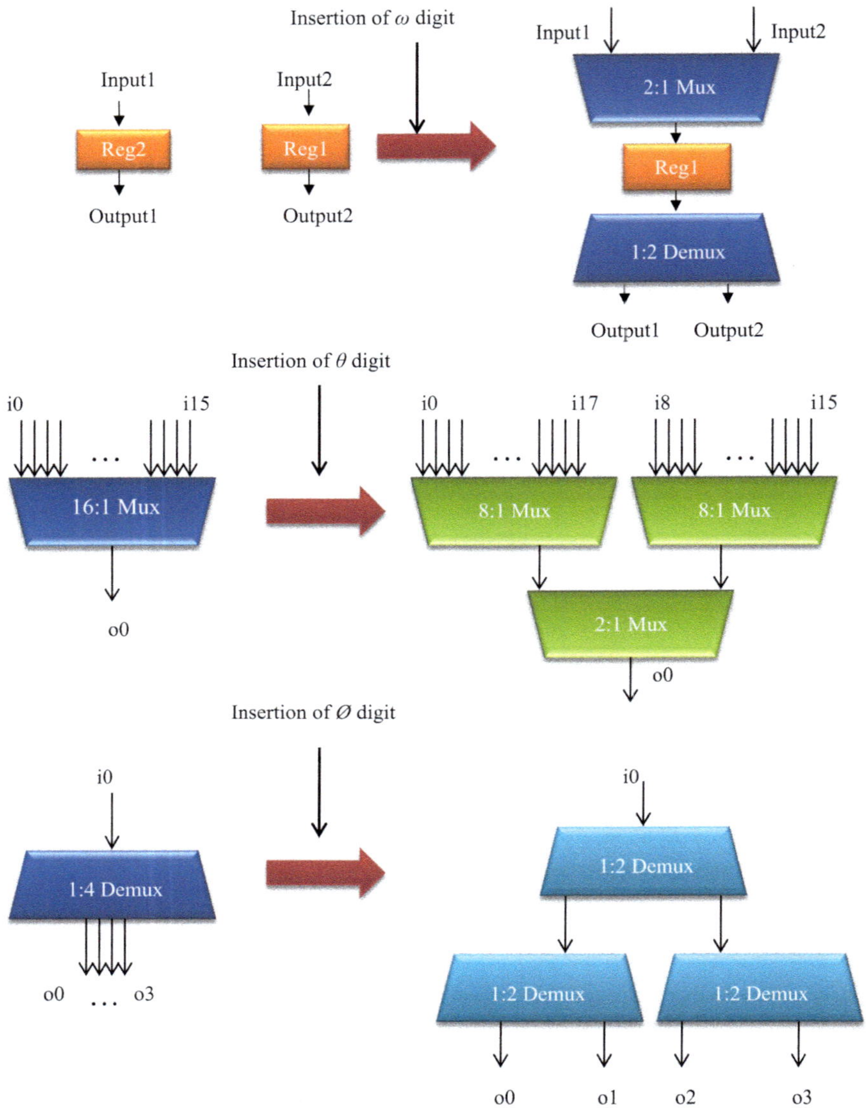

Figure 5.4 Demonstration of watermarking encoding rule (Sengupta et al., 2019)

5.4.2.3 Signature embedding process

The multi-level watermark embedding process at RTL and ESL for a DSP core is shown in Figure 5.5. Before embedding the multi-level watermark, CDFG representing the DSP application is scheduled based on the optimal resource configuration of the intended design. The optimal resource configuration is obtained

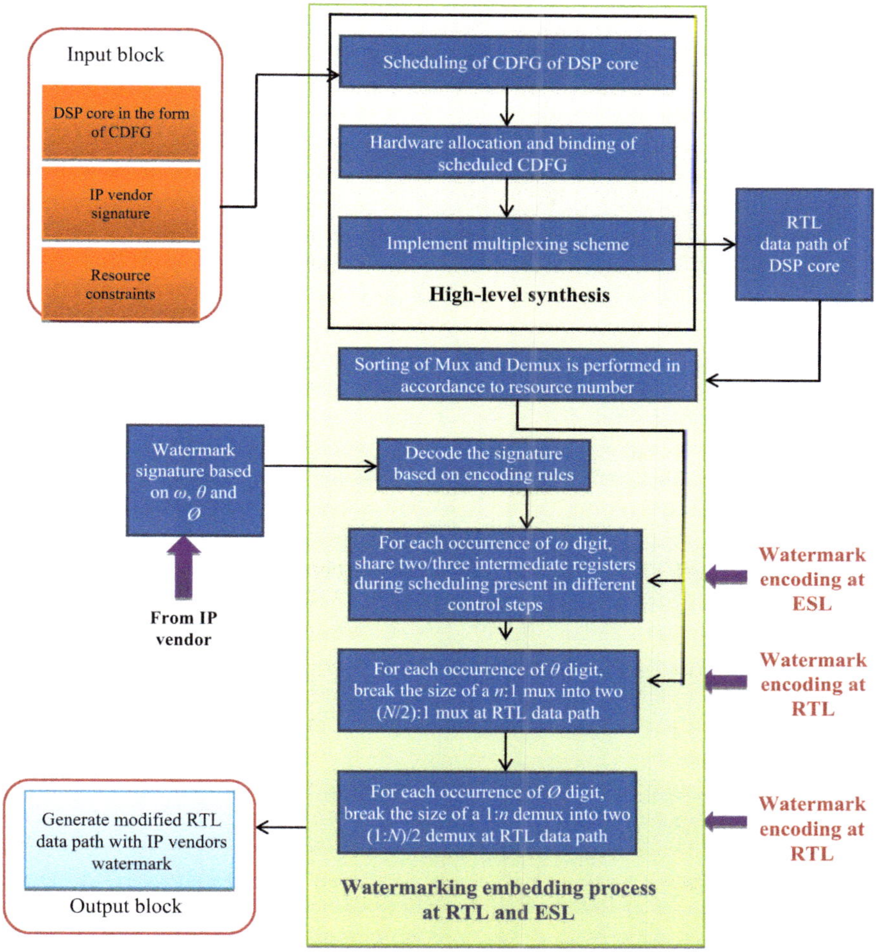

Figure 5.5 Multi-level watermark embedding process at RTL and ESL for DSP cores (Sengupta et al., 2019)

using a DSE framework. To perform DSE, an optimization framework such as particle swarm optimization (PSO) is employed. Once the scheduled and hardware (registers and resources) allocated CDFG of the DSP application is generated, certain steps are performed in order to implant the multi-level watermark:

1. Identify the number of registers (in scheduled CDFG) allocated in different control steps.
2. Divide the total number of such registers (identified in step (1)) into groups of two or three registers.

   Steps (1) and (2) are the pre-processing steps in order to implant signature constraints at ESL.

Table 5.2 Sorting of multiplexers and de-multiplexers based on corresponding hardware resources

Hardware resource	Multiplexers (mux)	De-multiplexers (demux)
Adder (A)	{mux for $A1$, mux for $A2,\ldots,$ mux for Au}	{demux for $A1$, demux for $A2,\ldots,$ demux for Au}
Multiplier (M)	{mux for $M1$, mux for $M2,\ldots,$ mux for Mv}	{demux for $M1$, demux for $M2,\ldots,$ demux for Mv}

Note: u, v denote the number of adders and multipliers, respectively, in the RTL design.

3. Corresponding to each hardware resource, the size of each multiplexer and de-multiplexer are determined in order to implement RTL design.
4. Based on the hardware resource number, all corresponding multiplexers and de-multiplexers are arranged in ascending order as shown in Table 5.2.

   Steps (3) and (4) are the pre-processing steps in order to implant signature constraints at RTL. The next steps can now be performed to embed the signature constraints simultaneously at both ESL and RTL:
5. Decode each variable of the IP owner/vendor signature from their corresponding encoding (stated in Table 5.1).
6. Each digit ω in the signature combination is implanted through sharing of two/three registers according to the encoding rule (stated in Table 5.1).
7. Each digit θ in the signature combination is implanted by fragmenting the size of a N:1 mux into two $N/2$:1 mux and one 2:1 mux, according to the encoding rule.
8. Each digit φ in the signature combination is implanted by fragmenting the size of a 1:N demux into two 1:$N/2$ demux and one 1:2 demux, according to the encoding rule.

However, while selecting the vendor signature θ digits must be kept at twice the number of φ digits. This is because there exists twice the number of multiplexers compared to the number of de-multiplexers in a DSP core RTL design.

5.5 Design process of a multi-level watermarked IP core

The process of employing a multi-level watermark in an IP core is demonstrated in this section using an FIR filter (Sengupta *et al.*, 2019). The FIR filter is a DSP core that is mainly used to eliminate the signal noise at the receiver end of the digital communication system. Additionally, it is widely exploited in audio, video and other signal processing applications in consumer electronics systems. Thus, an FIR filter IP core is a very significant DSP block and protection of such blocks against

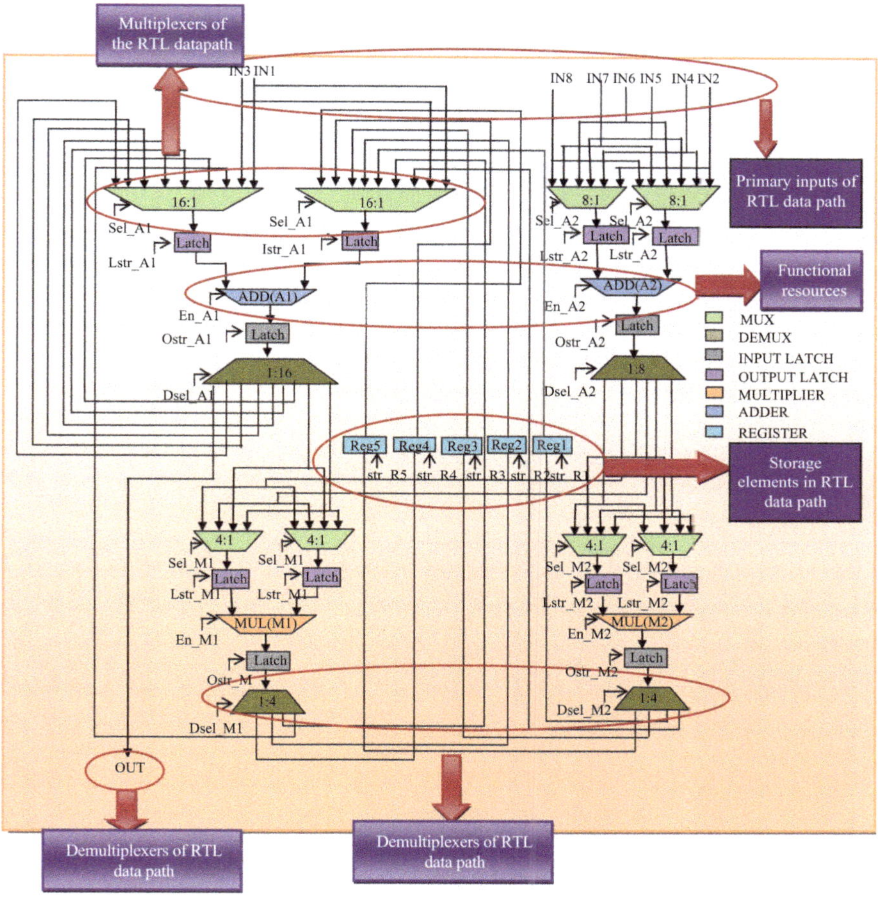

Figure 5.6 FIR filter RTL data path before watermark is embedded

fraudulent ownership claims is mandatory. An unprotected RTL data path of an FIR filter is shown in Figure 5.6.

In order to embed a multi-level watermark, a CDFG representing the FIR filter is first scheduled using the resource configuration obtained through PSO–DSE. Let us assume 2 adders (*A*1 and *A*2) and 2 multipliers (*M*1 and *M*2) are generated as the resource configuration through PSO–DSE. After performing hardware allocation, followed by binding of each hardware unit, the obtained CDFG of the FIR filter is shown in Figure 5.7. In the scheduled DFG, there are a total 15 addition operations (green node) and 8 multiplication operations (red node). As shown in the figure, operations 1, 3, 17, 18, 19, 20, 21, 22 and 23 are performed through adder *A*1, while

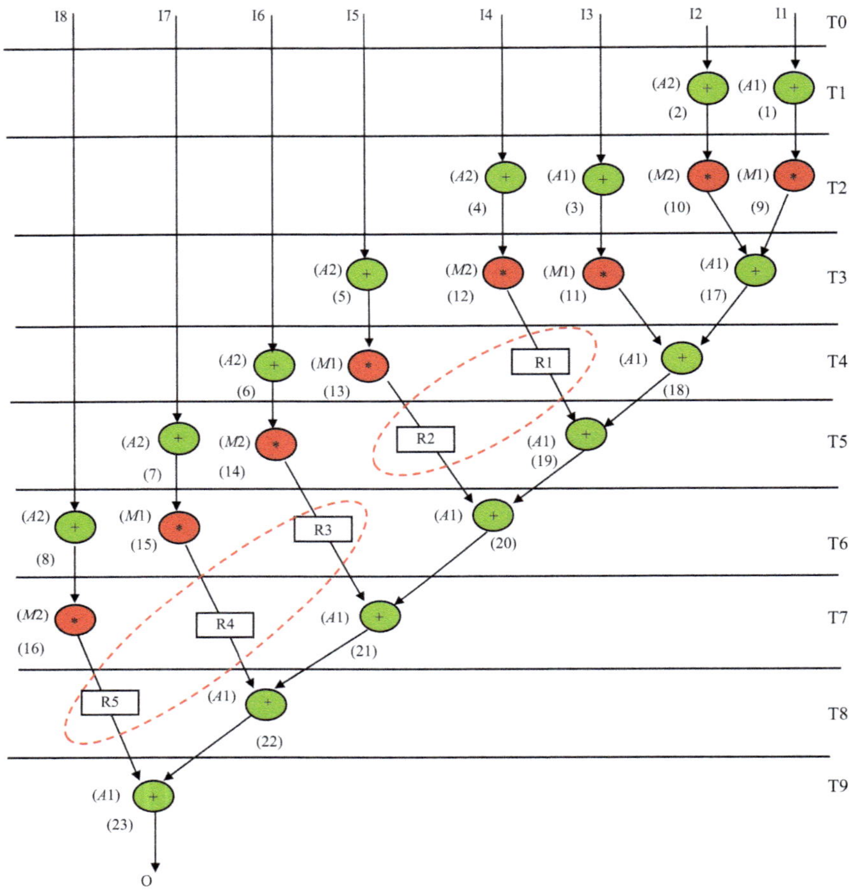

Figure 5.7 Scheduled DFG of FIR filter (Sengupta et al.*, 2019)*

operations 2, 4, 5, 6, 7 and 8 are performed through adder *A*2. Further, operations 9, 11, 13 and 15 are performed through multiplier *M*1, while operations 10, 12, 14 and 16 are performed through *M*2. The four steps of the multi-level watermark embedding process for the FIR filter are (Sengupta *et al.*, 2019):

1. The first step is to identify the total number of the registers allocated in different control steps: As shown in Figure 5.7, there are five intermediate registers (R1, R2, R3, R4 and R5) that have been allocated in different control steps (T4, T5, T6, T7 and T8, respectively).
2. These five registers are distributed into two groups each composed of two/three registers. One group contains R1 and R2 and another group contains R3, R4 and R5.

3. Corresponding to adder $A1$, 9 operations are being performed, which results into two 16:1 muxes at the inputs and one 1:16 demux at the output. Similarly, corresponding to adder $A2$, 6 operations are being performed, which results into two 8:1 muxes at the inputs one 1:8 demux at the output. Moreover, corresponding to multiplier $M1$, 4 operations are being performed, which results into two 4:1 muxes at the inputs one 1:4 demux at the output. Similarly, corresponding to multiplier $M2$, 4 operations are being performed, which results into two 4:1 muxes at the inputs one 1:4 demux at the output.
4. Assuming a vendor selected signature is $\theta\theta\theta\theta\theta\theta\theta\varphi\varphi\varphi\omega\omega$, due to the first θ digit, the first 16:1 mux of adder $A1$ is fragmented into two 8:1 muxes and one 2:1 mux. Further, due to the second θ digit, the second 16:1 mux of adder $A1$ is fragmented into two 8:1 muxes and one 2:1 mux. Similarly, the rest of the θ digits are encoded in the RTL design. Additionally, due to the first φ digit, the only 1:16 demux of adder $A1$ is fragmented into two 1:8 demuxes and one 1:2 demux. Similarly, the rest of the φ digits are encoded in the RTL design. Furthermore, due to the first ω digit, R1 and R2 are forced to be executed through a single shared register, while due to the second ω digit, R3, R4 and R5 are forced to be executed through another shared register. This process embeds the vendor-selected signature at the RTL and ESL stages of the FIR filter design. The RTL data path after embedding multi-level watermarking is shown in Figure 5.8, where the impact on the RTL data path due to watermark has been indicated in dotted red ovals.

The process of fragmenting in this approach is a covert way of inserting the vendor signature into the design because fragmenting designs into the next hierarchical level to reduce complexity is often a standard de-facto design practice in the industry. This covert way of inserting the signature thwarts an attacker from identifying the watermark, thus making it difficult to launch tampering or removal attacks.

5.6 Signature detection in a multi-level watermarked IP core

The signature detection process is performed through two sub-steps (Sengupta *et al.*, 2019):

1. *Design inspection:* The RTL design of the IP core is inspected to extract the structural properties of the data path circuit (available in the hardware description file of the top-level design) such as the arrangement of multiplexers, arrangement of de-multiplexers and arrangement of register sharing.
2. *Signature verification:* The signature of the vendor is decoded into its respective watermarking constraints and the presence of the constraints is verified in the extracted structural properties of the data path circuit (obtained in the previous step). If the constraints are found available in the extracted design post-verification, then the watermark is identified and the ownership is

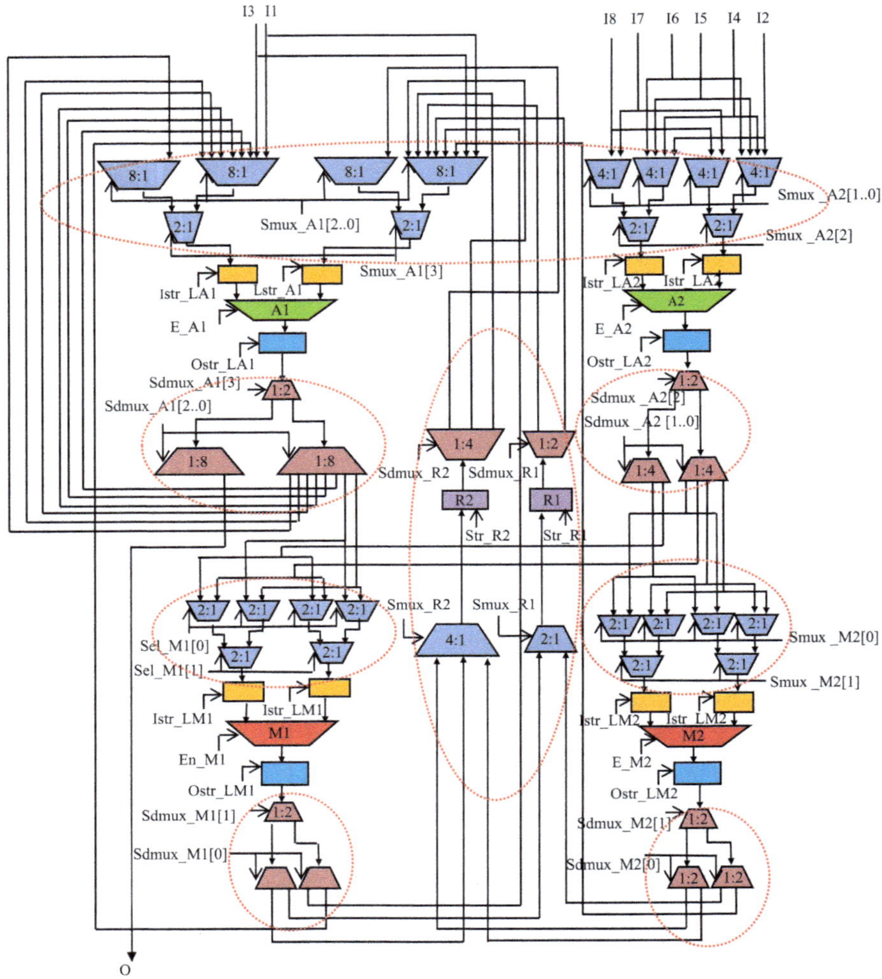

Figure 5.8 FIR filter RTL data path after embedding multi-level watermark (Sengupta et al.*, 2019)*

awarded to the actual vendor. In case some of the constraints are non-existent in the design, then it indicates that partial tampering may have occurred.

5.7 Analysis based on case studies

The multi-level watermarked FIR filter has been analysed in terms of security and design cost compared to the baseline design. Here, un-protected (without watermark) DSP design is termed as a baseline design. This analysis is followed by the comparative study with respect to an existing watermarking approach and baseline design.

5.7.1 Security and design cost analyses

The security of the multi-level watermarked FIR filter is assessed based on the tamper-tolerance ability of the design. An attacker may tamper with the vendor's signature by removing it partially or completely. However, in order to tamper, the attacker needs to know two parameters, namely the vendor's signature size and the number of variables required to encode it. Moreover, the encoding of each variable should be known to the attacker. Once these two parameters and encoding rules are known to the attacker, he/she may attempt to find the golden signature embedded into the intended DSP core. To do so, he/she guesses a signature of the same strength and decodes each of its digits. Thereafter, he/she detects the presence of each signature digit in the RTL data path of the DSP core. The process of guessing a signature and detecting its presence in the RTL data path is performed repeatedly until the original signature (with correct size, number of variables and their original encoding meaning) is not obtained. Therefore, the attacker is forced to perform a brute-force approach to find the original signature as shown in Figure 5.9. Only upon finding the correct signature can the attacker be able to tamper with the signature. A signature of a watermarked design is said to be highly tamper-tolerant if it is extremely difficult to find the correct signature among all possible combinations. The tamper-tolerance ability of a multi-watermarked design can be evaluated by calculating the number of all possible signatures using the equation (Sengupta *et al.*, 2019):

$$T_t = k^s \tag{5.1}$$

where, T_t denotes the tamper-tolerance metric, k denotes the number of encoding variables and s denotes the signature size. Table 5.3 shows the tamper-tolerance (using (5.1)) ability with respect to different signature size (s).

Furthermore, embedding a multi-level watermark into the design may result into extra overheads that could affect the design cost. Thus a design-cost analysis is also necessary. To analyse the design cost of a watermarked FIR filter, overheads due to the watermark are estimated with respect to the baseline (non-watermarked) design. Both the non-watermarked and watermark-embedded FIR filter are implemented in a field programmable gate array (FPGA) device using the Intel Quartus CAD synthesis tool. The overheads due to embedding the least and largest signature sizes are analysed in terms of logic elements (LE) with respect to a non-watermarked design. Table 5.4 provides a summary of device utilization of a non-watermarked FIR filter. Table 5.5 presents a summary of device utilization of a multi-level watermarked FIR filter for the least and largest possible signature sizes. Comparison of LE usage shows that a multi-level watermark incurs a nominal increase in overheads (3%–6%) compared to its baseline counterpart.

5.7.2 Comparative study

The existing contemporary watermarking approaches for DSP-based IP cores embed a secret mark only during ESL (Koushanfar *et al.*, 2005; Sengupta and Bhadauria, 2016; Sengupta and Roy, 2017). However, the multi-level watermarking

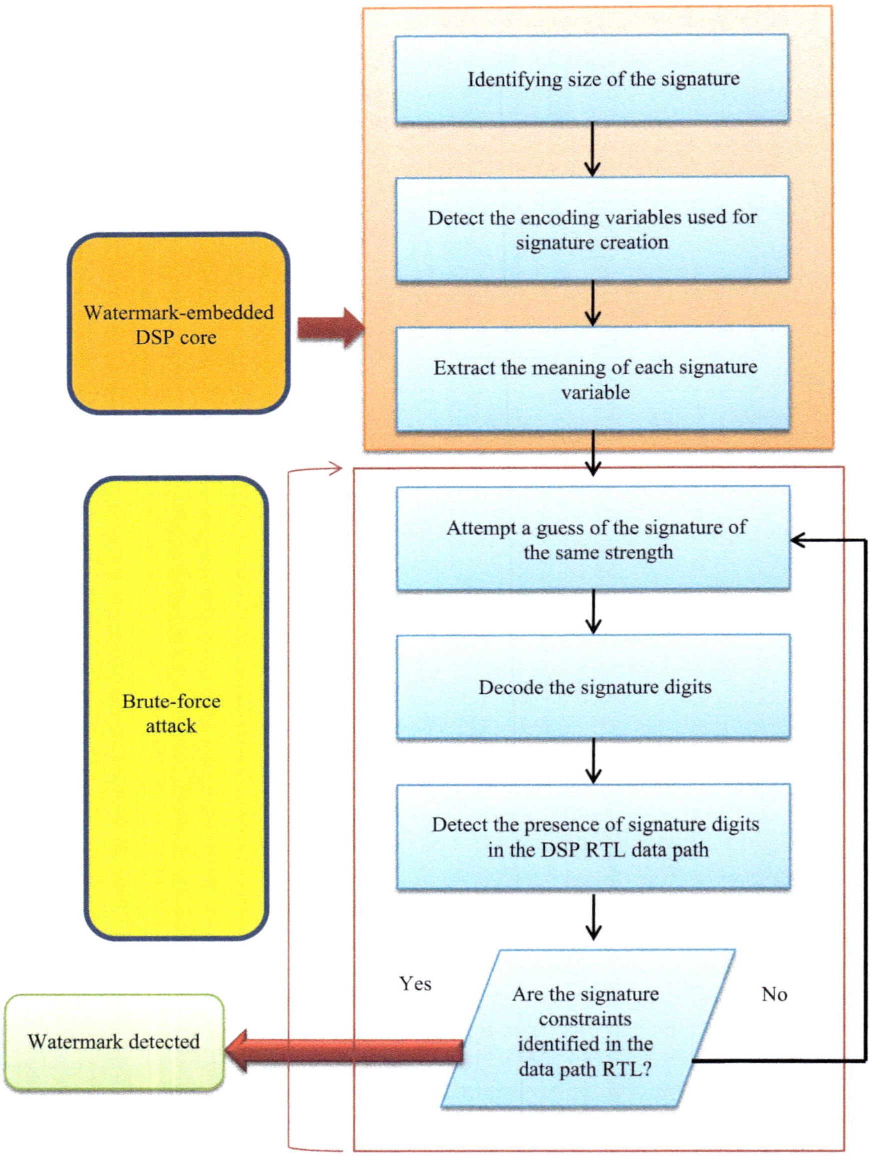

Figure 5.9 Signature detection in multi-phase watermarking using the brute-force approach

approach embeds the signature constraints in a DSP application at both ESL and RTL. So far, this is the only approach in the literature that leverages both ESL and RTL to employ watermarks. Therefore, the multi-level watermarking approach is compared with the baseline design (Tables 5.4 and 5.5). This comparison shows

Table 5.3 Ability of tamper-tolerance of multi-level watermarking for different signature sizes

Signature size	3	6	8	14
Signature combination	$\theta\theta\varphi$	$\theta\theta\theta\theta\varphi\varphi$	$\theta\theta\theta\theta\varphi\varphi\omega\omega$	$\theta\theta\theta\theta\theta\theta\theta\theta\varphi\varphi\varphi\varphi\omega\omega$
Tamper tolerance	27	729	6,561	4,782,972

Table 5.4 Device utilization summary of the eighth order FIR filter with no secret signature

Resource type	**Total logic elements**	**Total combina-tional function**	**Dedicated logic registers**	**Total registers**	**Total pins**
Resource usage	1,370/68,416 (2%)	1,369/68,416 (2%)	65/68,416 (<1%)	52	145/622 (23%)

Table 5.5 Summary of device utilization of FIR filter in the case of the least and largest possible size signature and comparison of LE overheads in contrast to the baseline FIR filter

Signature and resource details	**Details of maximum signature**	**Details of minimum signature**
Signature size	14	3
Signature digits	$\theta\theta\theta\theta\theta\theta\theta\theta\varphi\varphi\varphi\varphi\omega\omega$	$\theta\theta\varphi$
Total pin	145/622 (23%)	145/622 (23%)
Dedicated logic registers	65/68,416 (<1%)	65/68,416 (<1%)
Total register	52	52
Total combinational function	1,458/68,416 (2%)	1,417/68,416 (2%)
Total logic elements	1,461/68,416 (2%)	1,418/68,416 (2%)
Overhead	6%	3%

that the multi-level watermarking approach results into nominal additional overheads at higher security.

5.8 Conclusion

This chapter highlights the various abstraction levels of a design flow and discusses the significance of higher levels in the context of DSP or MP IP core protection. An approach of deploying protection (using a watermark) at two consecutive abstraction levels (ESL and RTL) was presented. Because of embedding watermark at more than one abstraction level, this approach is referred as multi-level watermarking. Like contemporary watermarking approaches for DSP–IP cores, signature constraints are also employed at ESL. However, it also leverages the RTL to implant a part of the signature constraints. Because of the presence of the watermark at two abstraction levels, detection and tampering of the signature by an adversary become highly complex. However, the multi-level watermark can easily be extracted during forensic detective control in the case of a false claim of ownership and IP piracy, counterfeiting or cloning. Additionally, a higher strength of watermark is achieved at nominal additional design overheads.

This chapter provides a deeper insight into multi-level watermarking. At the end of this chapter, the reader should understand the following concepts:

- Utility of the multi-level watermarking.
- Application of multi-level watermarking to protect IP core designs against various hardware threats such as IP piracy, counterfeiting, and false claim of ownership.
- Encoding rules of the signature variables.
- Process for implanting multi-level watermarking.
- Forensic detective control using multi-level watermarking to resolve the conflict of a fraudulent claim of ownership and detect piracy, counterfeiting or cloning.
- Design of a multi-level watermark-embedded IP core demonstrated using a real life DSP core application (FIR filter).
- Security and design cost analyses of the multi-level watermarking approach.
- Comparative study of a multi-level watermark-embedded DSP–IP core with respect to baseline design.

5.9 Questions and exercises

1. What is a multi-level watermark?
2. How is a multi-level watermark different from a multi-phase watermark?
3. Describe the encoding of multi-phase watermarking methodology.
4. In multi-phase watermarking, what is the role of the electronic-system level?
5. State the encoding rule of θ, and at which abstraction level is it implanted?
6. State the encoding rule of φ, and at which abstraction level is it implanted?
7. State the encoding rule of ω, and at which abstraction level is it implanted?
8. What is the rule of signature selection in multi-level watermarking?
9. Explain the process of signature detection in multi-level watermarking.
10. What is register sharing, and what hardware does it impact in the data path?
11. What is the role of PSO-based DSE in multi-level watermarking?
12. Define tamper-tolerance.
13. How is tamper-tolerance evaluated in multi-level watermarking?
14. Explain the flow chart of the signature detection process in a multi-level watermark through brute force.
15. State the other watermarking approaches used for DSP cores.

References

E. Castillo, U. Meyer-Baese, A. Garcia, L. Parilla, and A. Lloris, "IPP@HDL: Efficient intellectual property protection scheme for IP cores," IEEE Transactions on Very Large Scale Integration (VLSI) Systems, vol. 15, no. 5, Pages: 578–590, 2007.

F. Koushanfar, I. Hong, and M. Potkonjak, "Behavioral synthesis techniques for intellectual property protection," ACM Transactions on Design Automation of Electronic Systems, vol. 10, no. 3, Pages: 523–545, 2005.

B. Le Gal and L. Bossuet, "Automatic low-cost IP watermarking technique based on output mark insertions," Design Automation for Embedded Systems, vol. 16, no. 2, Pages: 71–92, 2012.

M. C. McFarland, A. C. Parker, and R. Camposano, "Tutorial on high-level synthesis," in DAC '88 Proceedings of the 25th ACM/IEEE Design Automation, vol. 27, no. 1, Pages: 330–336, 1988.

V. K. Mishra and A. Sengupta, "MO-PSE: adaptive multi-objective particle swarm optimization based design space exploration in architectural synthesis for application specific processor design," Advances in Engineering Software, vol. 67, Pages: 111–124, 2014.

S. M. Plaza and I. L. Markov, "Solving the third-shift problem in IC piracy with test-aware logic locking," IEEE Transactions on Computer-Aided Design of Integrated Circuits and Systems, vol. 34, no. 6, Pages: 961–971, 2015.

D. Roy and A. Sengupta, "Low overhead symmetrical protection of reusable IP core using robust fingerprinting and watermarking during high level synthesis," Future Generation Computing Systems, vol. 71, Pages: 89–101, 2017.

D. Roy and A. Sengupta, "Multilevel watermark for protecting DSP kernel in CE systems [hardware matters]," IEEE Consumer Electronics Magazine, vol. 8, no. 2, Pages: 100–102, 2019.

R. Schneiderman, "DSPs evolving in consumer electronics applications," IEEE Signal Processing Magazine, vol. 27, no. 3, Pages: 6–10, 2010.

A. Sengupta, "Intellectual property cores: Protection designs for CE products," IEEE Consumer Electronics Magazine, vol. 5, no. 1, Pages: 83–88, 2016.

A. Sengupta, "Hardware security of CE devices [hardware matters]," IEEE Consumer Electronics Magazine, vol. 6, no. 1, Pages: 130–133, 2017.

A. Sengupta and S. Bhadauria, "Exploring low cost optimal watermark for reusable IP cores during high level synthesis," IEEE Access, vol. 4, Pages: 2198–2215, 2016.

A. Sengupta and S. P. Mohanty, "Advanced encryption standard (AES) and its hardware watermarking for ownership protection," IP Core Protection and Hardware-Assisted Security for Consumer Electronics, Pages: 317–335, 2019a, e-ISBN: 9781785618000.

A. Sengupta and S. P. Mohanty, "IP core and integrated circuit protection using robust watermarking," IP Core Protection and Hardware-Assisted Security for Consumer Electronics, Pages: 123–170, 2019b, e-ISBN: 9781785618000.

A. Sengupta and D. Roy, "Antipiracy-aware IP chipset design for CE devices: A robust watermarking approach [hardware matters]," IEEE Consumer Electronics Magazine, vol. 6, no. 2, Pages: 118–124, 2017.

A. Sengupta and D. Roy, "Multi-phase watermark for IP core protection," in Proc. 36th IEEE International Conference on Consumer Electronics (ICCE), Pages: 1–3, 2018.

A. Sengupta, R. Sedaghat, and Z. Zeng, "A high level synthesis design flow with a novel approach for efficient design space exploration in case of multi-parametric optimization objective," Microelectronics Reliability, vol. 50, no. 3, Pages: 424–437, 2010.

A. Sengupta, D. Roy, and S. P. Mohanty, "Triple-phase watermarking for reusable IP core protection during architecture synthesis," IEEE Transactions on Computer-Aided Design of Integrated Circuits and Systems, vol. 37, no. 4, Pages: 742–755, 2018.

A. Sengupta, D. Roy, and S. P. Mohanty, "Low-overhead robust RTL signature for DSP core protection: new paradigm for smart CE design," in Proc. 37th IEEE International Conference on Consumer Electronics (ICCE), Pages: 1–6, 2019.

D. Ziener and J. Teich, "Power signature watermarking of IP cores for FPGAs," Journal of Signal Processing Systems, vol. 51, no. 1, Pages: 123–136, 2008.

Chapter 6

Protecting the rights of an IP buyer using cryptosystem-based multivariable fingerprinting

Anirban Sengupta[1]

This chapter describes a cryptosystem-based fingerprinting approach for IP cores that is employed during the register allocation phase of electronic-system level synthesis. The need for a buyer's fingerprint and its desirable features are highlighted. Further, a detailed elaboration on the fingerprint selection and encoding technique followed by the embedding process and, finally, implementation details on case studies are provided.

The chapter is organized as follows: Section 6.1 discusses the various challenges in electronic system design, followed by some fingerprint basics; Section 6.2 introduces some of the prior works in this domain; Section 6.3 presents an overview of the crypto-based fingerprinting process; Section 6.4 illustrates in detail the fingerprinting approach for digital signal processing (DSP) cores; and Section 6.5 discusses the security and design cost analyses using different case studies.

6.1 Introduction

'To satisfy time to market requirement' and 'reduction in design complexity' are two important challenges in a design process of electronic systems. These two challenges encourage the integration of reusable IP cores in a system on chip (SoC) to be employed in electronics devices. This not only reduces design complexity, but also helps to fulfil the raging demand. Further, the design of reusable IP cores such as joint photographic experts group (JPEG), moving picture experts group (MPEG), finite impulse response (FIR) and infinite impulse response (IIR) filter, among others (Schneiderman, 2010), through electronic-system level (ESL) synthesis (McFarland *et al.*, 1988; Martin and Smith, 2009; Sengupta *et al.*, 2010) is crucial in order to meet the industry requirements. Furthermore, the reusable IP cores not only increase the productivity of complex designs, but also lead to design-cost reduction. However, the unremitting applications of reusable IP cores in SoCs raise the need of their protection against several hardware threats such as IP forgery, infringement, fraudulent claims of ownership (Sengupta, 2016, 2017), etc. Such protection is important from

[1]Computer Science and Engineering, Indian Institute of Technology Indore, Indore, India

the perspective of the seller and buyer. For an IP owner or seller, deploying protection into IP cores is required in order to detect piracy/counterfeiting/cloning and secure ownership. This is required because the owner sends his/her IP core to a design house for the purpose of SoC integration or manufacturing. However, the design house may be deceitful, which raises the chances of forgery (Castillo *et al.*, 2007; Plaza and Markov, 2015; Sengupta and Roy, 2017). Further from the buyer's perspective, protection is deployed in order to provide him/her exclusive buyer's rights. This is required when an IP seller designs an IP core according to the buyer's specifications. Therefore providing exclusive buyer/user rights thwarts the reselling of the IP core to other users (Lach *et al.*, 2001; Roy and Sengupta, 2017).

Common protection techniques such as copyrights, patents, trademarks, etc., can also be used for intellectual properties. However, their limited applications render them unsuccessful to address the challenges. Therefore the need for a strong IP core protection measure from a buyer's perspective arises.

In order to ensure the exclusive buyer's rights over the IP cores, a 'buyer fingerprint' is a promising alternative. A buyer's fingerprint is embedded on top of the existing design without affecting the functionality. The secret fingerprinting constraints are obtained through a signature chosen by the buyer and a certain encoding scheme. These constraints are implanted covertly in addition to the usual design constraints. Thus the generated, reusable, IP core belongs to a specific buyer. The fingerprint is able to thwart the reselling or redistribution of illegal copies. In order to make IP rights protection schemes more secure, a cryptographic hashing algorithm can be integrated with the fingerprint scheme. The hashing algorithm is typically used to generate a secure hash digest (Sengupta *et al.*, 2019).

An optimal fingerprint should be able to satisfy certain criteria (Sengupta *et al.*, 2019):

1. An optimal fingerprint should result in a highly secure and low-cost solution that not only satisfies the user's constraints, but also additional security constraints.
2. The fingerprint must be permanently embedded onto the IP core design.
3. It must be unique for each buyer and hard to trace.
4. It should not interfere with the vendor IP rights protection mechanism.
5. It should incur minimal overhead post-embedding.
6. It should not affect the functionality of the IP core.

Further, the IP rights of a seller/vendor can also be protected using the well-known watermarking approach. A comparison of fingerprinting with the watermarking approach is highlighted in Section 6.1.1.

6.1.1 Fingerprinting vs watermarking

The comparison of fingerprinting and watermarking approach (Koushanfar *et al.*, 2005; Le Gal and Bossuet, 2012; Ziener and Teich, 2008; Sengupta and Bhadauria, 2016; Sengupta and Roy, 2018; Sengupta *et al.*, 2018; Roy and Sengupta, 2019; Sengupta and Mohanty, 2019a, 2019b) has been stated based on their threat model, objective and abstraction level of deployment as shown in Figure 6.1.

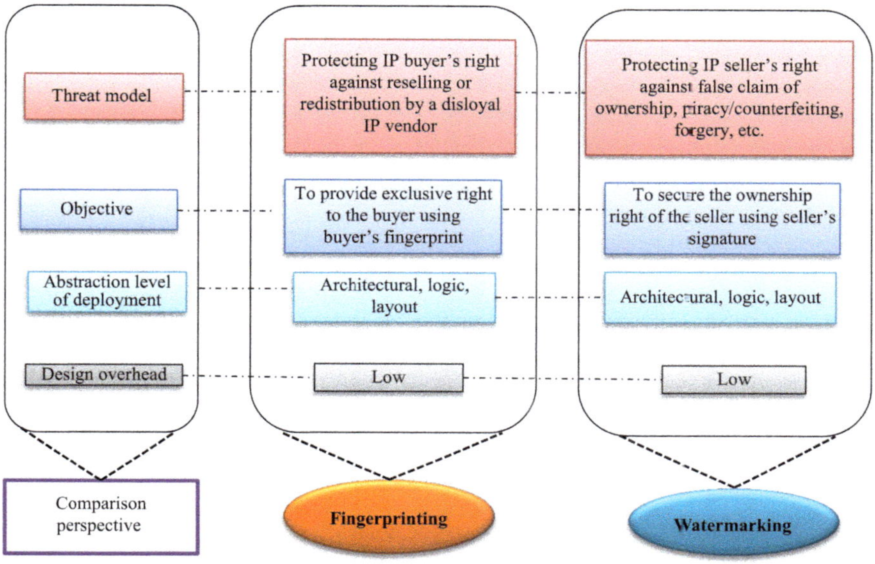

Figure 6.1 Comparison of fingerprinting with watermarking

6.2 Selected contemporary fingerprinting approach at architectural level

6.2.1 Overview, threat model and encoding process

At the architectural or ESL, Roy and Sengupta (2017) proposed a symmetric protection scheme to protect the IP seller's and buyer's rights over IP cores. The overview of this symmetric protection scheme (Roy and Sengupta, 2017) is shown in Figure 6.2. In this approach, the buyer's fingerprint is implanted into an IP core design along with the seller's watermark, thus providing symmetric protection. As shown in the figure, the buyer's fingerprint and seller's watermark are first decoded into respective security constraints. Thereafter, the security constraints of the buyer and seller are implanted covertly into the intended IP core during the architectural synthesis process. However, post-embedding the watermark and fingerprint, the functional behaviour of the IP core should remain intact. Post-architectural synthesis, a register-transfer level (RTL) description of the symmetrically protected (containing the buyer's fingerprint and seller's watermark) IP core is obtained.

The threat model of this approach targets the hardware threats with respect to the seller and buyer. This approach aims to protect the seller's IP rights in case of IP piracy/forgery/infringement and ownership fraud. Moreover, it promises to provide buyer's IP rights protection using a fingerprint to thwart the reselling of pirated copies of the IP core to a third user (without the knowledge of the genuine user/buyer). The buyer's fingerprint renders tracking illegally resold IP cores feasible, thus protecting exclusive user rights.

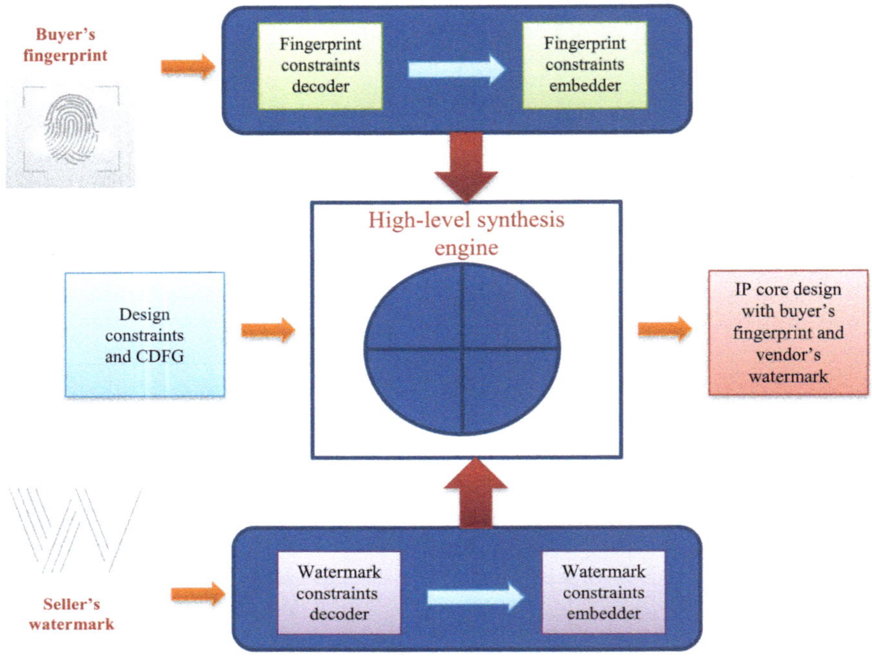

Figure 6.2 Overview of the fingerprinting and watermarking process – symmetrical IP core protection

Table 6.1 Buyer fingerprint and seller signature encoding

Buyer/seller	Variables	Encoding
Buyer's fingerprint	x	Force even operation to execute in odd control step. Scheduling conflict is resolved during scheduling phase
	y	Force odd operation to execute in even control step. Scheduling conflict is resolved during scheduling phase
	z	Encoded value of edge with node pair as (odd, odd) in coloured interval graph
Seller's signature	*i*	Encoded value of edge with node pair as (prime, prime)
	I	Encoded value of edge with node pair as (even, even)
	T	Encoded value of edge with node pair as (odd, even)
	!	Encoded value of edge with node pair as (0, any integer)

Further, careful encoding of the buyer's fingerprint and seller's signature are very significant in order to achieve higher protection and lower design overheads. The encoding rules of the buyer's fingerprint and seller's signature in this symmetric protection scheme (Roy and Sengupta, 2017) are shown in Table 6.1. The buyer's fingerprint is converted into constraints that are embedded into the design through scheduling and a graph colouring framework in an integrated fashion. For example, the *x* and *y* digits are embedded in the scheduling framework, while the *z* digit is implanted in the graph colouring framework. It is important to note that the fingerprint constraints comprise three unique variables that are independent of each other. This provides double-layered protection against tampering from an adversary. Additionally, implanting constraints in two independent frameworks of high-level synthesis (HLS) ensures that the constraints are covertly distributed throughout the design. However, while selecting the buyer's fingerprint, the buyer has to target a selection of more *x* and *y* digits than *z* digits. This is because *x* and *y* digits are embedded during scheduling the conflict resolution process, thereby not adding overheads onto the design. On the other hand, excessive *z* digits in a buyer fingerprint may result in register overheads. Thus a balanced selection of *x*, *y* and *z* digits ensure strong robustness and low overheads of the buyer's fingerprint.

In symmetrical IP protection, the buyer's fingerprint and seller's watermark are integrated simultaneously in the design. In Roy and Sengupta (2017), the seller's watermark is embedded after the buyer's fingerprint is implanted. In other words, the output of the buyer fingerprint acts as an input to the seller's watermark-embedding process. More explicitly, the coloured interval graph (CIG)-embedded fingerprint serves as an input to the watermarking process.

As shown in Table 6.1, the encoding rules of the seller's signature in this symmetric protection scheme (Roy and Sengupta, 2017) comprises four unique variables, viz. *i*, *I*, *T* and *!*. The seller's watermark constraints are embedded in the fingerprinted IP core design through the graph colouring framework. It is important to note that although watermarking constraints are also inserted in the register allocation phase using a graph colouring algorithm, the encoding rules are vastly different from the fingerprinting process. Inclusion of the variables in a watermark signature indicates that any number of secret additional edges can be added in the design by the vendor. Similar care must be exercised while selecting the watermark–signature size to perform a trade-off between high robustness and low overheads.

6.2.2 Embedding process

The data flow graph (DFG) representation of target IP core application acts as an input to the fingerprint and watermark embedding process. Scheduled DFG is obtained based on three primary inputs: (1) module library, (2) DFG and (3) resource constraints. Post-embedding the buyer's fingerprint and seller's watermark, the symmetrically protected IP core is generated at the output. Details of the process for obtaining a symmetrically protected IP core are shown in Figure 6.3. First, a buyer's fingerprint is selected and a sorted list of operations of the DFG is created. Then, while embedding the buyer's fingerprint, the *x* and *y* digits are inserted in sequence (based on the encoding rule) through operation conflict resolution during the scheduling phase. In case further fingerprint constraints are not available for embedding during conflict resolution during scheduling, conflict is broken based on sorted ordering. This results into a scheduled

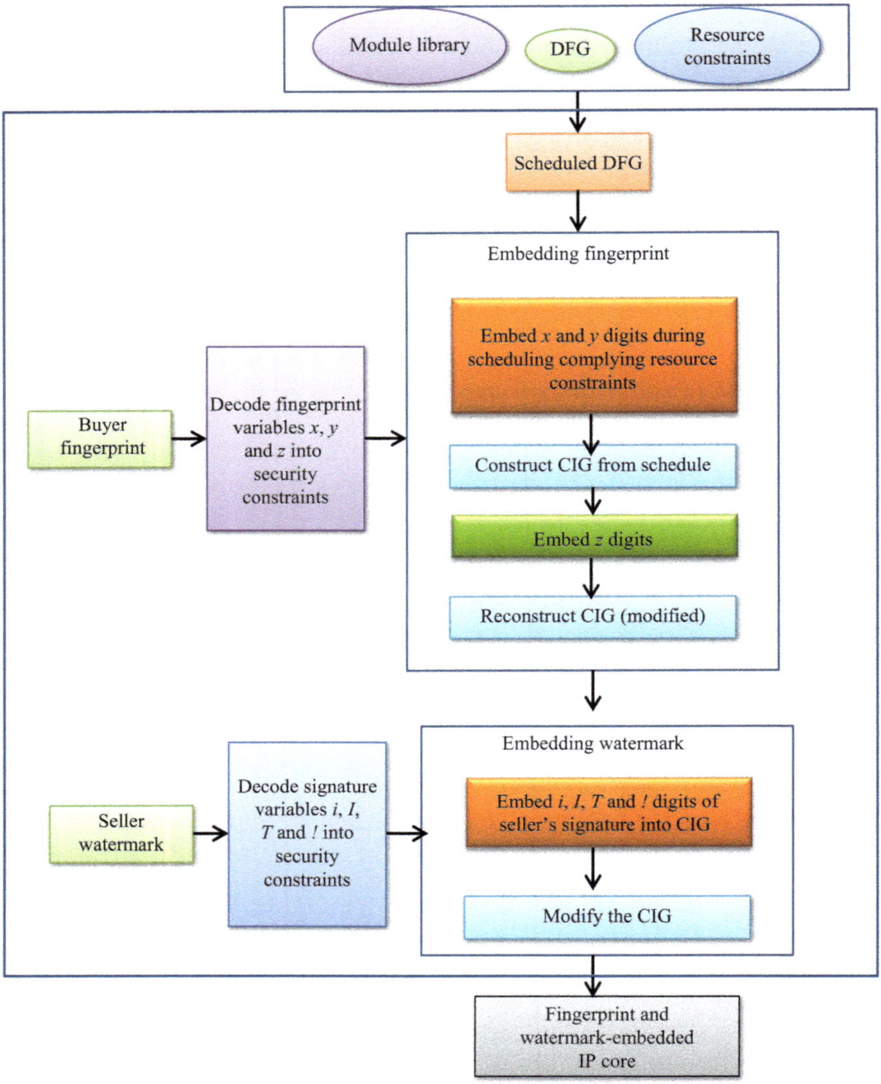

Figure 6.3 Details of fingerprinting and watermarking embedding process – symmetrical IP core protection (Roy and Sengupta, 2017)

fingerprint. The next step is to assign storage variables during generated scheduling and generate CIG. These storage variables are sorted in increasing order and stored in a list. Subsequently, the z digits are decoded based on the sorted list and embedded in the CIG (representing the register-allocation phase of HLS) as the second layer of the fingerprint constraints. In the next phase, the output of the modified CIG with the fingerprint is fed as an input to the watermarking module.

Again the storage variables are sorted in increasing order of their number. Subsequently, the seller's signature is selected. This signature is decoded based on the sorted list and embedded in the CIG (representing the register-allocation phase of HLS) as watermark constraints. The final modified version of the CIG contains both fingerprint and watermark constraints. This CIG and the schedule obtained earlier are then used for generating a symmetrically protected IP core data path and controller using an HLS framework.

6.3 Overview of cryptosystem-based multivariable fingerprinting

This chapter presents a multivariable fingerprinting approach for DSP–IP cores using a cryptographic hashing algorithm (Sengupta *et al.*, 2019). An overview of the presented approach is shown in Figure 6.4. Primary input to the presented fingerprinting approach is a DFG representing a DSP application (to be protected), resource constraints and module library. The final output generated from this approach is a buyer's fingerprint-protected IP core. With the aid of the buyer's fingerprint, all illegally resold pirated copies (by a dishonest vendor) of the IP core can be traced easily. Thus an embedded fingerprint thwarts the possibility to illegally resell IP cores.

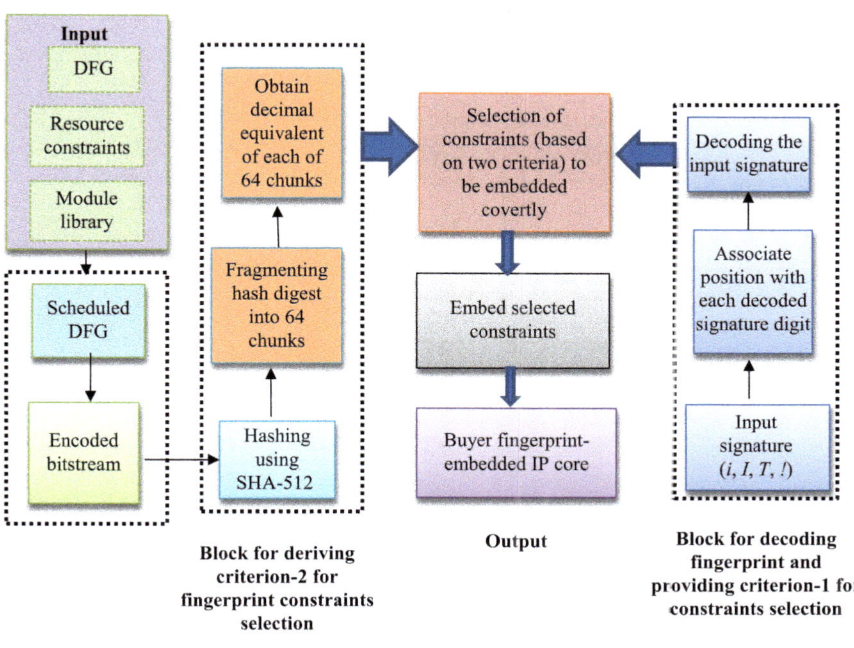

Figure 6.4 Overview of cryptosystem-based multivariable fingerprinting scheme for IP core protection (Sengupta et al., 2019)

In the cryptosystem-based multivariable fingerprinting approach the security constraints representing the buyer's fingerprint are implanted during the register-allocation phase of ESL synthesis. In order to implant constraints during the register-allocation phase, a CIG-based framework is employed. The complete process of generating a fingerprint-embedded IP core is accomplished through a number of steps, as shown in Figure 6.4. The fingerprint constraints to be embedded are selected based on two criteria. *Criterion-1* is the fingerprint size (total number of digits) in terms of variables *i*, *I*, *T* and *!*. Since each variable of a buyer's fingerprint carries an encoded meaning, it, therefore, needs to be decoded in terms of security constraints for implantation. However, all the constraints corresponding to all the variables are not chosen to be implanted into the design. Out of all the fingerprint digits, the potential digits to be embedded as security constraints are decided by selection criteria 1 and 2 jointly (Sengupta *et al.*, 2019).

Let us discuss the overview of the complete process step-by-step. In the first step, the DFG of the intended DSP core is scheduled based on resource constraints (multipliers and adders). The output of this step is a scheduled DFG complying with the resource constraints. In the next step, the scheduled DFG is encoded into a bitstream format according to certain encoding rules (described in Section 6.4.2.2). The next step performs a cryptographic hashing on the encoded bitstream of the design. This hashing transforms the bitstream into a secure hash-digest of 512 bits. The motive of converting the bitstream into a secure hash-digest is to obtain a unique and secure representation of the design. The collision resistance and pre-image resistance properties aid to fulfil the motive of employing the hashing scheme. Next, the secured, hashed, bitstream is fragmented into 64 chunks, each 8-bit in size. The decimal equivalents of 8-bit chunks play an important role to determine the fingerprint constraints to be embedded. Each value in a set of decimal equivalents (corresponding to 64 chunks of the hash digest), which is less than or equal to the size of the fingerprint, acts as the constraints selection *criterion-2*. The process of selecting fingerprint digits based on *criteria 1 and 2* is depicted in Figure 6.5. Only those digits of the fingerprint whose associated positions satisfy criterion-2 are eligible to act as security constraints.

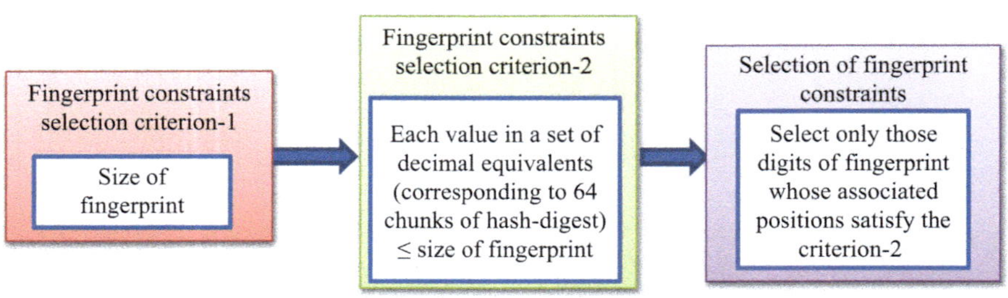

Figure 6.5 Selection process of fingerprint constraints based on two criteria (Sengupta et al., 2019)

Once the set of selected fingerprint digits has been determined, the corresponding security constraints are implanted into the CIG and the fingerprint-embedded DSP core is generated. The detailed procedure of selecting fingerprint constraints and embedding them into the DSP core are illustrated in Section 6.4.

6.4 Details of cryptosystem-based multivariable fingerprinting

This section illustrates the evaluation models used in the cryptosystem-based multivariable fingerprinting approach in detail followed by fingerprinting methodology (Sengupta *et al.*, 2019).

6.4.1 Evaluation models used

The cost of a fingerprint-embedded IP core is evaluated based on two models, namely the area model and delay model.

6.4.1.1 Area model

The total area of a fingerprint-embedded IP core is computed based on two parameters: (1) the total number of hardware resources such as number of functional units (adders, subtractors, multipliers), multiplexers and registers and (2) the unit area of each resource. The total area (A^t_{fp}) occupied is given as (adopted from Lach *et al.* (2001)):

$$A^t_{fp} = \sum_{k=1}^{n} A(R_k) * N(R_k) + A(mux) * N(mux) + A(reg) * N(reg) \tag{6.1}$$

where $N(R_k)$, $N(mux)$ and $N(reg)$ denote the number of the kth functional unit (multiplier/adder/subtractor), multiplexers and registers required, respectively. $A(R_k)$, $A(mux)$ and $A(reg)$ denote the unit area of the kth functional unit, multiplexer and one register, respectively.

6.4.1.2 Delay model

The total execution latency (L^t_{fp}) of a fingerprint-embedded IP core design is (adopted from Lach *et al.* (2001)):

$$L^t_{fp} = \sum_{T=1}^{n} \max(d(op_1), d(op_2), \ldots, d(op_k), d(reg)) \tag{6.2}$$

where $d(op_k)$ represents the delay of the kth operation in a given control step (T) and $d(reg)$ indicates the delay of a register.

Cost function

The cost of a fingerprint-embedded design evaluated from (6.1) and (6.2) is calculated by:

$$C_{fp} = \frac{w_1 \left(L^t_{fp}\right)}{L_m} + \frac{w_2 \left(A^t_{fp}\right)}{A_m} \tag{6.3}$$

where C_{fp} represents the cost of the design post-embedding the fingerprint, L^t_{fp} and A^t_{fp} denote the total execution latency and area of the design, respectively, L_m and A_m denote the maximum possible latency and area, while w_1 and w_2 represent the weights for latency and area, respectively. These weights are specified by the user and their values are in the range (0, 1). (*Note: To assign similar weightage to both latency and area during cost evaluation, both* w_1 *and* w_2 *are fixed at 0.5.*)

6.4.2 Fingerprinting methodology: pre-processing phase and demonstration

A buyer's fingerprint is embedded into a DSP core design during the register-allocation phase of ESL synthesis. In order to embed the fingerprint, the design needs to be transformed into a suitable format. This transformation phase is referred to as the pre-processing phase that is discussed through steps in Section 6.4.2.1. An explanation using an example is discussed in Section 6.4.2.2.

6.4.2.1 Pre-processing phase

1. Schedule the DFG representing the DSP application. The scheduling is responsible for performing time-stamping of the operations in various control steps based on the algorithm chosen. There are several scheduling algorithms that can be used for time-stamping: as soon as possible, as late as possible, list scheduling, etc. Here the LIST scheduling algorithm has been chosen because it allocates the critical operations first, resulting in minimization of latency or delay.
2. Encode the scheduled DFG into a bitstream format as per Table 6.2.
3. Generate 512-bit hash-digest of the bitstream using the SHA-512 algorithm (details of SHA-512 are discussed in Chapter 3).
4. Fragment the hash-digest into 64 chunks of data where each chunk of data is 8-bit in size.
5. Convert each 8-bit chunk of data into its corresponding decimal equivalent representation.

6.4.2.2 Demonstration on DCT core

A demonstration on discrete cosine transform (DCT) cores used in popular consumer electronics applications such as image or audio compression is discussed in this section. Considering a design constraint of two multipliers and one adder, the

Table 6.2 Encoding rules

Operation number (opn)	**Corresponding control step (T) number**	**Encoded bit**
Even	Even	0
Odd	Even	1
Even	Odd	0
Odd	Odd	0

scheduling of a DCT core is shown in Figure 6.6. As shown in the figure, there are 30 storage variables used for storing intermediate outputs and primary inputs/outputs. This scheduled design represents a baseline version as it does not embed the buyer's fingerprint yet. The next step is to obtain a graph colouring framework

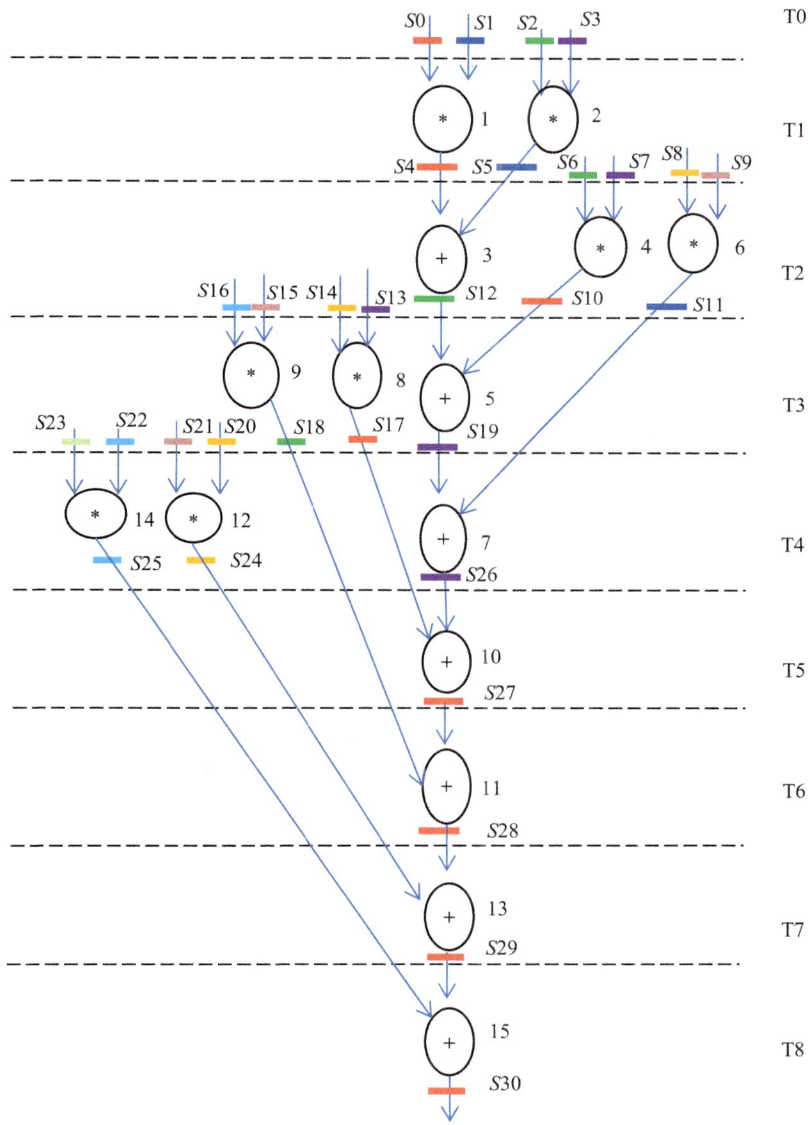

Figure 6.6 Scheduled and hardware allocated 8-point DCT using 1 (+) and 2 (*) (Sengupta et al., 2019)

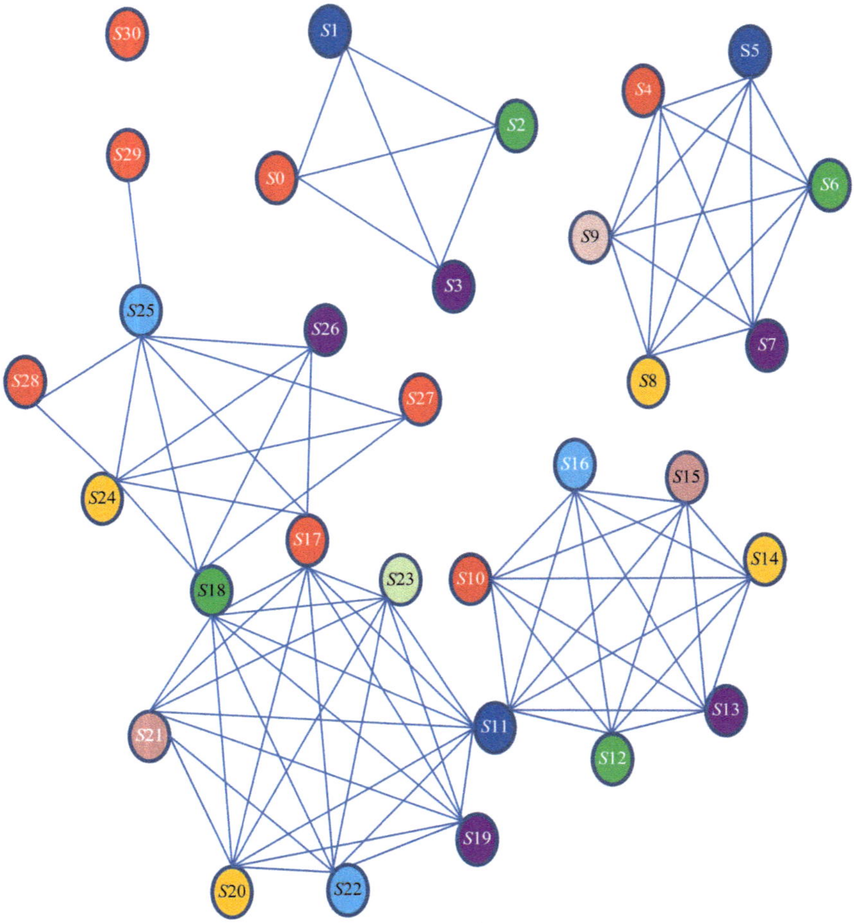

Figure 6.7 Coloured interval graph of baseline 8-point DCT core (without fingerprint constraints)

corresponding to the scheduling with storage variables. The corresponding CIG is shown in Figure 6.7. It is evident from the CIG that a maximum of eight registers are used to execute 30 storage variables (each node in the CIG represents a storage variable). The register allocation is provided in Table 6.3 (Sengupta *et al.*, 2019).

In the next phase, the respective bitstream format is generated using the scheduled design obtained before and the encoding rules shown in Table 6.2. For example, in the DCT core scheduling, since opn 1 is executed in control step 1 (T1), the encoded value (E_{opni}) of this operation is 0. The detailed encoded values of all operations are shown in Table 6.4 (Sengupta *et al.*, 2019). These encoded bits, upon appending, represent a bitstream format that is fed into the SHA-512 algorithm. This bitstream format is then converted into 1,024 bits using the

Table 6.3 *Register allocation table for baseline design of 8-point DCT (Note: T indicates the control step number and R0-R7 indicate the eight different registers)*

T	R0	R1	R2	R3	R4	R5	R6	R7
0	S0	S1	S2	S3	–	–	–	–
1	S4	S5	S6	S7	S8	S9	–	–
2	S10	S11	S12	S13	S14	S15	S16	–
3	S17	S11	S18	S19	S20	S21	S22	S23
4	S17	–	S18	S26	S24	–	S25	–
5	S27	–	S18	–	S24	–	S25	–
6	S28	–	–	–	S24	–	S25	–
7	S29	–	–	–	–	–	S25	–
8	S30	–	–	–	–	–	–	–

Table 6.4 *Encoded values of operations*

Operation number (opn)	Encoded bit (E_{opni})
1	0
2	0
3	1
4	0
5	0
6	0
7	1
8	0
9	0
10	0
11	1
12	0
13	0
14	0
15	1

standard appending rules of SHA. The SHA block is responsible for generating a 512-bit output that represents the design hash-digest. The next step is to generate 64 decimal values by dividing the 512 bits into groups of 8 bits each, as shown in Table 6.5 (Sengupta *et al.*, 2019). Each decimal value obtained will represent a position in the buyer's fingerprint to decide which digits of the buyer's fingerprint will serve as secret constraints to be embedded into the design.

Table 6.5   From the generated design digest and the 64 decimal equivalents are indicated (B: binary, D: decimal)

D	**151**	**124**	**167**	**96**	**57**	**38**	**190**	**63**
B	10010111	01111100	10100111	01100000	00111001	00100110	10111110	00111111
D	**39**	**246**	**226**	**195**	**165**	**143**	**115**	**172**
B	00100111	11110110	11100010	11000011	10100101	10001111	01110011	10101100
D	**132**	**227**	**166**	**22**	**220**	**48**	**109**	**85**
B	10000100	11100011	10100110	00010110	11011100	00110000	01101101	01010101
D	**254**	**68**	**77**	**79**	**150**	**31**	**234**	**53**
B	11111110	01000100	01001101	01001111	10010110	11101010	11101010	00110101
D	**173**	**150**	**63**	**85**	**143**	**224**	**100**	**185**
B	10101101	10010110	00111111	10001111	10001111	11100000	01100100	10111001
D	**221**	**227**	**133**	**114**	**206**	**30**	**106**	**2**
B	11011101	11100011	10000101	01110010	11001110	00011110	01101010	00000010
D	**131**	**117**	**128**	**103**	**63**	**28**	**252**	**84**
B	10000011	01110101	10000000	01100111	00111111	00011100	11111100	01010100
D	**82**	**197**	**40**	**201**	**117**	**200**	**13**	**142**
B	01010010	11000101	00101000	11001001	01110101	11001000	00001101	10001110

6.4.3 Fingerprinting methodology: selection of fingerprint and embedding process and demonstration

This subsection describes the selection and embedding process of the cryptosystem-based multivariable buyer's fingerprint followed by a demonstration using an example.

6.4.3.1 Choice of fingerprint and the decoding and embedding process

In the selection and embedding process of a fingerprint, the first step is to choose a buyer's signature with a fixed size consisting of four variables, viz. *i*, *I*, *T* and *!*. The encoding of each variable is shown in Table 6.6 (Sengupta *et al.*, 2019). According to the encoding rules, whenever digit *i* occurs in the fingerprint, it is decoded as an edge between node pair numbers (*Si*, *Sj*) of the form (prime, prime). Further, each occurrence of digits *I*, *T* and *!* are decoded as an edge between node pair numbers of the form (even, even), (odd, even) and (0, any integer), respectively. For example, let us consider a six-digit fingerprint 'IiiT!I'. The first digit *I* is decoded as an edge between the node pair (*S*2, *S*4). Similarly, second digit *i* is decoded as an edge between the node pair (*S*2, *S*3). Here, the third digit is the second occurrence of *i*, therefore it is decoded as an edge between the node pair (*S*2, *S*5). Likewise, other three digits are decoded as edges between (*S*1, *S*2), (*S*0, *S*1) and (*S*2, *S*6), respectively.

Once a fixed-size fingerprint is chosen, the next phase is to convert or decode it into constraints and implant the fingerprint constraints into a DSP core design. However, the fingerprint constraints corresponding to all signature digits are not implanted. Only those fingerprint digits which are selected based on selection criteria 1 and 2 (described in Figure 6.5) are eligible to be embedded as secret constraints. The embedding process is shown in Figure 6.8 (Sengupta *et al.*, 2019). As shown in the figure, once the eligible fingerprint constraints are extracted, they are implanted into the CIG of the design. In the CIG, two different storage variables (node pairs) executed in the same control step cannot be stored in the same register (colour). As a CIG represents the sharing of different registers (colours) among all the storage variables (node pairs), implanting an edge between any node pair imposes that the pair be assigned to distinct registers. If the addition of an edge results into a violation of the aforesaid rule, then it is resolved by swapping nodes within the same control step (T) so that both storage variables of this edge are

Table 6.6 Encoding the buyer's fingerprint

Variables of buyer's fingerprint	Encoding
i	Encoded value of edge with node pair as (prime, prime)
I	Encoded value of edge with node pair as (even, even)
T	Encoded value of edge with node pair as (odd, even)
!	Encoded value of edge with node pair as (0, any integer)

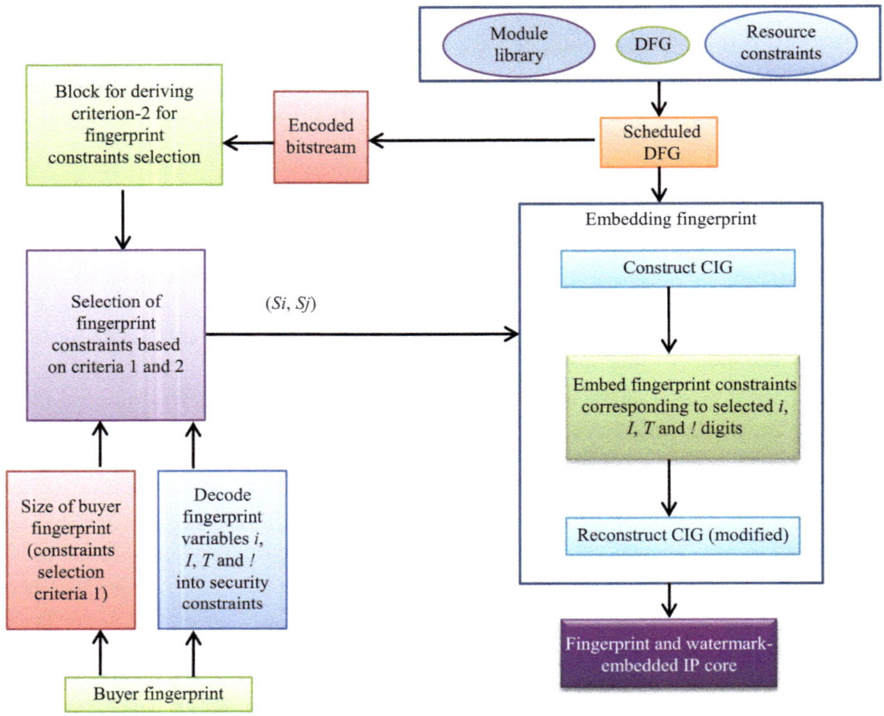

Figure 6.8 Fingerprinting embedding process for DSP core protection

assigned to different registers. Further, if the aforesaid resolution by swapping of nodes within the same T is not possible, then an extra register (another colour is added) is used to execute one of the nodes (storage variable) in the pair. Post-embedding all the selected secret constraints (edges) of the fingerprint, the CIG is modified and the corresponding modification in the register allocation table is performed.

6.4.3.2 Demonstration on DCT core

Let us assume an 80-digit selected fingerprint of an IP buyer is 'TII!iiTTii!TITiiiIIIiTTiiiIiIiii!!!ii!IT!!IiI!!!ITi!iiT!iiTT!iTiiTT!!!iTIII!Tii!'.

The respective position associated with each fingerprint digit is shown in Table 6.7 (Sengupta *et al.*, 2019). This buyer's fingerprint is intended to be embedded into the CIG (corresponding to DCT core). Now, based on the information available in Table 6.5 (which indicates the generated design digest), the set of all decimal values which are less than, or equal to, the fingerprint size is shortlisted for embedding (as it satisfies criteria 1 and 2). These two criteria jointly decide the number of fingerprint constraints eligible to be embedded. Out of 64 decimal values, only 16 digits satisfy both criteria, thus they are selected (shortlisted) as valid digits. The selected digits with their associated positions have been

Table 6.7 80-digit fingerprint with each digit associated with its respective position

1 T	**2 I**	3 I	4 !	5 i	6 i	7 T	8 T	9 i	10 i	11 !	12 T	**13 I**	14 T	15 i	16 i	17 i	18 I	19 I	20 I
21 i	**22 T**	23 T	24 i	25 i	26 i	27 I	**28 i**	29 I	**30 i**	**31 i**	32 i	33 !	34 !	35 !	36 i	37 i	**38 !**	**39 I**	**40 T**
41 !	42 !	43 I	44 i	45 I	46 !	47 !	**48 !**	49 I	50 T	51 i	52 !	**53 i**	54 i	55 T	56 !	**57 i**	58 i	59 T	60 T
61 !	62 i	**63 T**	64 i	65 i	66 T	67 T	**68 !**	69 !	70 !	71 i	72 T	73 I	74 I	75 I	76 !	**77 T**	78 i	**79 i**	80 !

Table 6.8 Decoded value of effective fingerprint constraints

Digit of signature	Decoded edge	Digit of signature	Decoded edge
57	($S5$, $S19$)	22	($S1$, $S12$)
38	($S0$, $S6$)	48	($S0$, $S11$)
63	($S1$, $S26$)	68	($S0$, $S15$)
39	($S2$, $S20$)	77	($S3$, $S6$)
79	($S7$, $S23$)	2	($S2$, $S4$)
31	($S3$, $S17$)	28	($S3$, $S11$)
53	($S5$, $S13$)	40	($S1$, $S16$)
30	($S3$, $S13$)	13	($S2$, $S8$)

highlighted in Table 6.7. Further, the selected digits and their corresponding decoded edges (representing fingerprint constraints) are shown in Table 6.8. These secret constraints (edges) are implanted into the CIG one-by-one. However, conflict may occur during the implantation of some of the constraint edges. For example, direct insertion of edge ($S3$, $S13$) corresponding to fingerprint digit 30, is not possible. This is because both nodes ($S3$ and $S13$) share the same register $R3$. This conflict is resolved by assigning $S13$ to register $R7$ in the second control step (T2). Likewise, conflict due to other constraint edges can be resolved. Thus all the constraint edges listed in Table 6.8 are implanted into the CIG one-by-one. Post-embedding fingerprint constraints, the modified CIG is shown in Figure 6.9 and the corresponding modification in the register allocation is presented in Table 6.9 (Sengupta *et al.*, 2019). Finally, the impact of the fingerprint constraints can also be observed in terms of changes in input–output connectivity of muxes and demuxes at RTL.

6.5 Analysis of case studies

The cryptosystem-based multivariable fingerprinting approach has been analysed in terms of security and design cost for various DSP applications (Sengupta *et al.*, 2019). Further, a comparison with a contemporary fingerprinting approach for DSP cores is presented in this section.

6.5.1 Security and design cost analysis

Security achieved through a multivariable fingerprint is assessed in terms of the probability of coincidence (P^c) metric given as (Sengupta and Roy, 2017, 2018):

$$P^c = \left(1 - \frac{1}{n}\right)^k \tag{6.4}$$

where P^c denotes the probability of coincidence, n denotes the number of colours used in the CIG before implanting fingerprint constraints and k denotes the number of additional constraints (edges) added or effective fingerprint size (potential

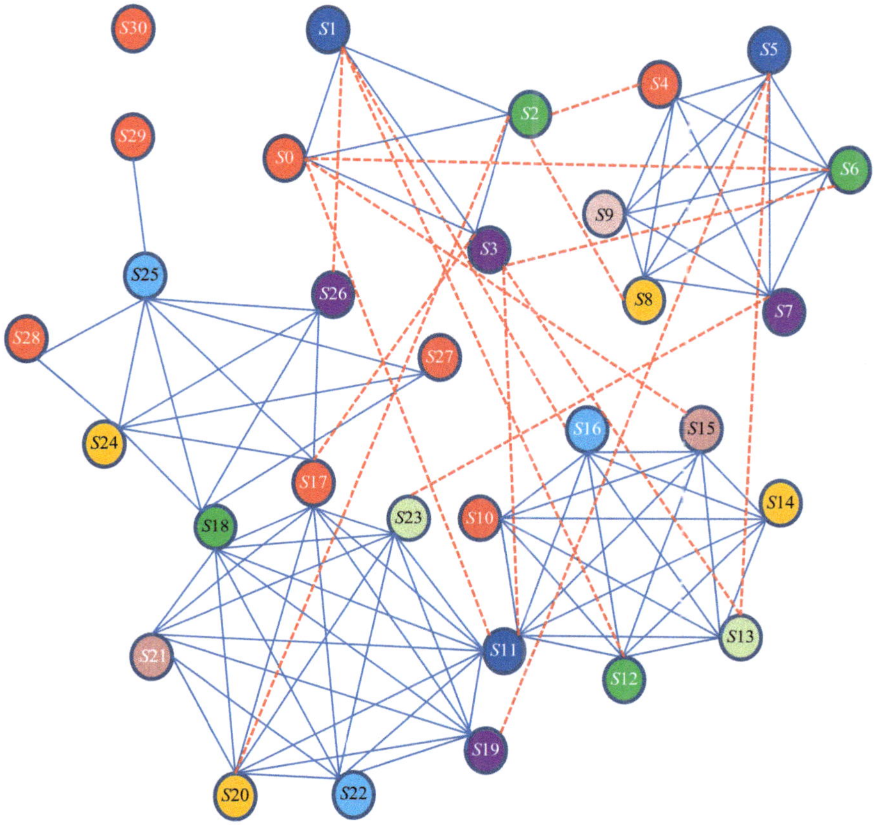

Figure 6.9 Coloured interval graph of 8-point DCT core with fingerprint-constraints (Sengupta et al., 2019) Note: Added constraints-edges have been shown with dotted red lines

Table 6.9 Modified register allocation table post-embedding fingerprint

T	R0	R1	R2	R3	R4	R5	R6	R7
0	S0	S1	S2	S3	–	–	–	–
1	S4	S5	S6	S7	S8	S9	–	–
2	S10	S11	S12	–	S14	S15	S16	S13
3	S17	S11	S18	S19	S20	S21	S22	S23
4	S17	–	S18	S26	S24	–	S25	–
5	S27	–	S18	–	S24	–	S25	–
6	S28	–	–	–	S24	–	S25	–
7	S29	–	–	–	–	–	S25	–
8	S30	–	–	–	–	–	–	–

Table 6.10 Security analysis of the cryptosystem-based multivariable fingerprinting approach

Applications	Number of registers	Fingerprint signature size = 80		Fingerprint signature size = 160		Fingerprint signature size = 240	
		Effective size (k)	P^c	Effective size (k)	P^c	Effective size (k)	P^c
DCT	10	17	0.167	37	0.020	48	0.0063
IDCT	12	20	0.175	35	0.047	48	0.0153
WDF	7	20	0.046	32	0.007	47	0.0007
MESA	6	16	0.054	27	0.007	31	0.0035
FFT	8	20	0.069	35	0.009	49	0.0014

fingerprint digits selected through criteria 1 and 2). Here, probability of coincidence signifies the proof of the buyer's right, which indicates a non-fingerprint solution carrying the fingerprint by coincidence. Thus the probability of coincidence metric reflects the strength of the fingerprint. As the probability of coincidence decreases, the strength of proof of the buyer's right increases. This is because as more fingerprint constraints are embedded into the CIG of the IP core, the likelihood of obtaining the same colouring solution by coincidence becomes less. According to (6.4), the value of P^c decreases with an increase in effective fingerprint constraints or edges (k) embedded into the CIG. The number of effective constraints essentially depends on the fingerprint size. The number of effective constraints increases with the increasing fingerprint size as shown in Table 6.10 (Sengupta *et al.*, 2019). This is because increasing the fingerprint size leads to an increase in the set of eligible decimal numbers (based on constraint selection criteria 1 and 2) obtained from the SHA output. The impact of three different fingerprint sizes ($Q = 80, 160, 240$) on probability of coincidence is shown in Table 6.10 (Sengupta *et al.*, 2019). Therefore to achieve a stronger proof of the buyer's right, a large fingerprint size should be selected.

Furthermore, the increase in fingerprint size may also affect the design cost of an IP core. However, the presented cryptosystem-based multivariable fingerprinting approach does not increase the design cost with an increase in fingerprint size. The design cost remains the same as that of the baseline version (design with no fingerprint) as shown in Table 6.11 (Sengupta *et al.*, 2019). The cost function parameters are computed based on a 15 nm technology scale and the design cost is assessed according to (6.3). It is evident from Table 6.11 that the presented fingerprinting approach incurs zero or nominal cost overheads and, hence, leads to a very low cost design solution (Sengupta *et al.*, 2019).

6.5.2 Comparative study

To assess the effectiveness of the cryptosystem-based multivariable fingerprinting approach (Sengupta *et al.*, 2019), it is compared with a contemporary fingerprinting

Table 6.11 Design cost analysis of the cryptosystem-based multivariable fingerprinting approach

Applications	Design cost of baseline design (no fingerprint)	Fingerprint signature size = 80		Fingerprint signature size = 160		Fingerprint signature size = 240	
		Effective size (k)	Cost	Effective size (k)	Cost	Effective size (k)	Cost
DCT	0.44	17	0.44	37	0.44	48	0.44
IDCT	0.33	20	0.33	35	0.33	48	0.33
WDF	0.61	20	0.61	32	0.61	47	0.61
MESA	0.64	16	0.64	27	0.64	31	0.64
FFT	0.59	20	0.59	35	0.59	49	0.59

Table 6.12 Comparative study of the cryptosystem-based multivariable fingerprinting approach with (Roy and Sengupta, 2017)

DSP applications	Effective number of fingerprint constraints	Design solution	Design cost (Sengupta *et al.*, 2019)	Design cost (Roy and Sengupta, 2017)	Percentage reduction in design cost
DCT	17	2A, 2M	0.44	0.46	4.35 %
IDCT	20	3A, 3M	0.33	0.37	10.81 %
WDF	20	2A, 2M	0.61	0.63	3.17 %
MESA	16	1A, 2M	0.64	0.66	3.03 %
FFT	20	4A, 4M	0.59	0.63	6.35 %

approach (Roy and Sengupta, 2017) for DSP cores. A comparison based on design cost of the fingerprint-embedded IP core is shown in Table 6.12. It is evident from the table that (Sengupta *et al.*, 2019) is capable to achieve a lower design cost (post-embedding fingerprint) compared to (Roy and Sengupta, 2017). This is because once the effective fingerprint size is obtained, the number of constraints to be embedded remain fixed in the cryptosystem-based multivariable fingerprinting approach. However, for the same effective fingerprint size, the contemporary fingerprinting approach (Roy and Sengupta, 2017) can have numerous possible combinations of fingerprint digits. Therefore the number of constraints to be embedded is variable and, hence, imposes different implications on the design cost. Furthermore, it is hard to estimate the optimal combination of fingerprint digits that would result into the lowest possible design cost post-embedding. Hence, the cryptosystem-based multivariable fingerprinting approach (Sengupta *et al.*, 2019) is more effective in comparison to the contemporary approach (Roy and Sengupta, 2017).

6.6 Conclusion

This chapter discusses the fingerprinting process for protection of the buyer's right in the context of DSP cores. Several fingerprinting approaches were discussed and special emphasis was made on cryptosystem-based multivariable fingerprinting. An embedded fingerprint helps to trace illegally resold IP cores by a disloyal seller, thus protecting the buyer's rights of the IP cores. The cryptosystem-based multivariable fingerprinting approach employs a hashing algorithm (SHA-512) to determine a unique set of constraints corresponding to the buyer's fingerprint. The register-allocation phase of ESL synthesis has been leveraged to embed the set of fingerprint constraints. Since in this approach a fingerprint is embedded at a higher abstraction level, the added constraints get distributed in the lower level throughout the design post-synthesis. Furthermore, the cryptosystem-based multivariable fingerprinting approach offers higher security at zero design and cost overheads compared to the contemporary fingerprinting approach (Roy and Sengupta, 2017).

This chapter provided a detailed discussion on the cryptosystem-based multivariable fingerprinting approach for DSP cores. At the end of this chapter, the reader is able to understand the following concepts:

- Utility of the fingerprinting approach.
- Application of the fingerprinting approach to protect the buyer's rights in the context of IP or DSP cores.
- Desirable features of a buyer's fingerprint.
- Comparison between watermarking and fingerprinting.
- Encoding rules of the fingerprint variables.
- Application of SHA-512 to generate unique fingerprint constraints in the multivariable fingerprinting approach.
- Demonstrating the design of the fingerprint-embedded IP core using a real-life DSP core application (DCT core).
- Security and design cost analysis of the fingerprinting approach with respect to the baseline version.
- Comparative study of a fingerprint-embedded DSP core with respect to a contemporary fingerprinting approach for several case studies.

6.7 Questions and exercises

1. Explain the difference between a watermark and fingerprint in the context of IP cores.
2. How is the threat model of IP fingerprinting different from IP watermarking?
3. In the context of DSP cores, what is the role of crypto block SHA-512 in generating a fingerprint?
4. Explain the encoding algorithm of control data flow graphs (CDFGs)s used in generating a fingerprint.
5. Explain the symmetrical protection of IP cores.
6. Why is this method is called 'symmetrical' protection?

7. Explain the encoding algorithm for fingerprinting in symmetrical IP core protection.
8. Explain the encoding algorithm for watermarking in symmetrical IP core protection.
9. Which incurs more chances of design and cost overheads, fingerprinting or watermarking in symmetrical IP core protection algorithm?
10. How is the multivariable fingerprinting process different from contemporary fingerprinting of DSP cores?
11. How are fingerprint constraints derived from the design digest?
12. Explain the decoding algorithm of IP fingerprinting.
13. Explain how input is created for the SHA-512 algorithm.
14. What is the impact of the fingerprint signature's strength on design cost?
15. What is the impact of the fingerprint signature's strength on the buyer's right?
16. What are storage variables, and how are they used in proving the buyer's right through a fingerprint?
17. State the different scheduling algorithms used in the HLS of DSP cores, and which one is advantageous in the context of cryptosystem-based multivariable fingerprinting of IP cores?
18. In symmetrical IP core protection, what is the input of the watermarking process?

References

E. Castillo, U. Meyer-Baese, A. Garcia, L. Parilla, and A. Lloris, "IPP@HDL: Efficient intellectual property protection scheme for IP cores," IEEE Transactions on Very Large Scale Integration (VLSI) Systems, vol. 15, no. 5, Pages: 578–590, 2007.

F. Koushanfar, *et al.*, "Behavioral synthesis techniques for intellectual property protection," ACM Transactions on Design Automation of Electronic Systems, vol. 10, no. 3, Pages: 523–545, 2005.

J. Lach, W. H. Mangione-Smith, and M. Potkonjak, "Fingerprinting techniques for field-programmable gate array intellectual property protection," IEEE Transactions on Computer-Aided Design of Integrated Circuits and Systems, vol. 20, no. 10, Pages: 1253–1261, 2001.

B. Le Gal and L. Bossuet, "Automatic low-cost IP watermarking technique based on output mark insertions," Design Automation for Embedded Systems, vol. 16, no. 2, Pages: 71–92, 2012.

G. Martin and G. Smith, "High-level synthesis: Past, present, and future," IEEE Design and Test of Computers, vol. 26, no. 4, Pages: 18–25, 2009.

M. C. McFarland, A. C. Parker, and R. Camposano, "Tutorial on high-level synthesis," in DAC '88 Proceedings of the 25th ACM/IEEE Design Automation, vol. 27, no. 1, Pages: 330–336, 1988.

S. M. Plaza and I. L. Markov, "Solving the third-shift problem in IC piracy with test-aware logic locking," IEEE Transactions on Computer-Aided Design of Integrated Circuits and Systems, vol. 34, no. 6, Pages: 961–971, 2015.

D. Roy and A. Sengupta, "Low overhead symmetrical protection of reusable IP core using robust fingerprinting and watermarking during high level synthesis," Elsevier Journal on Future Generation Computer Systems, vol. 71, Pages: 89–101, 2017.

D. Roy and A. Sengupta, "Multilevel watermark for protecting DSP Kernel in CE systems [hardware matters]," IEEE Consumer Electronics Magazine, vol. 8, no. 2, Pages: 100–102, 2019.

R. Schneiderman, "DSPs evolving in consumer electronics applications," IEEE Signal Processing Magazine, vol. 27, no. 3, Pages: 6–10, 2010.

A. Sengupta, "Intellectual property cores: Protection designs for CE products," IEEE Consumer Electronics Magazine, vol. 5, no. 1, Pages: 83–88, 2016.

A. Sengupta, "Hardware security of CE devices [hardware matters]," IEEE Consumer Electronics Magazine, vol. 6, no. 1, Pages: 130–133, 2017.

A. Sengupta and S. Bhadauria, "Exploring low cost optimal watermark for reusable IP cores during high level synthesis," IEEE Access, vol. 4, Pages: 2198–2215, 2016.

A. Sengupta and S. P. Mohanty, "Advanced Encryption Standard (AES) and Its Hardware Watermarking for Ownership Protection," IP Core Protection and Hardware-Assisted Security for Consumer Electronics, Pages: 317–335, 2019a, e-ISBN: 9781785618000.

A. Sengupta and S. P. Mohanty, "IP Core and Integrated Circuit Protection Using Robust Watermarking," IP Core Protection and Hardware-Assisted Security for Consumer Electronics, Pages: 123–170, 2019b, e-ISBN: 9781785618000.

A. Sengupta and D. Roy, "Antipiracy-aware IP chipset design for CE devices: A robust watermarking approach [hardware matters]," IEEE Consumer Electronics Magazine, vol. 6, no. 2, Pages: 118–124, 2017.

A. Sengupta and D. Roy, "Multi-phase watermark for IP core protection," in Proc. 36th IEEE International Conference on Consumer Electronics (ICCE), Pages: 1–3, 2018.

A. Sengupta, R. Sedaghat, and Z. Zeng, "A high level synthesis design flow with a novel approach for efficient design space exploration in case of multi-parametric optimization objective," Microelectronics Reliability, vol. 50, no. 3, Pages: 424–437, 2010.

A. Sengupta, D. Roy, and S. P. Mohanty, "Triple-phase watermarking for reusable IP core protection during architecture synthesis," IEEE Transactions on Computer-Aided Design of Integrated Circuits and Systems, vol. 37, no. 4, Pages: 742–755, 2018.

A. Sengupta, U. K. Singh, and P. K. Premchand, "Crypto based multi-variable fingerprinting for protecting DSP cores," in IEEE International Conference on Consumer Electronics (ICCE), Berlin, Sep 2019, IEEE. doi: 978-1-7281-2745-3/19/, available online at: https://edas.info/p25749.

D. Ziener and J. Teich, "Power signature watermarking of IP cores for FPGAs," Journal of Signal Processing Systems, vol. 51, no. 1, Pages: 123–136, 2008.

Chapter 7

Multi-phase obfuscation for fault-secured DSP circuits

Anirban Sengupta[1]

This chapter discusses a multi-phase obfuscation process for fault-secured intellectual property (IP) cores during electronic system level (ESL) synthesis. A detailed elaboration on the threat model for fault-secured IP cores is followed by the use of the multi-phase obfuscation process for digital signal processing (DSP) circuits and, finally, analyses of case studies.

The chapter is organized as follows: Section 7.1 discusses fault-secured IP cores and their needs, followed by different threats to fault-secured IP cores and how to solve them. Section 7.2 focuses on the differences between functional obfuscation and structural obfuscation. Section 7.3 presents the problem formulation for protecting fault-secured IP cores. Section 7.4 discusses selected contemporary structural obfuscation approaches used to date. Section 7.5 provides an overview of the multi-phase obfuscation approach followed by evaluation models used and details of the approach in the context of fault-secured DSP circuits. Section 7.6 demonstrates the multi-phase obfuscation approach on a fault-secured finite impulse response (FIR) filter. Section 7.7 presents analyses based on case studies. Section 7.8 concludes the chapter.

7.1 Introduction

In the current era of semiconductor technology, the trend in the design process of an integrated circuit (IC) includes globalization. More explicitly, fabless semiconductor companies are responsible only for designing the circuit, generating netlists and producing layouts. While for fabrication and manufacturing, the design netlists are transferred to off-shore foundries or manufacturing houses. Additionally, a system-on-chip integrator buys reusable IP cores from fabless design houses. This globalization in the supply chain has significantly increased design productivity and considerably reduced cost (Plaza and Markov, 2015; Yasin *et al.*, 2016). An IP core (such as a DSP kernel) may be susceptible to vulnerabilities such as single-event

[1]Computer Science and Engineering, Indian Institute of Technology Indore, Indore, India

upsets or single-event transients (SET) (Inoue *et al.*, 2011; Sengupta, 2016a; Sengupta and Kachave, 2017). These transients occur due to the emanation of α-particles from thorium and uranium impurities present in an IC package or when used in space applications under certain environmental radiations. The increase in the speed of electronic devices and substantial scaling in device geometry, has led to multi-cycle (temporal impact) and multi-unit (spatial impact) transient faults (Sengupta and Kachave, 2018). This challenge led to the design of fault-secured DSP circuits (Schneiderman, 2010). These fault-secured designs also need to be secured against hardware threats such as reverse engineering (RE) threats, overbuilding, Trojan insertion, etc. (Sengupta *et al.*, 2018).

Hardware threats can be realized by rouge elements (having malicious intents) present in the foundries. For example, a dishonest manufacturer may pirate an IP or IC netlist and sell them illegally (Sengupta, 2016b, 2017; Sengupta and Bhadauria, 2016). An IP or IC may suffer from various forms of infringements such as (1) after buying an IP from a design house, an adversary can make pirated copies and claim ownership, (2) an adversary can generate counterfeited copies of the original IP and earn illegal revenue by selling them under the same brand, (3) an adversary can sell cloned copies of the original IP under his/her own brand (different from original), (4) when an IP is sent to a manufacturing house for IC fabrication, a dishonest manufacturer may violate the terms of producing a fixed number of ICs and overbuild the ICs to gain personal benefit or (5) an adversary can reverse engineer (back engineer) (Torrance and James, 2009) the design netlist (GDS-II file), masks, packaged non-functional IC and functional IC to extract the designer's IP. Once the adversary extracts the original design, he/she can pirate, counterfeit or clone it or may insert malicious logic (called hardware Trojans) inside the design. Using Trojan insertion, the adversary can render the design function according to him/her, thus challenging the reliability and security of the IP core (Castillo *et al.*, 2007; Chakraborty and Bhunia, 2009; Colombier and Bossuet, 2015; Roy *et al.*, 2008). Furthermore, RE can also be performed to deduce the structure or functionality of the secret modules used in a chip (e.g., smart card, encryption key, etc.).

These hardware threats have drawn the attention of IP owners and designers towards the security of their IPs and ICs. Several protection mechanisms such as hardware metering (Koushanfar, 2012), IC camouflaging (Chow *et al.*, 2002; Rajendran *et al.*, 2013), obfuscation, etc., may be employed to address these challenges. However, obfuscation, in particular, has emerged as a strong countermeasure in the current era of IP core protection and is capable to thwart piracy, counterfeiting, cloning and RE attacks.

Obfuscation (Lao and Parhi, 2015; Sengupta *et al.*, 2017) is the art of making an object of interest unobvious to be understood by an attacker. In terms of electronics hardware, obfuscation is a mechanism of obscuring or hiding the architecture or description of a design through some modifications in order to cover its functionality. However, the functionality of the obscured design remains unaffected. Obfuscation makes the design unobvious for an attacker and thus renders it harder to reverse engineer. Obfuscation in electronics hardware is commonly performed in two ways, namely (1) logic or functional obfuscation and (2) structural obfuscation.

Logic obfuscation (Zhang, 2015; Yasin *et al.*, 2016; Sengupta *et al.*, 2019) is a mechanism of locking the functionality of a design using some additional gates that can be inserted randomly or through an algorithm. Thus a obtained logically locked or functionally obfuscated design reveals its functionality only upon applying a valid key. The additionally inserted gates are associated with key-inputs, therefore, they are referred to as key gates. The valid key is stored in a tamper-evident memory such that it can be prevented from being accessed by an attacker. However, logic obfuscation suffers from some key-based attacks such as key sensitization (Yasin *et al.*, 2016), Boolean satisfiability (SAT)-based attack (Subramanyan *et al.*, 2015), etc. Therefore a keyless obfuscation method can be used as an alternative to logic obfuscation. Structural obfuscation (Lao and Parhi, 2015; Sengupta *et al.*, 2017) is a keyless technique of obscuring the functionality of a design. Structural obfuscation renders the functionality and original architecture unobvious by introducing intelligent modifications into the structure of the design. Thus the obtained structurally obfuscated design is functionally equivalent to the original; however it is significantly harder to reverse engineer.

This chapter discusses a multi-phase obfuscation approach to protect the design of fault-secured DSP circuits against hardware threats (Sengupta *et al.*, 2018). This multi-phase obfuscation approach employs obfuscation at two different phases of ESL synthesis (McFarland *et al.*, 1988; Sengupta *et al.*, 2010) to protect the fault-secured design. Further, a dual modular redundant (DMR) design technique to generate fault-secured designs against k-cycles transient faults has also been discussed in this chapter. The multi-phase obfuscation approach conceals the functionality of the fault-secured design by modifying its architecture or structure without affecting the original behaviour. Additionally, checkpoints insertion through applying multi-cuts in the DMR design further enhances the fault-secured capability. Currently, the multi-phase obfuscation proposed by Sengupta *et al.* (2018) is the only approach in the literature that offers obfuscation of fault-secured DSP designs against RE threats.

7.2 Functional obfuscation vs. structural obfuscation

Target threats are similar for functional obfuscation and structural obfuscation; however the two approaches differ in key aspects such as security protocol, abstraction level of deployment, activation mechanism, vulnerability to different attacks and resulting design overheads. A detailed comparison is shown in Figure 7.1. Security through functional obfuscation essentially depends on the encryption keys. Therefore functional obfuscation may be vulnerable to key-based attacks (key sensitization, SAT) and removal attack on locking logic (comprising key gates). Further, incorporation of key gates into the design incurs cost overheads. Structural obfuscation, on the other hand, is a keyless approach which renders it non-vulnerable to key-based attacks. Moreover, the basic idea of this

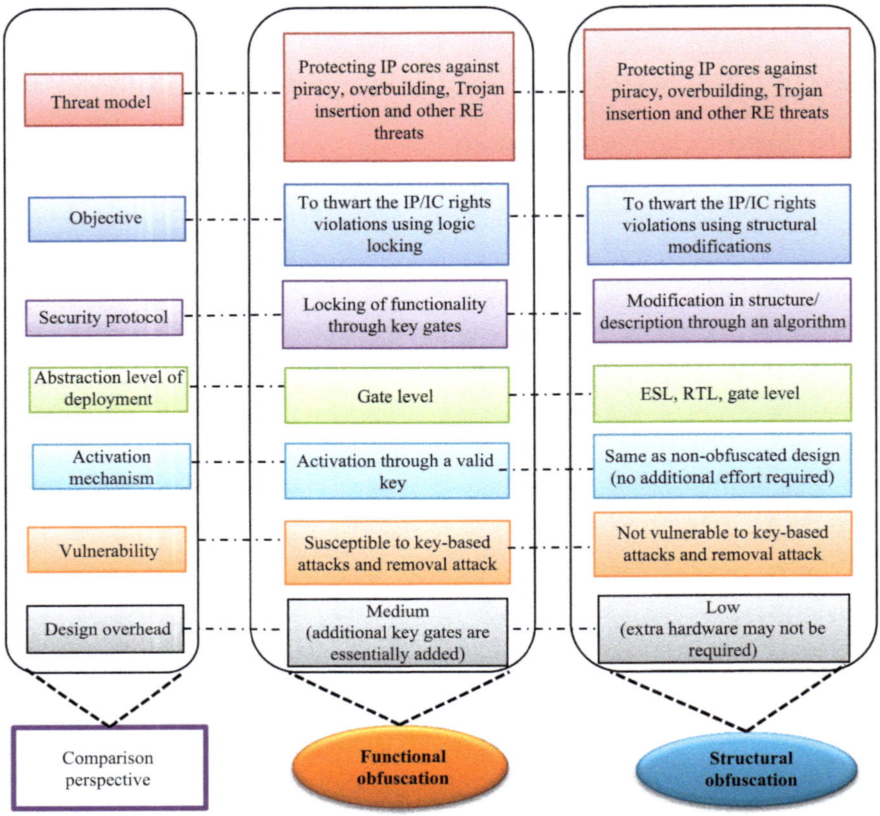

Figure 7.1 Comparison of functional and structural obfuscation

obfuscation approach is to modify the circuit structure of the design in order to make it unobvious to an attacker. Therefore structural obfuscation does not add extra hardware (or nominal extra hardware) into the design and is hence capable of providing a low-cost solution. Further, structural obfuscation may be implemented at different abstraction levels such as ESL, register transfer level (RTL) and gate level. Nevertheless, for highly complex designs such as DSP circuits, structural obfuscation is more beneficial when implemented at a higher abstraction level such as ESL and RTL. This is because if structural obfuscation is implemented at ESL and RTL, then overhead control is better, complexity is less and design objectives (such as area, power and delay) can be optimized through a design space exploration (DSE) framework (Mishra and Sengupta, 2014) resulting in lower design cost. However, functional obfuscation is always implemented at gate level, therefore it is less feasibility to optimize the design overheads and complexity.

7.3 Problem formulation

The aim of this chapter is to present a security mechanism of transient, fault-secured DSP circuits using multi-phase structural obfuscation (Sengupta *et al.*, 2018). A control data flow graph (CDFG) or C-code representing a DSP application is transformed into an obfuscated fault-secured design using multi-phase obfuscation, DMR design technique and ESL synthesis framework. The employed multi-phase obfuscation should be able to modify the structure of the fault-secured design significantly so that an attacker would not easily understand the functionality of the design, thus thwarting RE. To obtain a multi-cycle, transient, fault-secured design, the maximum possible transient-fault strength ($K^c = 2$, which is equivalent to 2,000 ps @ 1 GHz) is usually considered as the specification. The goal is that subjected to transient-fault strength ($K^c = 2$ control steps), the employed obfuscation on a fault-secured DSP circuit should incur minimal design cost while, simultaneously, being capable of enhancing the strength of obfuscation. The design cost of an obfuscated, fault-secured DSP core is calculated in terms of hardware area (A^T_{Ob+FS}) and execution latency (L^T_{Ob+FS}) (Sengupta *et al.*, 2018).

7.4 Discussion on selected approaches

This section discusses the techniques of employing structural obfuscation during ESL synthesis. During ESL synthesis, structural obfuscation has been deployed for (1) IP cores (DSP circuits) and (2) fault-secured IP cores (DSP circuits). Related approaches that employed structural obfuscation for DSP circuits are Lao and Parhi (2015), Sengupta *et al.* (2017) and Sengupta and Roy (2017). The IP core protection of DSP circuits using obfuscation was proposed for the first time in the literature by Lao and Parhi (2015). The structural obfuscation employed by Lao and Parhi (2015) is based on the idea of high-level transformations (HLTs). Lao and Parhi (2015) leveraged HLTs of iterative data-flow graphs (representing DSP application) to perform trade-offs between area power and speed. Lao and Parhi (2015) used HLTs in order to perform obfuscation on DSP circuits while simultaneously meeting the same trade-off. The HLTs are capable of modifying the structure of a DSP circuit without compromising the original functionality. The HLTs result into unobvious architectures. Thus functionality can be hidden using HLTs by exploiting the fact that circuits with identical structures can exhibit different functionality and circuits with different structures can exhibit identical functionality. Lao and Parhi (2015) performed the HLTs using folding in order to achieve structural obfuscation. The folding transformation approach is the reverse of unfolding transformation. In folding transformation, folded variants are produced according to a folding set. Since an appropriate folding order results in a structure with lower power and area, the choice of the folding order or set plays a crucial role in determining the performance of the folded structure. For assistance to HLTs, Lao and Parhi (2015) also leveraged a key-based finite state machine (FSM) and a reconfigurator to obfuscate DSP circuits. When a valid key is provided to the FSM, it activates the reconfigurator. A reconfigurator has been used to reconfigure the filter (such as a finite impulse response filter) order in multiple modes of operation to employ obfuscation. The purpose of this obfuscation approach is to conceal the functionality of the DSP circuits so that an attacker cannot discern it. Thus Lao and Parhi (2015) secured DSP circuits by rendering them harder to reverse engineer. Since an attacker does not know the initialization key to the FSM and the reconfigured states, this approach is capable to protect DSP circuits against piracy and overbuilding. A genuine vendor knows the key and reconfigured states, therefore only he/she can make the IC functional after fabrication or provide the key to the buyer. This renders the overbuilt ICs (without the keys) useless. Further, an attacker can identify the functionality of the design in two ways, namely (1) by examining structure of the netlist in order to detect and isolate the obfuscated design or (2) by performing simulation-based RE. However, the obfuscation of Lao and Parhi (2015) secures the design against the attack based on structural examination. Figure 7.2 shows the structural obfuscation approach proposed by Lao and Parhi (2015). In this approach, Lao and Parhi (2015) used a key-based FSM. The key monitors the states of the FSM. Upon applying a valid key to the FSM, the reconfigurator is enabled. Further, the different functionality modes of the DSP circuit are configured by the reconfigurator. Several equivalent DSP circuits are possible using HLTs. This leads to structural obscurity in the circuit. Additionally, for a DSP circuit, similar data paths with different controllers can be designed using HLTs. Thus to perform obfuscation, a designer can insert different variation modes into the DSP circuit. Some modes produce functionally invalid output, nevertheless these may represent valid output under different circumstances. However, obfuscation for DSP circuits by Lao and Parhi (2015) has some limitations such as (1) resulting in significant design overheads due to not considering an optimal design solution as well as not exploring an optimal folding factor, (2) it does not exploit the use of successive HLTs and (3) it does not consider the obfuscation of fault-secured DSP circuits. Sengupta and Roy (2017) and Sengupta *et al.* (2017) leveraged multi-stage (compiler based), HLTs such as redundant operation elimination (ROE), logic transformation (LT) and tree-height transformation (THT) to employ structural obfuscation in DSP circuits. However, Sengupta and Roy (2017) did not consider loop-based or control-intensive DSP circuits. Further, Sengupta *et al.* (2017) exploited two more transformations called loop unrolling (LU) and loop invariant code motion in addition to ROE, LT and THT. Thus the approach by Sengupta *et al.* (2017) is also capable of handling loop-based DSP circuits and exploring optimal LU factors. Additionally, Sengupta *et al.* (2017) obtain lower design cost by exploring a low-cost design solution using a particle swarm optimization–based DSE framework. However, Sengupta and Roy (2017) and Sengupta *et al.* (2017) did not employ obfuscation for fault-secured DSP circuits. The multi-phase obfuscation approach proposed by Sengupta *et al.* (2018) is the only approach in the literature to date that employs structural obfuscation on fault-secured DSP circuits.

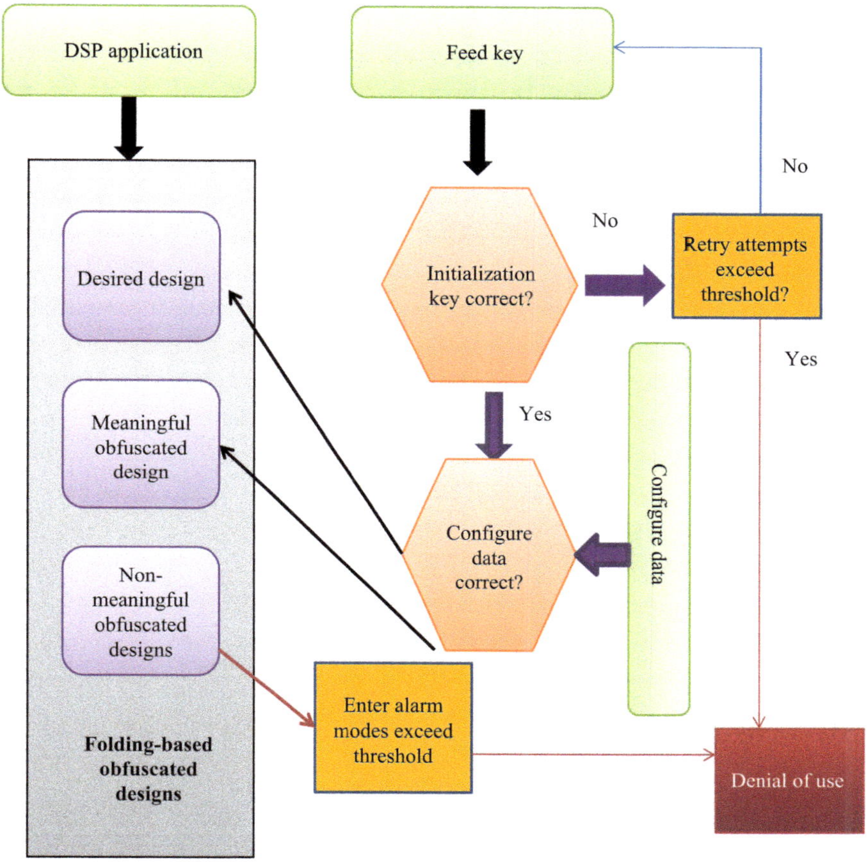

Figure 7.2 Structural and functional obfuscation using Lao and Parhi (2015)

7.5 Structural obfuscation of fault-secured DSP circuits

This section presents structural obfuscation of fault-secured DSP circuits. A fault-secured design generated through a DMR technique is secured by employing obfuscation at two different phases (Sengupta *et al.*, 2018) of ESL synthesis. Phase-1 obfuscation is employed before pre-generating fault-secured designs. Once the obfuscated (using phase-1 obfuscation) fault-secured design is obtained, then phase-2 obfuscation is employed. The application of multi-phase structural obfuscation not only hinders RE threats by rendering the circuit architectures unobvious, but is also capable of optimizing the delay overhead (using multiple cut-insertion) and enhance fault security (using multiple checkpoints). Details of this approach followed by evaluation models are discussed from Section 7.5.1 onward (Sengupta *et al.*, 2018).

7.5.1 Overview

An overview of the multi-phase obfuscation approach for fault-secured DSP circuits is shown in Figure 7.3. As shown in the figure, primary inputs for this approach are (1) C/C++ code or CDFG representing a DSP application, (2) module library, (3) resource constraints and (4) transient-fault strength (i.e., how many cycles will it affect in the worst-case scenario). An obfuscated, transient, fault-secured DSP circuit is generated at the output. The two phases of the multi-phase obfuscation approach for protecting a fault-secured DSP core are illustrated in Figure 7.4. In order to secure a fault-secured DSP circuit against RE threats, first phase-1 obfuscation is performed on the DSP application. During phase-1 obfuscation, three compiler-based HLTs and a functional-unit (FU) transformation (FUT) are performed. The three HLTs are (Sengupta *et al.*, 2018): (1) ROE, (2) LT and (3) THT. These three HTLs are performed in sequence, followed by an FUT (Sengupta *et al.*, 2018) process. Once the phase-1 obfuscation is completed, a fault-secured design algorithm is applied to obtain security (detection) against *k*-cycles transient fault. This algorithm is executed in two steps, namely (1) constructing the DMR design and (2) scheduling of the DMR design based on resource constraints (obtained using a DSE system). Output of this

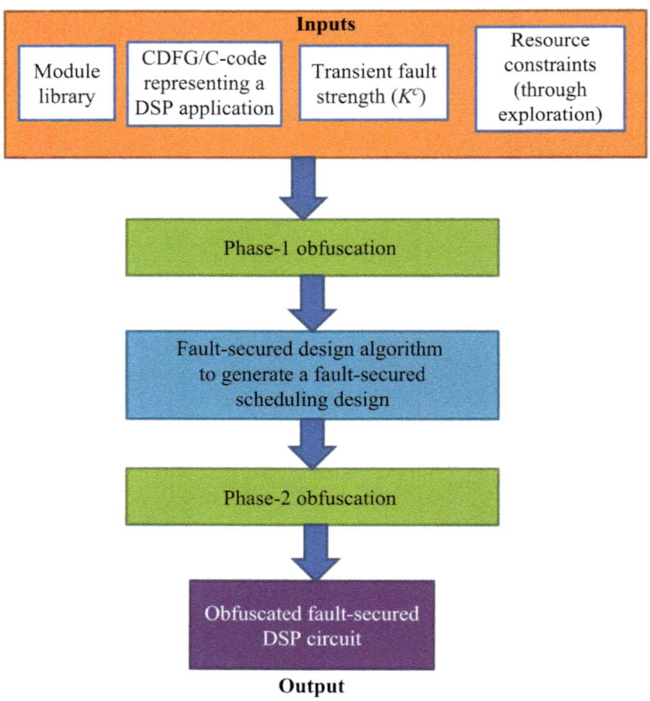

Figure 7.3 Overview of multi-phase obfuscation of fault-secured DSP circuits (Sengupta et al., 2018)

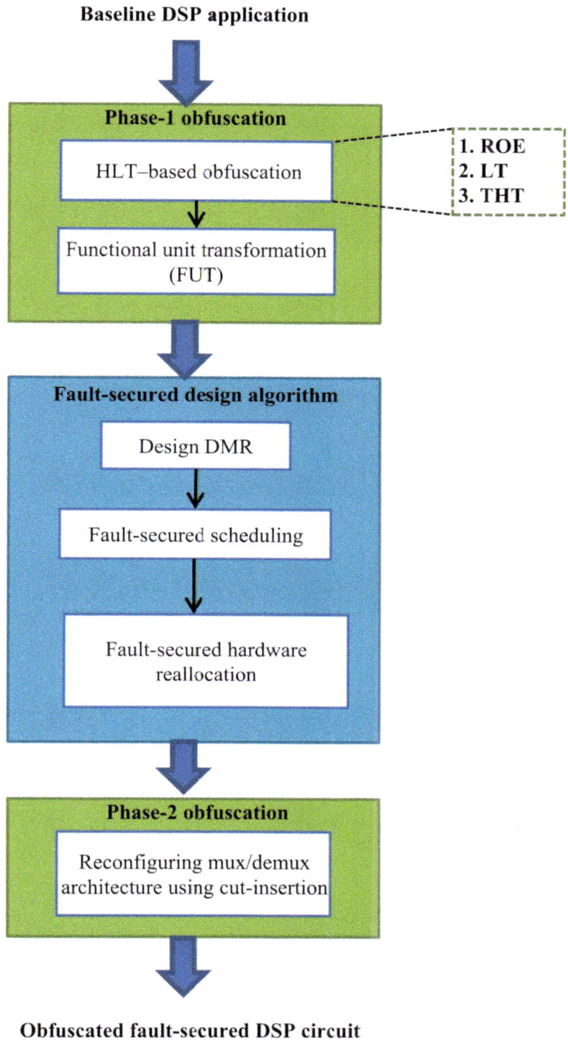

Figure 7.4 The two phases of multi-phase obfuscation for a fault-secured DSP core (Sengupta et al., 2018)

algorithm is a fault-secured scheduling and re-allocation. The next step is to employ phase-2 obfuscation, which not only enhances the obfuscation strength, but also enhances the strength against transient faults and optimizes execution latency. In this step, multiple cuts are inserted into the fault-secured DMR scheduled design in order to further structurally obfuscate the design enabled through the modification of interconnectivity and reconfiguration of mux/demux sizes/architectures.

The inserted multi-cuts also enhance the strength of transient-fault security (Sengupta *et al.*, 2018).

7.5.2 Evaluation models

The cost of a multi-phase obfuscated fault-secured IP core is evaluated based on the area model and delay model (Sengupta *et al.*, 2018):

7.5.2.1 Area model

The total area of an obfuscated, fault-secured design is computed based on two parameters: (1) the total number of hardware resources such as the number of FUs (adders, subtractors and multipliers), multiplexers and registers, (2) unit area of each resource. The total area (A^T_{Ob+FS}) occupied is given as (Sengupta *et al.*, 2018):

$$A^T_{Ob+FS} = \sum_{k=1}^{n} A(R_k) \times N(R_k) + A(mux) \times N(mux) + A(reg) \times N(reg) \tag{7.1}$$

where $N(R_k)$, $N(mux)$ and $N(reg)$ denote the number of extracted *k*th FU (multiplier, adder, subtractor), multiplexers and registers required. $A(R_k)$, $A(mux)$ and $A(reg)$ denote the unit area of the *k*th FU, multiplexer and unit register, respectively. The number of extracted *k*th FUs refers to the maximum number of type-*k* FUs utilized in the control step (T) of the obfuscated, scheduled DMR design.

7.5.2.2 Delay model

The total execution latency (L^T_{Ob+FS}) of an obfuscated fault-secured design is evaluated by

$$L^T_{Ob+FS} = \text{Number of control steps } (T) \tag{7.2}$$

where a control step has been considered to be 100 ps and the total number of control steps consumed for completing a given task is the total execution latency (L^T_{Ob+FS}).

7.5.2.3 Cost function

The design cost of a fault-secured, obfuscated design is evaluated based on (7.1) and (7.2) as

$$C_{Ob+FS} = w_1 \times \left(\frac{L^T_{Ob+FS}}{L^m_{Ob+FS}}\right) + w_2 \times \left(\frac{A^T_{Ob+FS}}{A^m_{Ob+FS}}\right) \tag{7.3}$$

where C_{Ob+FS} represents the cost of the obfuscated, fault-secured design, L^T_{Ob+FS} and A^T_{Ob+FS} denote the total execution latency and area of the design, respectively, L^m_{Ob+FS} and A^m_{Ob+FS} denote the maximum possible latency and area, respectively. And w_1 and w_2 represent the designer-specified weights for latency and area, respectively. These weights range between (0, 1) (*Note: To assign* *similar weightage to both latency and area during cost evaluation, w_1 and w_2 are fixed at 0.5*).

7.5.3 Details of multi-phase obfuscation for fault-secured DSP circuits

The multi-phase obfuscation approach for fault-secured DSP circuits is illustrated in detail (Sengupta *et al.*, 2018).

7.5.3.1 Phase-1: Structural obfuscation using HLT and FUT

As shown in Figure 7.4, phase-1 structural obfuscation performs HLT (ROE, LT, THT)–based transformation, followed by FUT, which both affect its equivalent gate-level design netlist. These transformation-based obfuscations are (Sengupta *et al.*, 2018):

1. *ROE*: In this process, redundant nodes (representing operations) in a CDFG of a DSP application are eliminated without affecting the functionality of the design. When two such nodes exist in the CDFG whose operation and inputs are same, then one of the nodes is considered to be redundant. The process of ROE is accomplished through a number of steps. In the first step, each node is inspected with respect to the remaining nodes to check whether its operation and inputs are same as the other nodes. In this way, all the nodes are inspected in ascending order. In the next step, all the redundant nodes are enlisted where each row contains the nodes with same redundancy. Then, all the redundant nodes except the node with the least operation number are eliminated from each row. This elimination results into changes in the inputs of the remaining operations. However, this elimination of redundant nodes and corresponding changes in the inputs of the remaining nodes should not affect the functionality of the design. This transformation affects the gate-level design netlist in terms of interconnectivity and circuit structure, while preserving the functional behaviour.
2. *LT:* LT is a process of altering nodes logically in a CDFG in order to introduce some structural changes. However, altering nodes in the graph should not affect the functionality. Sengupta *et al.* (2018) leverage the first layer of operations in the graph to apply LT. Application of LT results in changes to the operation types of potential nodes. However, the number of nodes in the graph is increased as a result of LT. This transformation also affects the gate-level design netlist in terms of interconnectivity and circuit structure, while preserving the functional behaviour.
3. *THT:* THT is another HLT approach used by Sengupta *et al.* (2018) during phase-1 obfuscation. In the THT process, the height of the CDFG is intentionally reduced to modify the tree structure. In this process, the critical-path dependency is divided into temporary sub-computations. These sub-computations are executed in parallel, while ensuring the correct functionality of the design. Thus the THT results in changes to resource (operations) inputs and mux/demux configuration, however functionality remains intact.

4. *FUT:* This type transformation is a kind of operator amalgamation in which two FUs (operations) are merged into a single unit. This newly generated FU represents a customized resource that has a new input configuration and mux/demux connectivity. The custom-function unit is denoted by symbol '$'. This custom-function unit performs an 'addition followed by storage, followed by multiplication' operation (Sengupta *et al.*, 2018). It is notable that the FUT is feasible only for nodes having independent inputs. FUT is performed whenever an addition operation is followed by a multiplication operation in a CDFG. This transformation also affects the gate-level design netlist in terms of interconnectivity and circuit structure, while preserving the functional behaviour.

7.5.3.2 Fault-secured design process

Transient faults are such faults that upset the device momentarily (transiently). A resource affected by a transient fault generates erroneous output only for a very short moment. The main cause of a transient fault is the strike of α-particles emanating from thorium or uranium impurities present in the IC and environmental radiations in space applications. The α-particles' strike leads to SET in DSP cores integrated into an IC. The transient time of a fault depends upon the fault-strength (K^c) which, in the worst case, is typically 2,000 ps. During the execution of operations, this may affect more than single-control steps (lasting up to 2,000 ps) caused by the high-energy particles. Considering the fact that the effect of a transient fault (due to particles with high-linear energy transfer) may typically last between 1,000 ($K^c = 1$ control step) and 2,000 ps ($K^c = 2$ control steps) considering 1 GHz frequency, the fault-secured design was therefore generated for this specification. Once generated, phase-2 obfuscation can be employed (Sengupta *et al.*, 2018).

1. *DMR design:* Once phase-1 obfuscation is accomplished, the obfuscated CDFG is subjected to a DMR design process. To construct a DMR design, the obfuscated graph (post-phase-1 obfuscation) is duplicated. The original obfuscated graph and its duplicate are jointly called DMR design.
2. *Fault-secured scheduling and resource hardware reallocation:* Obfuscated DMR design is initially scheduled based on given resource constraints. In the scheduling process, operations of original and duplicate units are concurrently scheduled. After performing scheduling using the given resource constraints, information about the number of maximum resources (FUs) of a particular type being executed in a control step is extracted. This process is called resource extraction and is performed for each particular FU resource. However, the effect of transient fault strength (K^c) has not yet been considered. In order to generate the transient fault-secured design, the effect of the fault-security constraint K^c is incorporated into the DMR design. While incorporating the effect of the transient-fault strength, the DMR design may require to be rescheduled and hardware resources would need to be reallocated accordingly. Considering the strength of transient fault $K^c = 2$ control steps, the following

rules are employed in order to obtain the K^c-cycle, transient, fault-secured, DMR design (Sengupta *et al.*, 2018):

(i) If the original and duplicate sister operations have a control step (delay) difference of greater than $K^c = 2$ control steps, then the same FU resource is allocated to both sister operations of the DMR design.

(ii) If the original and duplicate sister operations have a control step difference of less than, or equal to, $K^c = 2$ control steps, then distinct resources of the same FU are allocated to both sister operations of the DMR design.

(iii) If the original and duplicate sister operations have a control step difference of less than, or equal to, $K^c = 2$ control steps and distinct resources of the same FU are not available, then the duplicate sister operation is shifted down by 1 control step. This shifting of duplicate operations by 1 control step is performed until the control-step difference between two sister operations becomes greater than $K^c = 2$ control steps. Then, both sister operations can be allocated to the same FU resource.

7.5.3.3 Phase-2: structural obfuscation using FU reallocation and mux/demux reconfiguration

Once the obfuscated (using phase-1 obfuscation) fault-secured design is generated, phase-2 structural obfuscation is performed (Sengupta *et al.*, 2018). In this phase, the obfuscated fault-secured DMR design is again obfuscated using mux/demux reconfiguration and additional comparator hardware. In order to accomplish phase-2 obfuscation, the cut-insertion technique is applied on the fault-secured DMR design. In this technique, cuts are applied on some data dependency edges in the graph which lead to the shifting of corresponding operations by at least one control step upward with respective changes in operation inputs. This shifting of operations results in FU reallocation, mux/demux reconfiguration and comparator hardware insertion. Further, through cut insertions, multiple checkpoints are deployed in the graph to enhance the fault security. Maximum possible cuts can be applied to maximize the security (checkpointing). Additionally, phase-2 obfuscation through cut-insertion also helps to optimize the delay overheads caused by constructing the fault-secured DMR design. Further, incorporating multiple cuts requires inserting a number of comparators resulting in more structural transformation. Although it also results in nominal area overheads, the reduction in delay overhead (due to cut-insertions) balances the overall design cost.

7.6 Demonstration of structural obfuscation of fault-secured DSP circuits

In this section, structural obfuscation of fault-secured designs is demonstrated using examples of DSP applications (Sengupta *et al.*, 2018). Figure 7.5 shows the flow of the multi-phase structural obfuscation approach through phase-1 obfuscation,

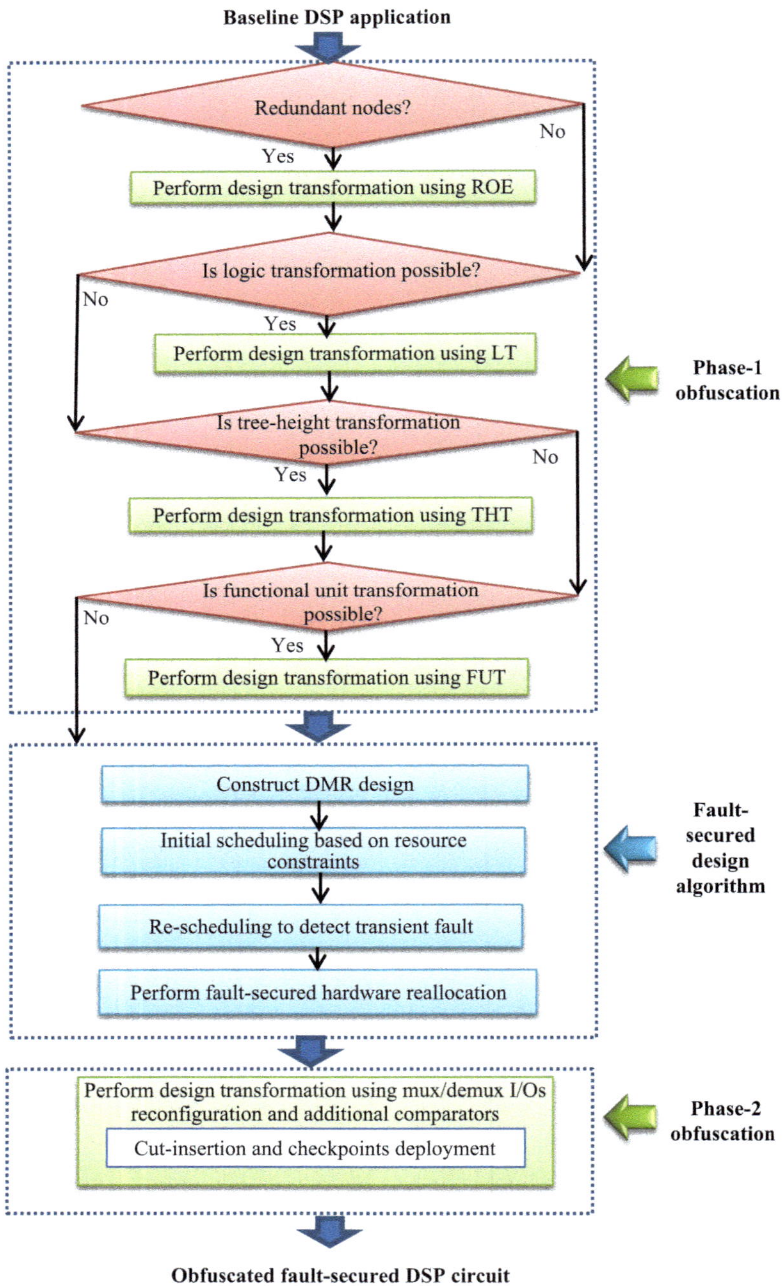

Figure 7.5 Flow of multi-phase obfuscation approach to obfuscate a fault-secured DSP circuit (Sengupta et al., 2018)

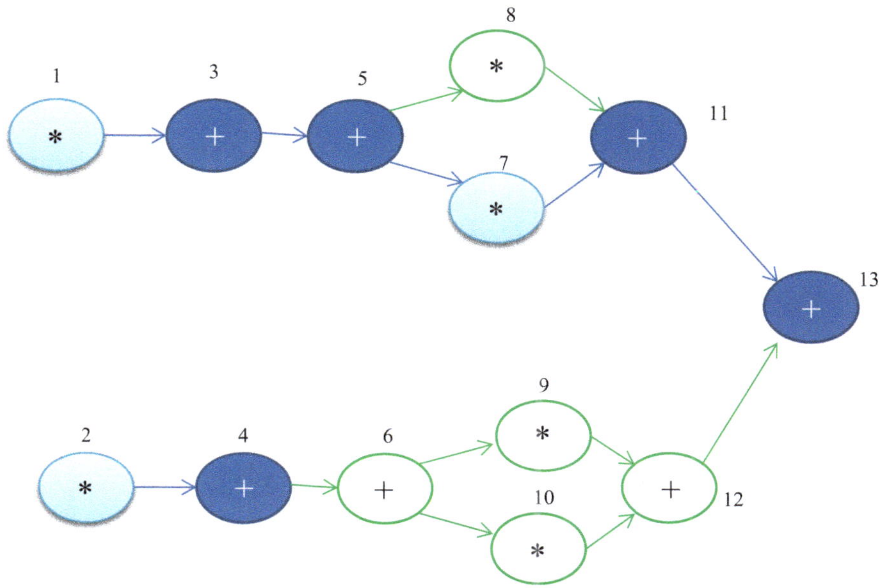

Figure 7.6 Redundant operation removal–based obfuscation (Sengupta et al., 2018)

construction of a fault-secured DMR design and phase-2 obfuscation. Demonstration of the multi-phase obfuscation approach is as follows (Sengupta *et al.*, 2018).

7.6.1 Demonstration of phase-1 obfuscation

As shown in Figure 7.5, the first step of phase-1 obfuscation is to check whether a CDFG representing a sample application has redundant nodes. For example, as shown in Figure 7.6, the CDFG has five redundant nodes (green outline). Once redundant nodes are eliminated, only shaded nodes will remain in the graph. The process of redundant operations (nodes) elimination is as follows. First, each node is inspected with respect to other nodes in the graph. Post-inspection, it is observed that nodes 5 and 6 have same operation (opn) and inputs, so they are put in a row containing the list of redundant nodes. As opn 5 has the smallest opn number, all other nodes with higher opn numbers are eliminated. After eliminating operation 6, inputs of its children nodes 9 and 10 are now fed from 5 instead of 6. Next opns 7, 8, 9 and 10 are also put in the row containing the list of redundant nodes. Except for opn 7, all other nodes (8, 9 and 10) in this row are eliminated, which results into a change in inputs of their children nodes. Similarly, opns 11 and 12 are also put in a row and opn 12 is eliminated as a redundant node. The CDFG with only shaded nodes represents the transformation-based obfuscated graph after ROE. How this transformation helps the DSP circuit in structural obfuscation is discussed in Section 7.7.1.

Once ROE is performed, the next step of phase-1 obfuscation is to check whether LT is feasible in the CDFG. For example, in the CDFG representing infinite impulse response (IIR) filter application, LT can be performed. Figure 7.7 represents the IIR filter application. The upper half of the figure shows the original graph and the lower half shows the obfuscated graph post–LT. As shown in the figure, the first layer of the original graph can be leveraged to perform LT. Assuming inputs of opns 1 and 2 are 4 and *i*, both operations (multiplications) are

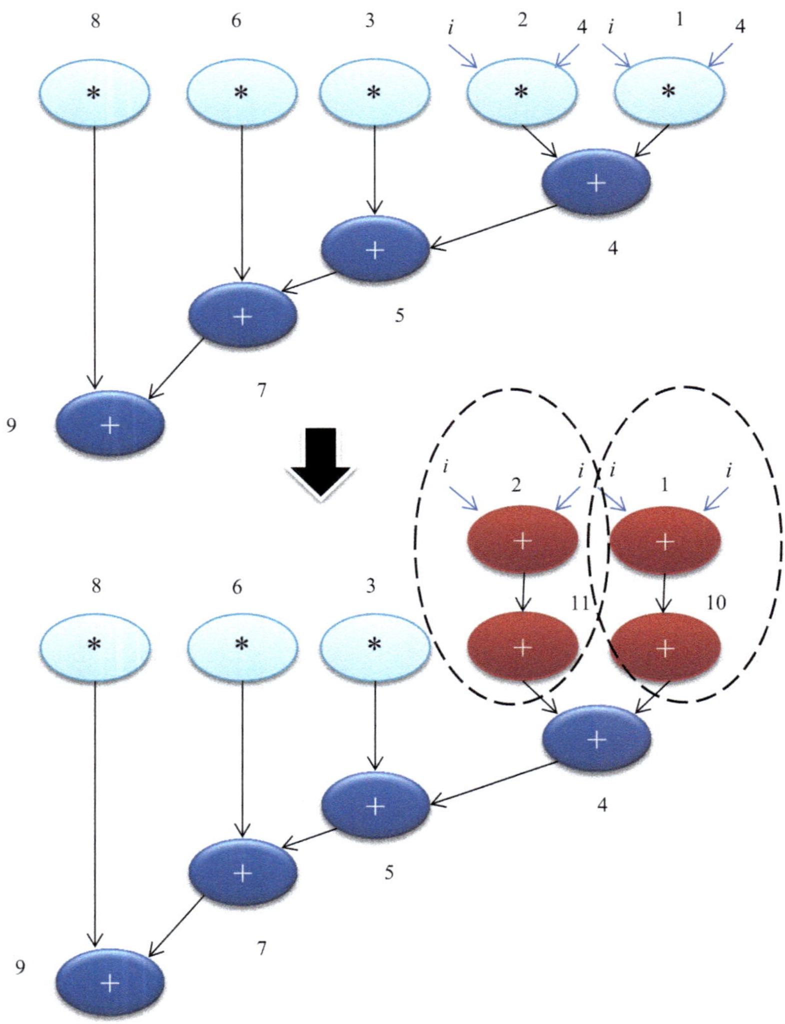

Figure 7.7 Logic transformation–based obfuscation of an IIR filter (Sengupta et al., 2018)

substituted by two addition operations. This results in change in the input of some operations in the graph. Modified or newly added operations (nodes) have been highlighted in red. Post–LT, the obfuscated graph performs the same functionality (produces same output) as the original. How this transformation helps a DSP circuit in structural obfuscation is discussed in Section 7.7.1.

Further, as shown in Figure 7.5, the next step of phase-1 obfuscation is to perform THT, if possible. For example, THT has been performed on a CDFG representing an FIR filter application as shown in Figure 7.8. The upper half of the figure shows the CDFG before THT and the lower half shows the obfuscated graph

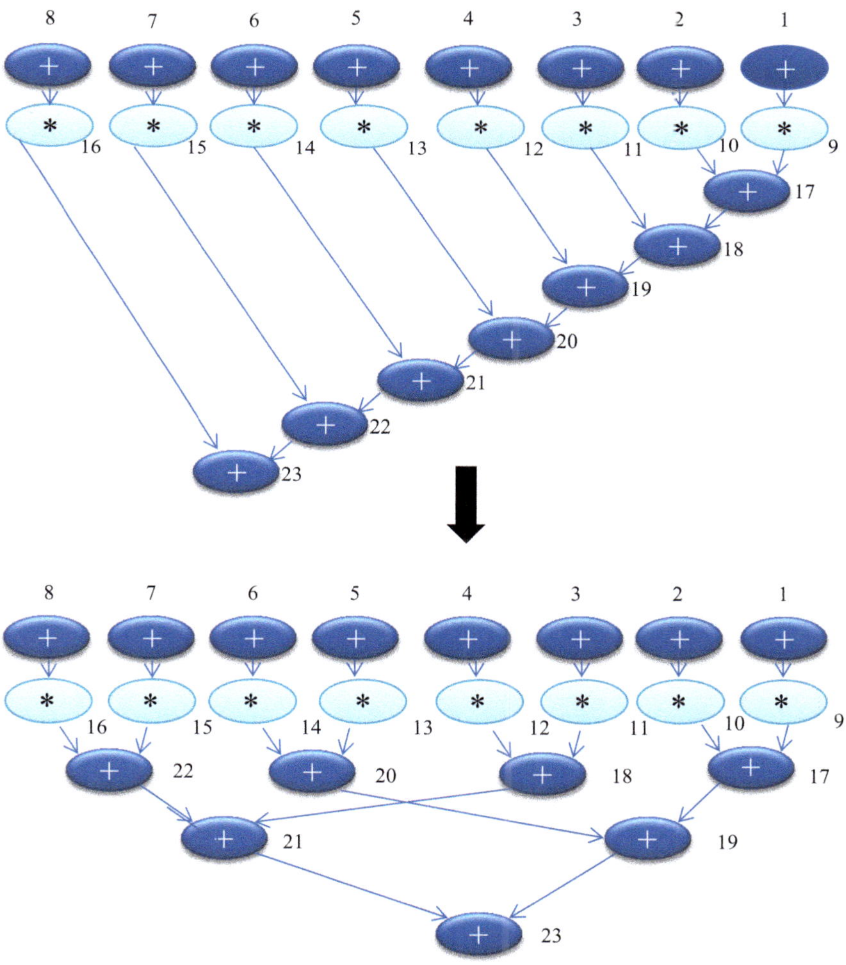

Figure 7.8 Tree-height transformation–based obfuscation of an FIR filter (Sengupta et al., 2018)

post–THT. In the original graph, addition operations (17, 18, 19, 20, 21, 22 and 23) are executed sequentially. Obfuscation using THT has been performed by executing some addition operations concurrently by shifting them up without affecting the correct functionality. Operations 18, 20, 21 and 22 have been shifted up in order to execute 18, 20 and 22 concurrently with 17 and execute 21 concurrently with 19 as shown in Figure 7.8. Owing to these shifts, 23 also shifts up.

Further, the next step of phase-1 obfuscation is to perform FUT. However, FUT is only possible if the execution of the addition operations is followed by multiplication. For example, FUT can be performed on CDFG representing an FIR filter application, as shown in Figure 7.9. The upper half of the figure shows the CDFG before FUT and the lower half shows the obfuscated graph post–FUT. To perform obfuscation using FUT, all addition operations followed by multiplications are replaced by a custom operation (FU) that executes the same functionality as before. The new FU has been labelled with '$' and highlighted with red in Figure 7.9. The design of the new custom resource (custom FU) is shown in Figure 7.10.

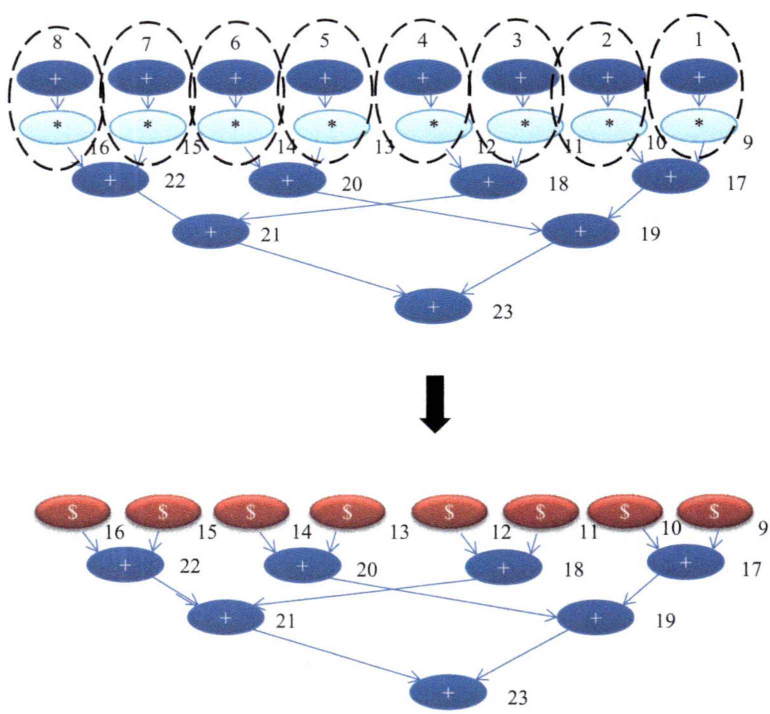

Figure 7.9. Resource transformation–based obfuscation of an FIR filter (Sengupta et al., 2018)

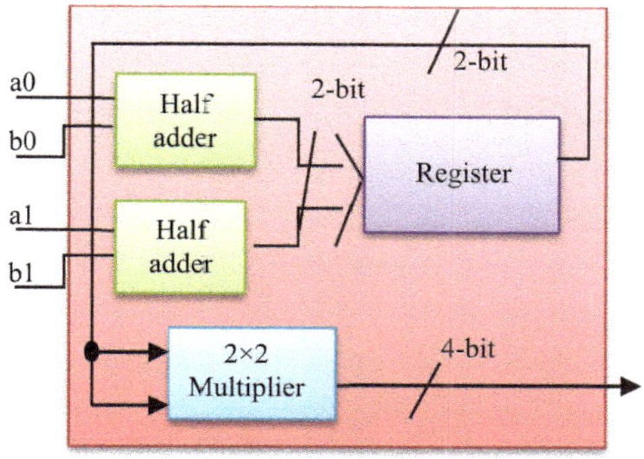

Figure 7.10 Custom functional unit used for resource transformation (Sengupta et al., 2018)

Thus phase-1 obfuscation is accomplished by performing three HLTs (ROE, LT and THT) followed by FUT. After performing phase-1 obfuscation on the FIR filter application, the fault-secured DMR design is constructed, followed by phase-2 obfuscation. Figure 7.9 shows the phase-1 obfuscated FIR filter application.

7.6.2 Demonstration of fault-secured design process

The first step of the fault-secured design process is to construct a DMR design. The DMR design of an FIR filter (post-phase-1 obfuscation) is shown in Figure 7.11. To construct the DMR design, each operation is replicated, as shown in the figure. A checkpoint using a comparator is deployed to detect the transient fault. The DMR design is initially scheduled based on the given resource constraints (4A, 4$). Further, the DMR design is rescheduled to accommodate transient-fault strength ($K^c = 2$ control steps). This obtained scheduled DMR design is secured (detectable) against the k-cycle transient fault. The transient fault-secured scheduling of the DMR design with hardware reallocation is shown in Figure 7.12. Rescheduling and resource (FU) reallocation is required to satisfy the transient fault security constraint and the resource constraints simultaneously.

7.6.3 Demonstration of phase-2 obfuscation

Phase-2 obfuscation is performed through maximal cut-insertion in the phase-1 obfuscated fault-secured scheduled DMR. This phase-2 obfuscation process is responsible for obtaining reconfigured mux and demux units (indicating different inputs–output and connectivity) and additional comparator units. In the fault-secured FIR design, the authors inserted the maximum 10 cuts to obtain the maximum possible obfuscation through reconfigured muxes, demuxes and comparators in the RTL architecture. It is important to note that multiple cuts result in reduced delay of the controller (FSM) due to less control steps. This also contributes to the obfuscation process as reduced control steps means reduced states, thus indicating less flip-flops being used in the FSM of the design. These two contributing factors

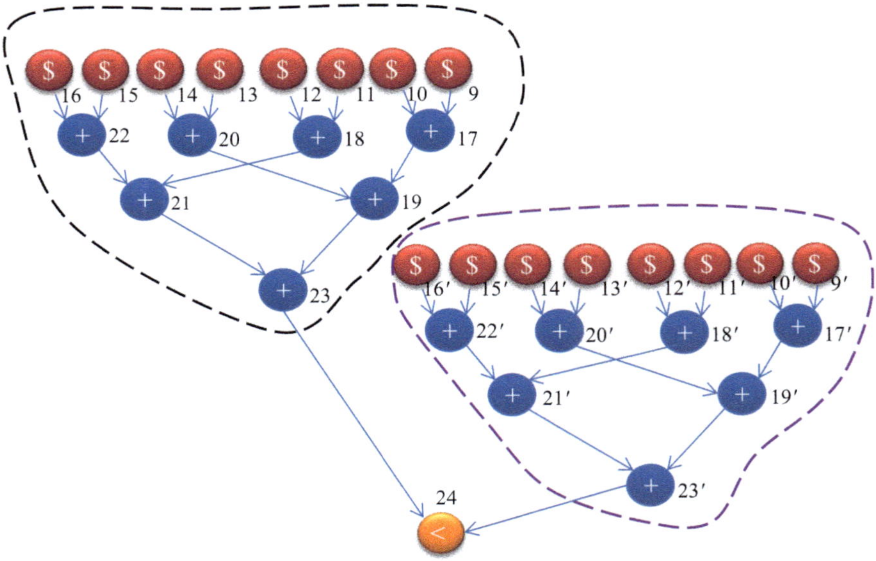

Figure 7.11 Phase-1 obfuscated design of an FIR filter (DMR version) (Sengupta et al., 2018)

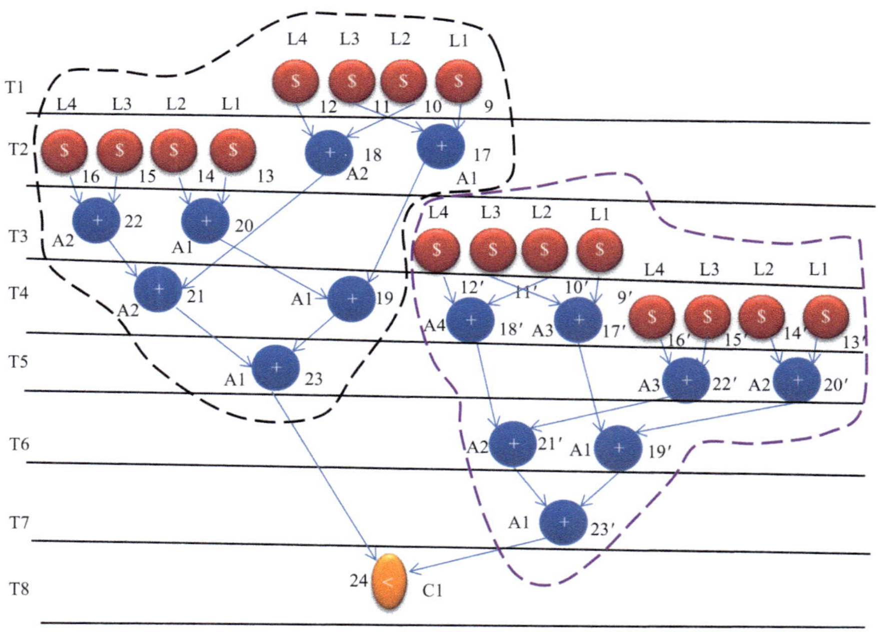

Figure 7.12 Phase-1 obfuscated design of a fault-secured FIR filter scheduled by using 4L and 4A resources

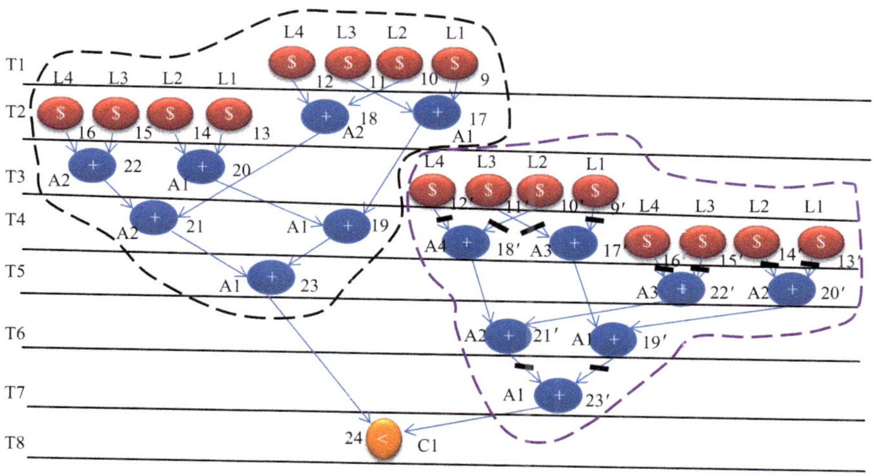

Figure 7.13 Phase-1 obfuscated design of a fault-secured FIR filter with cut insertions (indicated in bold black lines) at the output of 9′, 10′, 11′, 12′, 13′, 14′, 15′, 16′, 19′, 21′

of structural obfuscation result in modification in the gate-level structure in the DSP circuit. In Figure 7.13, the following 10 cuts have been inserted: 9′, 10′, 11′, 12′, 13′, 14′, 15′, 16′, 19′ and 21′. Due to these additional cut insertions, the RTL design of the fault-secured FIR filter now utilizes 11 comparator units, as shown in Figure 7.14. This also further structurally obfuscates the gate structure of the final netlist without affecting the functionality. For example, as shown in Figure 7.14, comparator C1 accepts the output of operations 9 (original unit) and 9′ (duplicate unit), post-cut insertion. Similarly, comparator C2 accepts the output of operations 10 (original unit) and 10′ (duplicate unit), post-cut insertion. A similar process is followed for rest of the comparators from C3 to C10. Comparator C11 is the default comparator that existed in the design prior to the cut insertion. Therefore in the un-obfuscated design of a fault-secured FIR filter only a single comparator C11 exists that compares the output from operations 23 (original unit) and 23′ (duplicate unit). However, in the final obfuscated design of the fault-secured FIR filter, there exist 10 additional comparators due to 10 cut insertions. Figure 7.15 shows the non-obfuscated fault-secured FIR filter design and Figure 7.14 shows its equivalent multi-phase obfuscated fault-secured FIR filter design. Modifications in the scheduling and hardware allocation of the nodes (after performing multi-phase obfuscation) can be observed by comparing the two figures. Further, the differences in the control steps and changes in hardware resources are also evident from both figures. These observations show the structural modifications (without change in functionality) owing to performing multi-phase structural obfuscation. Furthermore, the obfuscated RTL data path of the fault-secured FIR filter circuit with structural (and interconnectivity) changes (marked in green) is shown in Figure 7.16.

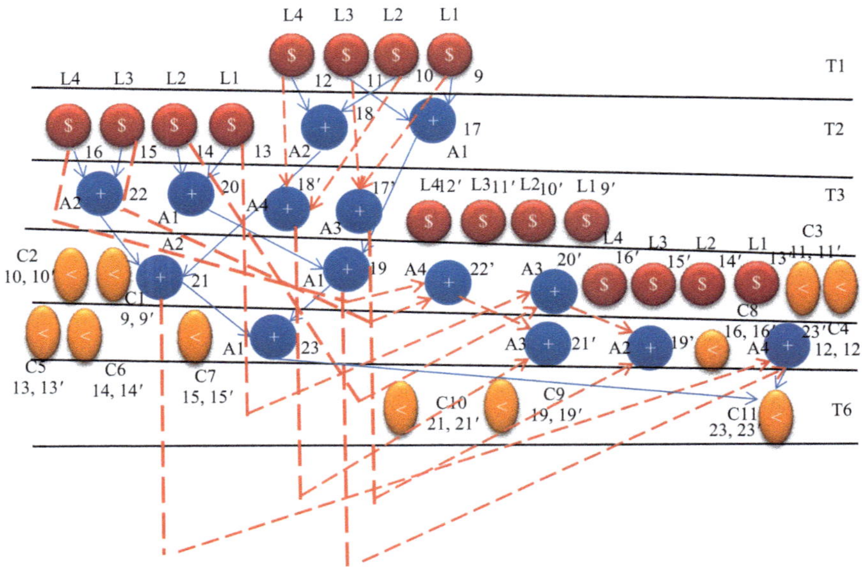

Figure 7.14 Multi-phase (phases 1 and 2) obfuscated design of a fault-secured FIR filter with additional comparators (C1–C11) (Sengupta et al., 2018)

7.7 Analyses of case studies

The multi-phase obfuscation approach (Sengupta *et al.*, 2018) for fault-secured designs has been analysed for various DSP applications in terms of security and design cost. In order to assess the effectiveness of the multi-phase obfuscation approach for fault-secured designs, the security and design cost analyses is followed by a comparative study with the baseline design and other existing approaches. *However, no other approaches have performed obfuscation on fault-secured design, therefore a comparison has been done with an obfuscation approach on a non-fault–secured DSP circuit (Sengupta et al., 2017) and an un-obfuscated fault-secured approach (Inoue et al., 2011).*

7.7.1 Security analysis

The security using multi-phase obfuscation is assessed in terms of strength of obfuscation (S_{ob}^{MP}). The strength of obfuscation is a security metric that is used to assess the structural difference between the obfuscated design and its original counterpart. The strength of obfuscation is defined as (Sengupta *et al.*, 2018)

$$S_{ob}^{MP} = \frac{\Sigma_{i=1}^{t}\text{Number of unique nodes modified}}{\text{Number of nodes before obfuscation}} \tag{7.4}$$

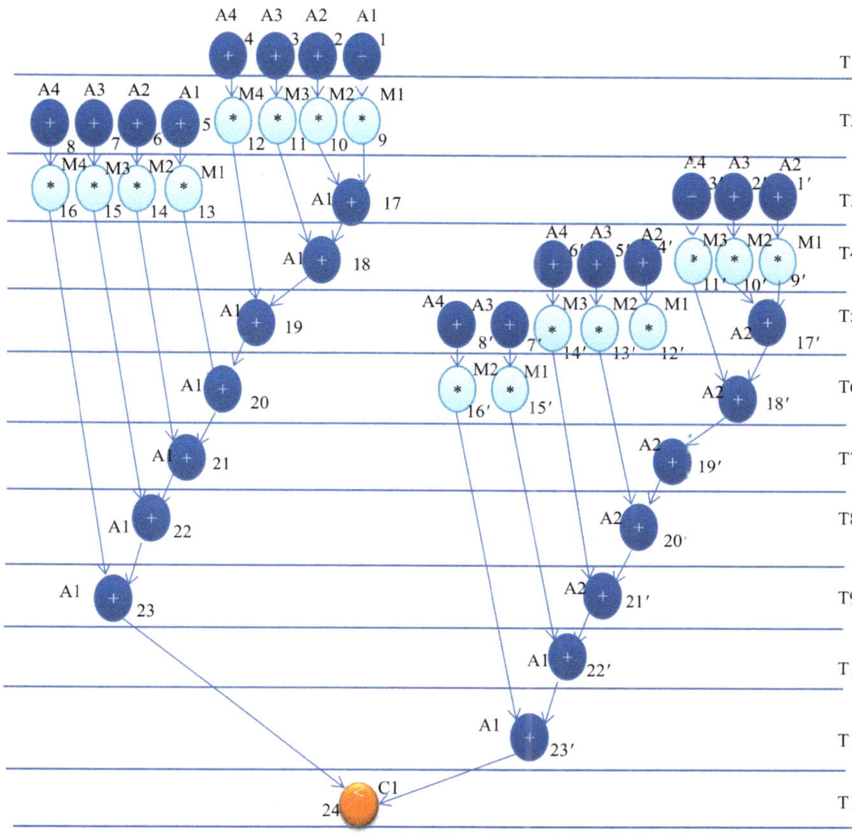

Figure 7.15 Non-obfuscated, fault-secured, scheduled (based on 4A, 4M) FIR filter design

where *t* denotes the number of transformations used in obfuscation. The strength of obfuscation reflects the security achieved through obfuscation. To obtain higher security, the strength of obfuscation $\left(S_{ob}^{MP}\right)$ value must essentially be high. According to (7.4), high S_{ob}^{MP} is achieved when a large number of nodes in the graph of the intended application is modified. Whether a node in the graph is modified or not can be decided based on certain conditions (Sengupta *et al.*, 2018).

1. A node is considered to be modified if primary input to this node in the obfuscated graph is distinct from the original counterpart.
2. A node is considered to be modified if its child in the obfuscated graph is distinct from the original counterpart.

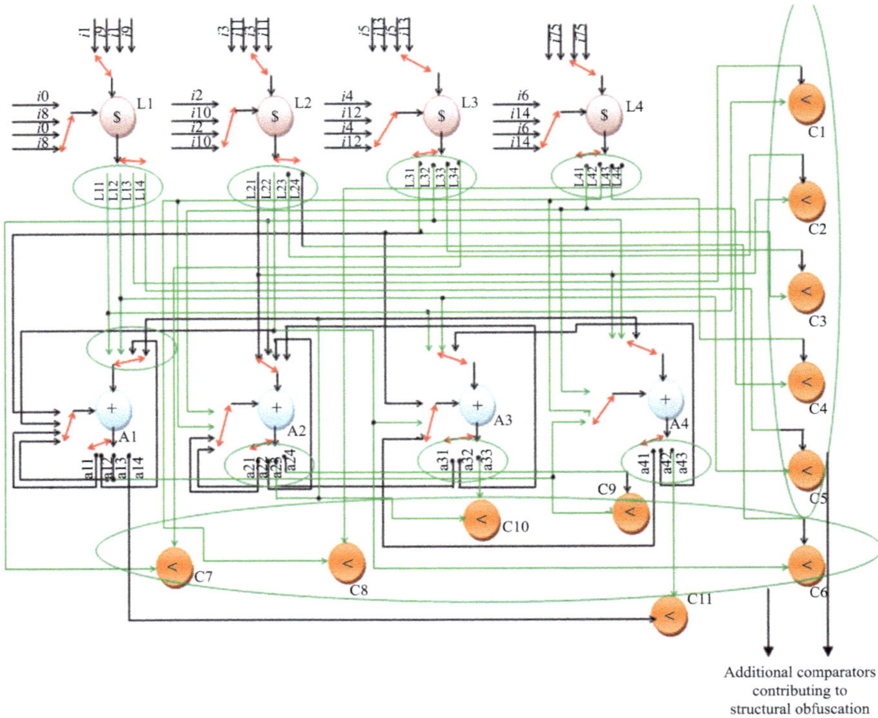

Figure 7.16 Multi-phase, obfuscated, fault-secured DSP circuit of FIR filter with an enhanced fault security feature. Note: the green coloured wires and encircled units indicate structurally obfuscated hardware

3. A node is considered to be modified if its type of FU in the obfuscated graph is altered post-obfuscation.
4. A node of the original graph is considered to be modified if it does not exist in the obfuscated graph post-obfuscation.

Comparison of S_{ob}^{MP} of the multiphase obfuscation approach (Sengupta *et al.*, 2018) with another obfuscation approach (Sengupta *et al.*, 2017) has been presented in Table 7.1. The comparison has been presented based on the number of nodes modified through the individual obfuscation phases. Table 7.1 also indicates the number of gates transformed (affected) using multi-phase obfuscation (Sengupta *et al.*, 2018). The larger the number of gates affected or modified, the stronger is the obfuscation achieved, resulting in higher protection from an adversary. A seen in Table 7.1, multi-phase obfuscation (Sengupta *et al.*, 2018) achieves stronger strength of obfuscation compared to Sengupta *et al.* (2017). The comparison with Sengupta *et al.* (2017) has been performed because Sengupta *et al.* (2017) also implemented HLT–based obfuscation on DSP circuits. However, besides

Table 7.1 Comparative study of multi-phase obfuscation (Sengupta et al., 2018) with (Sengupta et al., 2017) in terms of structural obfuscation strength

DSP applications	**Sengupta *et al.* (2018)**						**Sengupta *et al.* (2017)**	**Enhancement in strength of obfuscation (%)**
	REO (unique nodes)	**LTO (unique nodes)**	**THT (unique nodes)**	**ALU (unique nodes)**	**Number of gates affected**	S_{ob}^{MP}		
FIR	–	–	12	11	4,288	1	0.5	100
IIR	–	3	3	–	3,040	0.666667	0.33333	100
DWT	–	10	–	–	6,688	0.588235	0.52941	11.11
BPF	5	6	2	2	960	0.517241	0.44827	15.38
ARF	10	6	–	–	2,400	0.571429	0.42857	33.33

Sengupta *et al.* (2018) there is no work that has performed obfuscation on fault-secured DSP circuits.

7.7.2 Design cost analysis

This section reports the design cost analysis of the multi-phase obfuscation approach (Sengupta *et al.*, 2018). More explicitly, results have been reported for fault-secured DSP circuits (at $K^c = 2$) after phase-1 obfuscation, in terms of functional resources utilized (number of adders, ALUs, comparators), register area, area of function resources, area of multiplexers and design delay. The detailed results are shown in Table 7.2. Further, Table 7.3 highlights the results of un-obfuscated, non-fault–secured design (baseline version) in terms of functional resource utilized (number of adders, ALUs, comparators), register area, area of function resources, area of multiplexers and design delay. A comparison of Tables 7.2 and 7.3 clearly shows that the design cost overheads due to obfuscation and fault security is nominal compared to the baseline design. The reason for not attaining higher overheads is because in the multi-phase obfuscation approach, the series of HLTs applied causes a reduction in the latency and, sometimes, mux sizes. Further, in multi-phase obfuscation (Sengupta *et al.*, 2018), delay reduction is also achieved due to multi-cuts inserted.

A comparative analysis of multi-phase obfuscated designs (after phase-1 obfuscation) with un-obfuscated designs (Inoue *et al.*, 2011) for fault-secured DSP circuits is highlighted in Table 7.4. In both approaches, DMR logic followed by certain fault-security rules has been used to achieve the fault security value. However, Inoue *et al.* (2011) have not employed hybrid transformation-based obfuscation unlike Sengupta *et al.* (2018). Nevertheless, the comparison reported here for multi-phase obfuscation (Sengupta *et al.*, 2018) is only after phase-1 obfuscation (and does not include the obfuscation achieved through multi-cuts). Therefore the reduction of obfuscated design cost achieved in Sengupta *et al.* (2018) compared to Inoue *et al.* (2011) is due to hybrid transformations and applied fault-security rules that allow optimization in design cost.

The results and analysis of the multi-phase obfuscation approach for fault-secured DSP circuits (after phase-1 and phase-2, and after both) has been shown in Table 7.5. A comparative view of multi-phase obfuscation with only phase-1 obfuscation or a single checkpoint (Table 7.4) vs. multi-phase obfuscation with phase-1 and phase-2 or multiple checkpoints (Table 7.5), shows the reduction in latency for multi-phase obfuscation (Sengupta *et al.*, 2018) with multiple checkpoints. The number of cuts applied for obfuscation also result in delay reduction, as highlighted in Table 7.5. It is clearly evident from Table 7.5 that the reduction in obfuscated design cost after phase-2 obfuscation is due to the reduction in delay through multi-cut insertions.

Table 7.2 Results of multi-phase obfuscation (Sengupta et al., 2018) methodology for obfuscated fault-secured DSP designs

DSP applications	K^c	**Resources**	**FU area (μm²)**	**Mux area (μm²)**	**Register area (μm²)**	**Total area (μm²)**	**Latency (ns)**	**Cost**
FIR	2	4A,4ALU,1C	473.433	34.5047	3.14573	511.083	8	0.35022
IIR	2	3A,3M,1C	300.418	25.559	6.29146	332.268	8	0.50893
DWT	2	5A,1M,1C	187.171	63.8976	7.07789	258.147	15	0.49000
BPF	2	5A,2M,1ALU,1C	357.827	75.3992	9.43718	442.663	13	0.50526
ARF	2	5A,6M,1C	564.659	77.9551	10.2236	652.838	13	0.52853

Table 7.3 Results of un-obfuscated, non-fault–secured (baseline) DSP circuits

DSP applications	K^c	Resources	FU area (μm^2)	Mux area (μm^2)	Total area (μm^2)	Latency (ns)	Cost
FIR	0	4A,4M	377.4884	28.1149	405.603	9	0.34422066
IIR	0	1A,3M	245.3674	6.38976	251.757	5	0.35439387
DWT	0	3A,4M	358.614	19.1693	377.783	10	0.57481188
BPF	0	3A,4M	358.6134	34.5047	393.118	8	0.40932128
ARF	0	3A,6M	509.609	34.5047	544.114	8	0.40416636

7.8 Conclusion

In the beginning of this chapter, the need for fault-secured IP cores in the current semiconductor era was discussed in brief. This chapter highlights the necessity of protecting fault-secured IP cores against various threats (such as piracy and RE) and discusses a few well-known solutions, particularly the use of multi-phase obfuscation. Multi-phase obfuscation aims to obfuscate the IP design in such a way that it becomes unobvious for an attacker and harder to reverse engineer. In this chapter, multi-phase obfuscation has been performed on fault-secured DSP circuits using two phases during ESL synthesis. Phase-1 obfuscation is performed before constructing the fault-secured design. All design transformations (HLT–based and FUT) during phase-1 obfuscation are performed on the CDFG (of DSP cores) before the scheduling process of ESL synthesis. Further, phase-2 obfuscation is performed on the phase-1 obfuscated fault-secured scheduled design using multiple-cut insertions. It was discussed from the literature how phase-2 obfuscation not only enhances the obfuscation strength, but is also capable of improving the fault-security strength owing to multiple checkpoints. It was also reviewed how various transformations applied on the design using multi-phase obfuscation achieved higher obfuscation strength at low design cost.

This chapter provides deep insight into the multi-phase obfuscation of fault-secured DSP circuits. At the end of this chapter, the reader is able to understand the following concepts:

- Need of the fault-secured IP cores.
- Necessity of protecting fault-secured IP cores against piracy and RE threats.
- Comparative study between functional obfuscation and structural obfuscation.
- Explanation of various HLTs (compiler based) applicable on DSP applications.
- Design transformations during phase-1 of multi-phase obfuscation.
- Transient fault-secured design process.
- Design transformations during phase-2 of multi-phase obfuscation.
- Demonstration of the multi-phase obfuscation approach using a real-life DSP core application (IIR and FIR filters).
- Security and design cost analyses on several case studies (different DSP applications).

Table 7.4 Results of the multi-phase obfuscated fault-secured approach (Sengupta et al., 2018) (without additional checkpointing) and the un-obfuscated fault-secured approach (Inoue et al., 2011)

DSP applications	**Sengupta *et al.* (2018) (without additional checkpointing)**				**Inoue *et al.* (2011)**			
	Resources	**Area (μm^2)**	**Latency (ns)**	**Cost**	**Resources**	**Area (μm^2)**	**Latency (ns)**	**Cost**
FIR	4A,4ALU,1C	511.083	8	0.35022486	8A,8M,1C	837.945	11	0.5213008
IIR	3A,3M,1C	332.268	8	0.50893418	3A,6M,1C	545.195	7	0.6548789
DWT	5A,1M,1C	258.147	15	0.4900046	3A,3M,1C	350.946	13	0.5781495
BPF	5A,2M,1ALU,1C	442.663	13	0.5052612	6A,4M,1C	545.686	10	0.5559024
ARF	5A,6M,1C	652.383	13	0.52853461	7A,7M,1C	747.013	10	0.5422742

Table 7.5 Results of the multi-phase obfuscated fault-secured approach (Sengupta et al., 2018) (including multi-checkpointing)

DSP applications	K^c	Number of checkpoint(s)	Resources	FU area (μm^2)	Mux area (μm^2)	Register area (μm^2)	Area (μm^2)	Latency (ns)	Cost
FIR	2	10	4A,4ALU,11C	646.448	30.6708	9.43718	686.556	6	0.35180240
IIR	2	1	3A,3M,2C	317.719	30.6708	6.29146	354.681	7	0.4979806
DWT	2	4	5A,1M,5C	256.377	74.1212	10.2236	340.722	12	0.55391034
BPF	2	4	5A,2M,1ALU,5C	427.033	86.9007	11.7965	525.73	11	0.55074730
ARF	2	4	5A,6M,5C	633.865	72.8433	11.7965	718.505	10	0.52647227

7.9 Questions and exercises

1. What is fault security?
2. What is transient fault?
3. What is multi-phase obfuscation?
4. What is high-level transformation?
5. Describe ROE.
6. Describe THT.
7. How is FU transformation performed?
8. What is checkpointing?
9. What is the trade-off involved in inserting multi-cut insertions?
10. What is the role of comparator hardware in fault security?
11. What is the role of DMR design in fault security?
12. How do multi-cut insertions play a role in structural obfuscation?
13. How does phase-2 obfuscation affect design cost?
14. What is the strength of obfuscation metric?
15. How does phase-1 obfuscation affect the gate-level structure?
16. How does phase-2 obfuscation affect the gate-level structure?
17. Define the transient fault-security rules.
18. How does THT contribute towards structural obfuscation?
19. Explain the difference between structural and functional obfuscation.
20. What is a fault-secured DSP circuit? How is it different from a regular DSP circuit?

References

E. Castillo, U. Meyer-Baese, A. Garcia, L. Parilla, and A. Lloris, "IPP@HDL: Efficient intellectual property protection scheme for IP cores," IEEE Transactions on Very Large Scale Integration (VLSI) Systems, vol. 15, no. 5, Pages: 578–590, 2007.

R. S. Chakraborty and S. Bhunia, "HARPOON: An obfuscation-based SoC design methodology for hardware protection," IEEE Transactions on Computer-Aided Design of Integrated Circuits and Systems, vol. 28, no. 10, Pages: 1493–1502, 2009.

L. W. Chow, J. P. Baukus, and W. M. Clark, "Integrated circuits protected against reverse engineering and method for fabricating the same using an apparent metal contact line terminating on field oxide," U.S. Patent 7 294 935, 2002.

B. Colombier and L. Bossuet, "Survey of hardware protection of design data for integrated circuits and intellectual properties," IET Computers & Digital Techniques, vol. 8, no. 6, Pages: 274–287, 2015.

T. Inoue, H. Henmi, Y. Yoshikawa, and H. Ichihara, "High-level synthesis for multi-cycle transient fault tolerant datapaths," in Proc. 17th IEEE Int. Online Test. Symp., Pages: 13–18, 2011.

F. Koushanfar, "Hardware Metering: A Survey," Springer, New York, NY, USA, Pages: 103–122, 2012.

Y. Lao and K. K. Parhi, "Obfuscating DSP circuits via high-level transformations," IEEE Transactions on Very Large Scale Integration (VLSI) Systems, vol. 23, no. 5, Pages: 819–830, 2015.

M. C. McFarland, A. C. Parker, and R. Camposano, "Tutorial on high-level synthesis," in DAC '88 Proceedings of the 25th ACM/IEEE Design Automation, vol. 27, no. 1, Pages: 330–336, 1988.

V. K. Mishra and A. Sengupta, "MO-PSE: Adaptive multi-objective particle swarm optimization based design space exploration in architectural synthesis for application specific processor design," Advances in Engineering Software, vol. 67, Pages: 111–124, 2014.

S. M. Plaza and I. L. Markov, "Solving the third-shift problem in IC piracy with test-aware logic locking," IEEE Transactions on Computer-Aided Design of Integrated Circuits and Systems, vol. 34, no. 6, Pages: 961–971, 2015.

J. Rajendran, M. Sam, O. Sinanoglu, and R. Karri, "Security analysis of integrated circuit camouflaging," in Proc. ACM/SIGSAC Conf. Comput. Commun. Secur, Pages: 709–720, 2013.

J. A. Roy, F. Koushanfar, and I. L. Markov, "EPIC: Ending piracy of integrated circuits," in Proc. Design Autom. Test Europe, Pages: 1069–1074, 2008.

R. Schneiderman, "DSPs evolving in consumer electronics applications," IEEE Signal Processing Magazine, vol. 27, no. 3, Pages: 6–10, 2010.

A. Sengupta, "Intellectual property cores: Protection designs for CE products," IEEE Consumer Electronics Magazine, vol. 5, no. 1, Pages: 83–88, 2016a.

A. Sengupta, "Soft IP core design resiliency against terrestrial transient faults for CE products [hardware matters]," IEEE Consumer Electronics Magazine, vol. 5, no. 4, Pages: 129–131, 2016b.

A. Sengupta, "Hardware security of CE devices [hardware matters]," IEEE Consumer Electronics Magazine, vol. 6, no. 1, Pages: 130–133, 2017.

A. Sengupta and S. Bhadauria, "Exploring low cost optimal watermark for reusable IP cores during high level synthesis," IEEE Access, vol. 4, Pages: 2198–2215, 2016.

A. Sengupta and D. Kachave, "Particle swarm optimisation driven low cost single event transient fault secured design during architectural synthesis," The Journal of Engineering, vol. 2017, no. 6, Pages: 184–194, 2017.

A. Sengupta and D. Kachave, "Spatial and temporal redundancy for transient fault-tolerant datapath," IEEE Transactions on Aerospace and Electronic Systems, vol. 54, no. 3, Pages: 1168–1183, 2018.

A. Sengupta and D. Roy, "Protecting an intellectual property core during architectural synthesis using high-level transformation based obfuscation," IET Electronics Letters, vol. 53, no. 13, Pages: 849–851, 2017.

A. Sengupta, R. Sedaghat, and Z. Zeng, "A high level synthesis design flow with a novel approach for efficient design space exploration in case of multi-parametric optimization objective," Microelectronics Reliability, vol. 50, no. 3, Pages: 424–437, 2010.

A. Sengupta, D. Roy, S. P. Mohanty, and P. Corcoran, "DSP design protection in CE through algorithmic transformation based structural obfuscation," IEEE Transactions on Consumer Electronics, vol. 63, no. 4, Pages: 467–476, 2017.

A. Sengupta, S. P. Mohanty, F. Pescador, and P. Corcoran, "Multi-phase obfuscation of fault secured DSP designs with enhanced security feature," IEEE Transactions on Consumer Electronics, vol. 64, no. 3, Pages: 356–364, 2018.

A. Sengupta, D. Kachave, and D. Roy, "Low cost functional obfuscation of reusable IP cores used in CE hardware through robust locking," IEEE Transactions on Computer-Aided Design of Integrated Circuits and Systems, vol. 38, no. 4, Pages: 604–616, 2019.

P. Subramanyan, S. Ray, and S. Malik, "Evaluating the security of logic encryption algorithms," in IEEE International Symposium on Hardware Oriented Security and Trust (HOST), Pages: 137–143, 2015.

R. Torrance and D. James, "The state-of-the-art in IC reverse engineering," Canadian Healthcare Engineering Society, vol. 5747, Pages: 363–381, 2009.

M. Yasin, J. J. Rajendran, O. Sinanoglu, and R. Karri, "On Improving the Security of Logic Locking," IEEE Transactions on Computer-Aided Design of Integrated Circuits and Systems, vol. 35, no. 9, Pages: 1411–1424, 2016.

J. Zhang, "A practical logic obfuscation technique for hardware security," IEEE Transactions on Very Large Scale Integration (VLSI) Systems, vol. 24, no. 3, Pages: 1193–1197, 2015.

Chapter 8
Hologram-based structural obfuscation for DSP cores

Anirban Sengupta[1] *and Mahendra Rathor*[1]

This chapter highlights an alternative paradigm for the design of structurally obfuscated digital signal processing (DSP) cores inspired by security hologram. Discussion on different structural-obfuscation techniques have been included in this chapter with emphasis on the algorithms used for obfuscation during design synthesis of high-level synthesis (HLS).

The chapter is organized as follows: Section 8.1 introduces the importance of obfuscation for IP cores during the system-on-chip design process; Section 8.2 discusses the background on security hologram; Section 8.3 highlights the possible use of applying the hologram concept for structural obfuscation; Section 8.4 explains the hologram-based obfuscation methodology for DSP cores; Section 8.5 presents illustrative examples for hologram-based obfuscation on DSP cores; Section 8.6 discusses the process of determination of gate count for un-obfuscated and obfuscated designs; Section 8.7 presents a demonstration and examples of high-level transformation–based obfuscated DSP circuits; Section 8.8 presents design examples of un-obfuscated (baseline) DSP circuits; Section 8.9 presents analysis on case studies; Section 8.10 concludes the chapter.

8.1 Introduction

A DSP kernel inside a digital signal processor executes DSP algorithms to facilitate specific applications, such as de-noising, compression–decompression, correlation, transformation. For example, finite impulse response (FIR) and infinite impulse response (IIR) filters are used for the purpose of filtering or de-noising intended signals; discrete cosine transform (DCT), inverse discrete cosine transform (IDCT) are used underneath the joint photographic experts group (JPEG) compressor–decompressor to facilitate image compression or decompression; discrete wavelet transform is used for compression and de-noising; fast Fourier transform (FFT) is used for fast computation of discrete Fourier transform and DCT, etc. Owing to

[1]Computer Science and Engineering, Indian Institute of Technology Indore, Indore, India

huge applications of DSP kernels in contemporary smart electronic devices, their security perspective has become very important (Schneiderman, 2010; Sengupta, 2016, 2017). Structural obfuscation (Lao and Parhi, 2015; Sengupta and Roy, 2017; Sengupta *et al.*, 2017a) is one of the major security mechanisms that can be employed to secure DSP designs against reverse engineering (RE) (Torrance and James, 2009) threats resulting into attacks such as Trojan insertion, piracy, etc. (Sengupta *et al.*, 2017b, 2017c).

8.2 Background on hologram

A hologram is a special feature whereby multiple images can be overlapped with each other, while simultaneously being capable of being displayed at the same time. The overlapping process is similar to concept of camouflaging. Different images can be viewed at different angles in a hologram. The different angles of viewing can be horizontal to vertical (in both directions) or top to bottom (in both directions). At different angles of viewing, distinct images evolve through switch elements (Matrix Technologies, Hologram Features, 2016). This phenomenon can be a useful tool for performing structural obfuscation of DSP circuits as well (Sengupta and Rathor, 2019). This is explained in Section 8.3.

8.3 Applying hologram concept for structural obfuscation

Similar to the hologram concept where two or more images are camouflaged, hologram-based obfuscation has been formulated such that at least two DSP applications are integrated together in a camouflaged fashion (Sengupta and Rathor, 2019). These camouflaged DSP applications (in circuit form) are possible to be viewed and activated at a specific switching value. In other words, individual DSP circuit architecture is functional based on the application of a specific bit combination. Similar to holograms, these bit patterns are equivalent to the switching element used for viewing different images. Sengupta and Rathor (2019) have introduced a hologram-based obfuscation methodology where the concept of camouflaging is inspired from image hologram phenomenon.

A conceptual view of hologram-based obfuscation for IP core protection is shown in Figure 8.1. Similar to other structural-obfuscation approaches, hologram-based obfuscation proposed in Sengupta and Rathor (2019) also targets the same threat model. More explicitly, hologram-based obfuscation designs provide the necessary camouflaging to thwart RE attacks as it becomes un-obvious (post obfuscation) for an attacker to identify the architecture of the application. Hologram-based obfuscated designs enhance the complexity of understanding the functionality of the architecture, thus hindering RE attacks.

One of the most important aspects of an obfuscation approach is to perform obfuscation at minimal design overheads. Hologram-based obfuscation satisfies this property as it obfuscates two applications simultaneously into a single-design

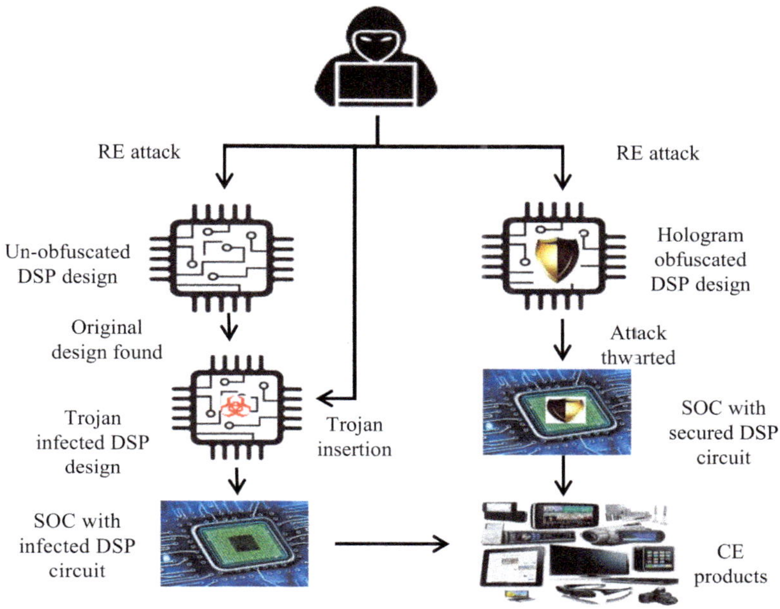

Figure 8.1 Schematic representation of protecting DSP designs using hologram-based obfuscation

architecture. A discussion on some of the contemporary obfuscation approaches used for DSP cores as well as a detailed description of functional vs. structural obfuscation are provided in Chapter 7.

8.4 Hologram-based obfuscation methodology

8.4.1 Features of hologram-based obfuscation (Sengupta and Rathor, 2019)

1. Performs structural obfuscation during data path and controller-generation phases of HLS (McFarland *et al.*, 1988; Sengupta *et al.*, 2010) for DSP applications.
2. Utilizes hologram camouflaging concept for structural obfuscation.
3. Integration of two DSP applications into a single-design architecture using multiple-obfuscation algorithms.
4. Capable to achieve massive savings in utilization of resources owing to obfuscating two applications simultaneously compared to two individually obfuscated designs.

8.4.2 Threat model

Hologram-based obfuscation is capable of thwarting RE attacks that are used to implant malicious logics into the design, perform cloning/piracy or fraudulent claims of ownership. This form of structural obfuscation enhances the complexity of launching RE attacks by hindering an attacker from identifying the actual functional architecture of the design. In other words, it confuses an adversary by making the design architecture un-obvious (Sengupta and Rathor, 2019).

8.4.3 Overview

DSP cores, such as FIR, IIR, DCT, IDCT and FFT, are widely used in electronic systems to facilitate several applications, such as video or audio processing, image processing and de-noising. Since the standard design architecture of DSP cores are well known, an adversary can analyse and understand the design of a DSP–IP core through RE and may insert Trojan hardware (malicious logic) inside the design or pirate/clone it. Therefore to ensure the security and reliability of DSP cores, the security against RE attack is crucial. This chapter presents a structural obfuscation approach to secure DSP cores by leveraging the hologram feature. The aim of hologram-based obfuscation is to camouflage two DSP cores into one another by integrating their architectures into a single data path. Since, the integration of DSPs into a single register transfer level (RTL) data path architecture renders the design un-obvious to an attacker, RE attacks are thwarted using hologram-based obfuscation (Sengupta and Rathor, 2019).

The inputs to this approach are the scheduled data flow graphs (DFGs) of two DSP applications and output is a hologram-obfuscated design that is a camouflaged integrated RTL data path of both applications. Hologram-based obfuscation is performed in an HLS framework using three rules as shown in Figure 8.2. According to the first rule, only a sub-set of inputs of both DSP applications are multiplexed. According to the second rule, all inputs of both DSP applications are multiplexed, while according to the third rule, only inputs of similar components of both DSP applications are multiplexed. Depending on the design architecture of two selected DSP applications, these rules are applied for generating a hologram-obfuscated design (Sengupta and Rathor, 2019).

8.4.4 Details of hologram-based obfuscation

The complete process of generating a hologram-based obfuscated design is described in this section (Sengupta and Rathor, 2019). Figure 8.3 depicts the process which is centred around the three rules for employing hologram-based obfuscation, namely:

1. *Rule-1: Multiplexing sub-set of inputs of two DSP applications*

   If a portion of the scheduled DFGs of two different DSP applications is structurally identical (identical portions of both applications require similar resources with similar input–output connectivity), then the following two actions are performed:
   (i) The inputs of only similar portions of both applications are multiplexed.
   (ii) The outputs of both applications are multiplexed.
   (iii) The switching between two designs is performed using multiplexers (muxes) each of size 2:1. At select line $H = 0$, one DSP design activates (gets inputs through switching muxes) while at $H = 1$, the other DSP design is activated.

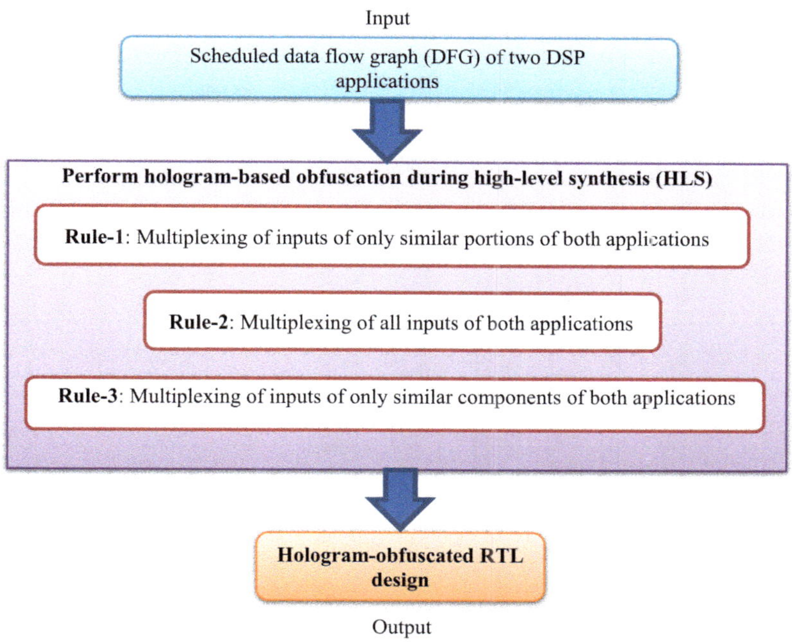

Figure 8.2 An overview of the hologram-based structural obfuscation approach (Sengupta and Rathor, 2019)

2. *Rule-2: Multiplexing of all inputs of two DSP applications*
   If two scheduled DFGs representing two different DSP applications are identical (i.e., both applications require the same number of inputs and outputs and similar resources with similar input–output connectivity), then the following action is performed:
   (i) All inputs of both applications are multiplexed.
   (ii) At $H = 0$, one DSP design gets functionally enabled while at $H = 1$, functionality of the other DSP design gets enabled.
3. *Rule-3: Multiplexing of inputs of similar components in two DSP applications*
   This rule is applied when two applications have some similar components in their data paths; however, the paths of inputs of those components are different. When this condition is satisfied, multiplexing of inputs of only similar components of both applications is performed. Components of one application accept inputs at $H = 0$ while components of the other application accept inputs at $H = 1$.

   The process of applying these rules on DSP applications is shown in Figure 8.3. When these rules are applied on the scheduled DFGs of two DSP applications and processed through an HLS framework, then a hologram-obfuscated RTL design is generated. In the hologram-obfuscated design, the architecture of two DSP applications are camouflaged into a single RTL data path. The obfuscated design is a common data path for both DSP applications.

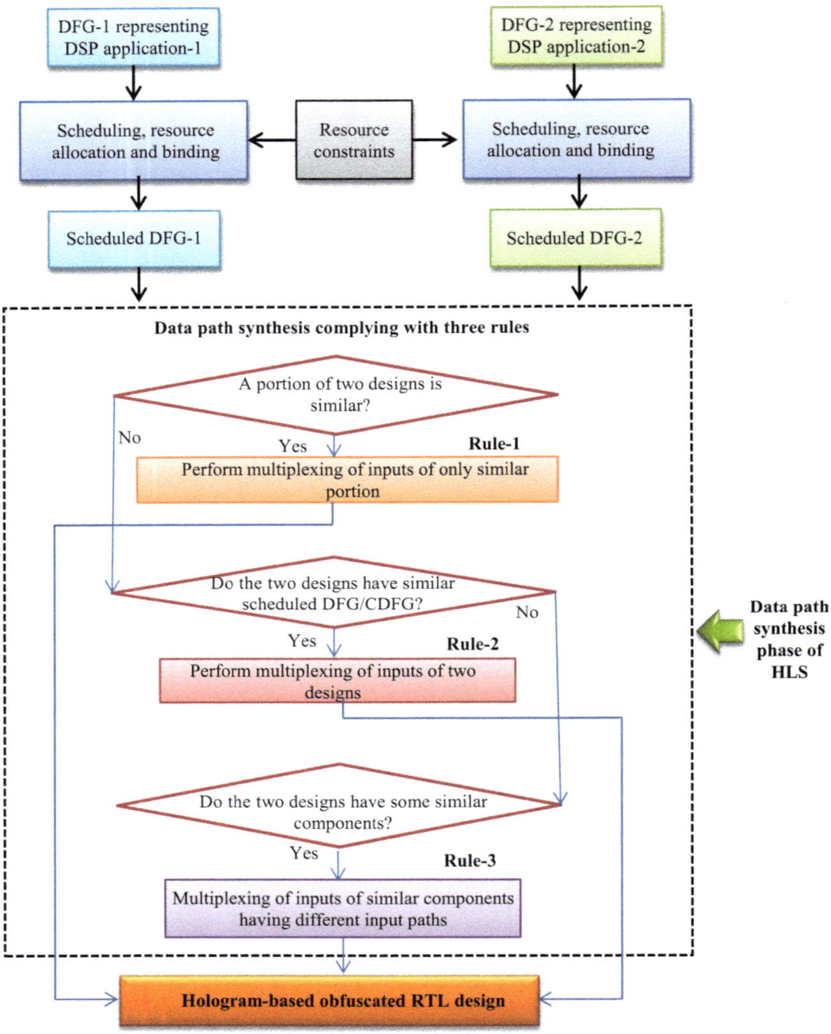

Figure 8.3 Flow of generating a hologram-obfuscated DSP design

8.5 Illustrative examples of hologram-based obfuscation for DSP cores

This section demonstrates the hologram-based obfuscation approach (Sengupta and Rathor, 2019) based on rule-1 and rule-2, respectively, for two case studies. In the first case study, the process of generating IIR filter–FIR filter obfuscated design has been illustrated based on rule-1. In the second case study, the process of generating 4-point DCT–4-point IDCT obfuscated design has been illustrated based on rule-2.

8.5.1 Case study of IIR–FIR filter obfuscated design

IIR and FIR filters are widely used in modern electronic systems to facilitate applications such as telecommunications, speech processing, removal or attenuation of selected frequencies, noise suppression of signals stored in analogue media, bandwidth limiting, etc. Due to their wide utilization, security against RE is very important. In order to secure the IIR filter and FIR filter designs simultaneously at low cost, hologram-based structural obfuscation is a promising solution. The process of obtaining IIR–FIR obfuscated design based on rule-1 is illustrated in this sub-section (Sengupta and Rathor, 2019). The generic equation of IIR and FIR filters are:

$$\begin{aligned}\textbf{IIR}: Y[n] &= b0 \times X[n] + b1 \times X[n-1] + b2 \times X[n-2] \\ &\quad + b3 \times X[n-3] - a1 \times Y[n-1] \\ &\quad - a2 \times Y[n-2] - a3 \times Y[n-3] \end{aligned} \tag{8.1}$$

$$\begin{aligned}\textbf{FIR}: Y'[n] &= h0 \times X'[n] + h1 \times X'[n-1] + h2 \times X'[n-2] \\ &\quad + h3 \times X'[n-3] \end{aligned} \tag{8.2}$$

where $Y[n]$ and $Y'[n]$ represent the output of IIR and FIR filters, respectively. Further,

$\{a1, a2, a3, b0, b1, b2, b3\}$ set of input coefficients of IIR filter.
$\{h0, h1, h2, h3\}$ set of input coefficients of FIR filter.
$\{X[n], X[n-1], X[n-2], X[n-3]\}$ set of previous values of input of IIR filter.
$\{Y[n-1], Y[n-2], Y[n-3]\}$ set of previous values of output of IIR filter.
$\{X'[n-1], X'[n-2], X'[n-3]\}$ set of previous values of input of FIR filter.

In the process of generating a hologram-obfuscated design of IIR and FIR filters, the first step is to create DFGs of both DSP applications. The DFGs representing IIR and FIR filter are created using (8.1) and (8.2), respectively. Further, the scheduling is performed on both DFGs based on resource (functional unit) constraints of one adder and one multiplier ($1A$, $1M$). Figures 8.4 and 8.5 show the scheduled DFGs of IIR and FIR filters, respectively. In the figures, T indicates a control step and $M1$ and $A1$ indicate the allocated hardware resource of the multiplier and adder units, respectively. Multiplication and addition operations are highlighted in orange and violet, respectively.

Once the scheduled DFGs of IIR and FIR filters are obtained, they are fed to the data path synthesis phase of HLS in order to generate a common obfuscated RTL of both DSP applications. Here, hologram-based obfuscation is employed using rule-1. This is because, structurally, a portion of IIR filter is identical to the FIR filter. As per rule-1, inputs of only similar portions of both designs are multiplexed through a number of 2:1 muxes and the outputs of both designs are also multiplexed through a single 2:1 mux. Based on the value of select line H, either of the design gets activated. Application of the specific-bit combination in the hologram design results into the manifestation of the respective data path architecture. For example, at $H = 0$, the IIR design gets functionally enabled while at $H = 1$, the

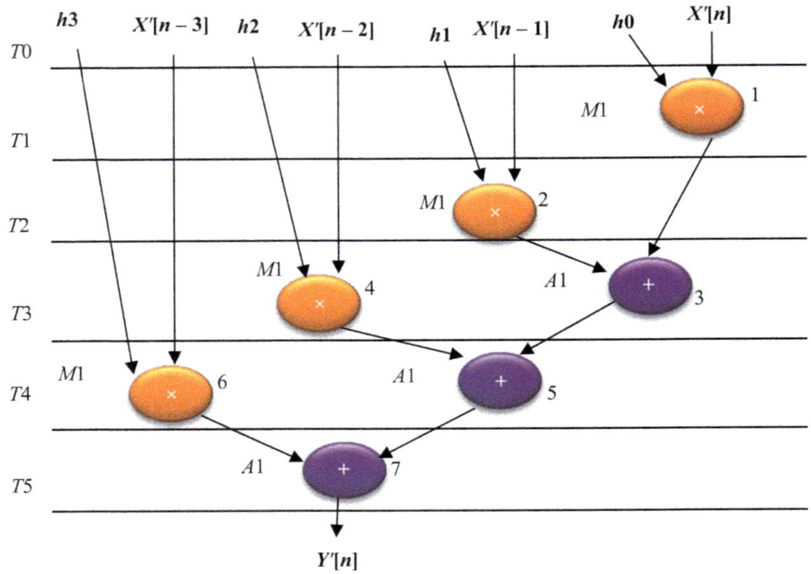

Figure 8.4 Scheduled DFG of FIR filter based on one adder and one multiplier

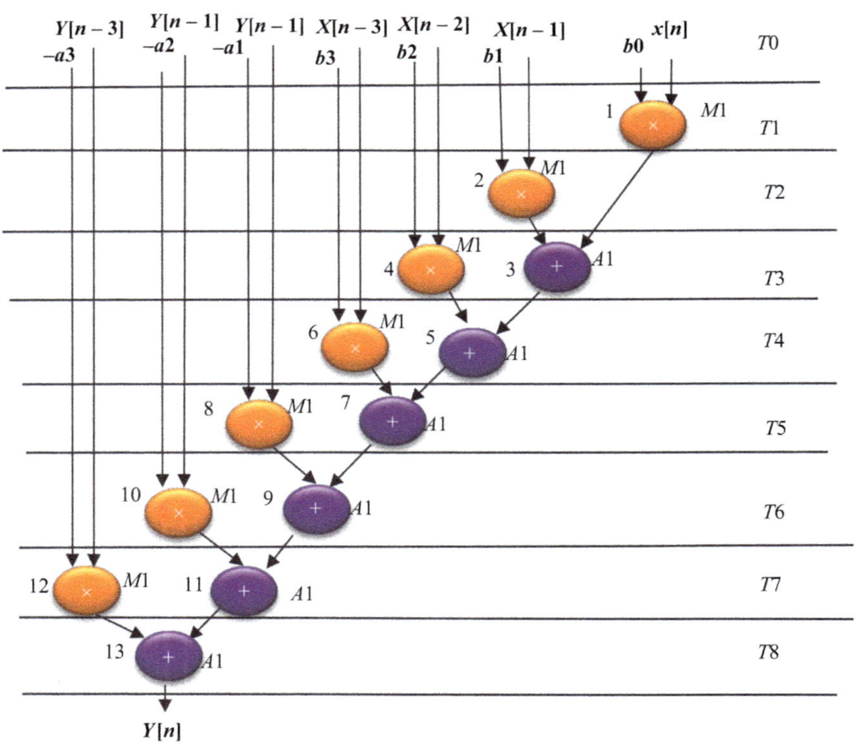

Figure 8.5 Scheduling of IIR filter based on resource constraints (1A, 1M)

FIR design gets activated. Here the switching element of the hologram-obfuscated design is the select control of the 2:1 muxes, which can be directly controlled by the IP vendor or designer. Complying with rule-1, the data path synthesis phase of HLS generates a hybridized obfuscated RTL data path of both designs (IIR and FIR) simultaneously. This is shown in Figure 8.6.

Resemblance of an image hologram with a hologram-obfuscated design (Sengupta and Rathor, 2019): Switching or flipping elements integrated in a hologram is responsible for switching between two (or more) images at different viewing angles. Likewise in a hologram-obfuscated design, a number of additional

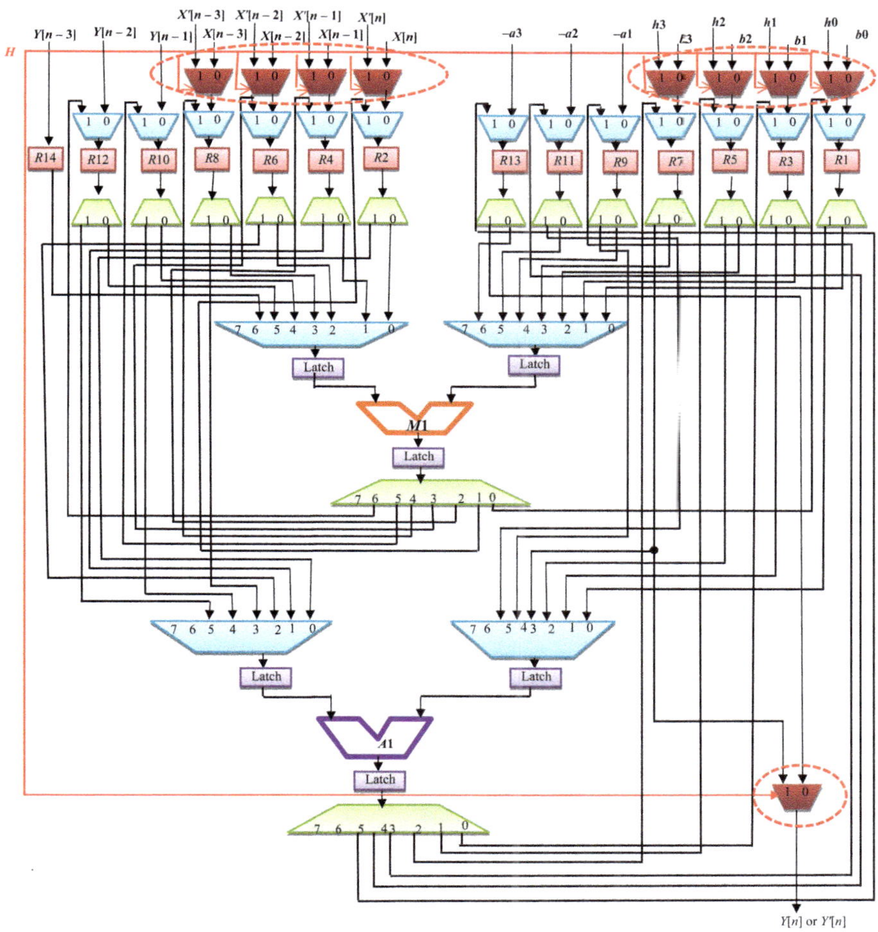

Figure 8.6 Hologram-obfuscated RTL data path of IIR–FIR filter integrated design. Note: switching muxes of the hologram are circled in red-dotted ovals

multiplexers (acting as switching elements) have been used for mimicking the hologram features. The switching muxes of a hologram-obfuscated design have been shown in red in the integrated RTL data path (select/control lines of switching muxes are also been shown in red).

8.5.2 Case study of 4-point DCT–IDCT obfuscated design

Consumer electronics systems such as digital cameras, etc., use a JPEG core to facilitate image compression. The DCT is utilized underneath the JPEG to transform the image pixels from time to frequency domain. The process of obtaining 4-point DCT–IDCT obfuscated design based on obfuscation rule-2 is illustrated in this sub-section (Sengupta and Rathor, 2019). The generic equation of 4-point DCT and 4-point IDCT are

$$\textit{4-Point DCT}: X[0] = k1 \times x[0] + k2 \times x[1] + k3 \times x[2] + k4 \times x[3] \tag{8.3}$$

$$\textit{4-Point IDCT}: x'[0] = k'1 \times X'[0] + k'2 \times X'[1] + k'3 \times X'[2] + k'4 \times X'[3] \tag{8.4}$$

where $X[0]$ and $x'[0]$ represent the first output value of 4-point DCT and 4-point IDCT, respectively. Further,

$\{k1, k2, k3, k4\}$ set of generic values of input coefficients of 4-point DCT.
$\{k'1, k'2, k'3, k'4\}$ set of generic values of input coefficients of 4-point IDCT.
$\{x[0], x[1], x[2], x[3]\}$ set of input values of 4-point DCT.
$\{X'[0], X'[1], X'[2], X'[3]\}$ set of input values of 4-point IDCT.

In the process of generating a hologram-obfuscated design of 4-point DCT–IDCT cores, the first step is to create DFGs of both DSP applications. The DFGs representing 4-point DCT and 4-point IDCT are created using (8.3) and (8.4), respectively. Further, the scheduled DFGs of 4-point DCT and 4-point IDCT (based on resource constraints 1*A*, 1*M*) are shown in Figures 8.7 and 8.8, respectively. Once the scheduled DFG of 4-point DCT and 4-point IDCT are obtained, they are fed to the data path synthesis phase of HLS in order to generate a common obfuscated RTL of both DSP applications. Here, hologram-based obfuscation is employed based on obfuscation rule-2 because, structurally, a 4-point DCT is identical to a 4-point IDCT. As per rule-2, inputs of both designs are multiplexed through a number of 2:1 muxes and output of both design are also multiplexed through a single 2:1 mux. The select line H determines which design will manifest in the hologram. At $H = 0$, the 4-point DCT design is functionally enabled while at $H = 1$, the 4-point IDCT design is activated. Complying with rule-2, the data path synthesis phase of HLS generates an obfuscated RTL data path of both designs (4-point DCT and 4-point IDCT) simultaneously. This is shown in Figure 8.9.

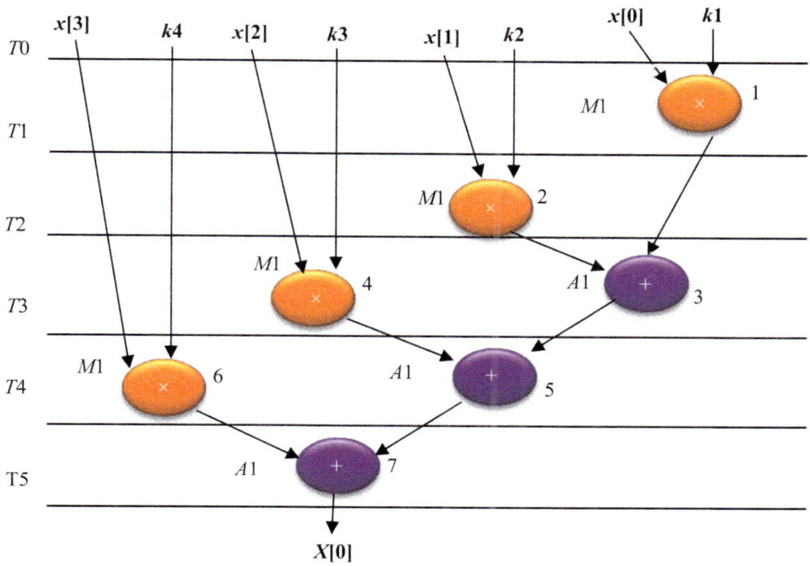

Figure 8.7 Scheduled DFG of 4-point DCT based on one adder and one multiplier (Sengupta and Rathor, 2019)

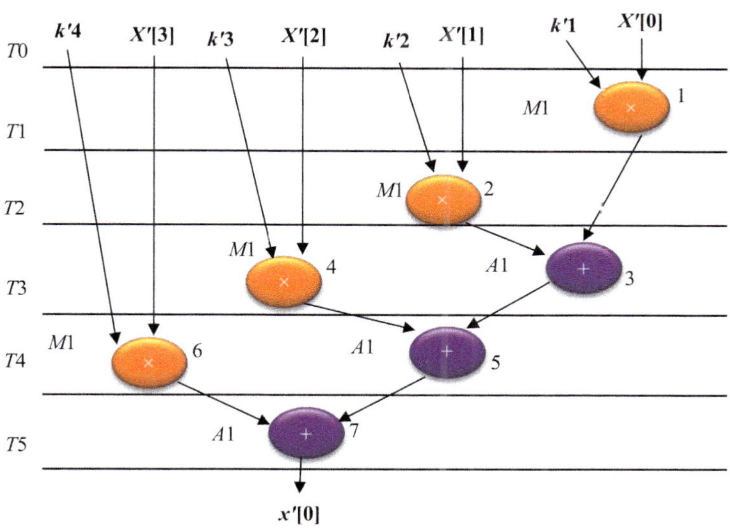

Figure 8.8 Scheduled DFG of 4-point IDCT based on one adder and one multiplier (Sengupta and Rathor, 2019)

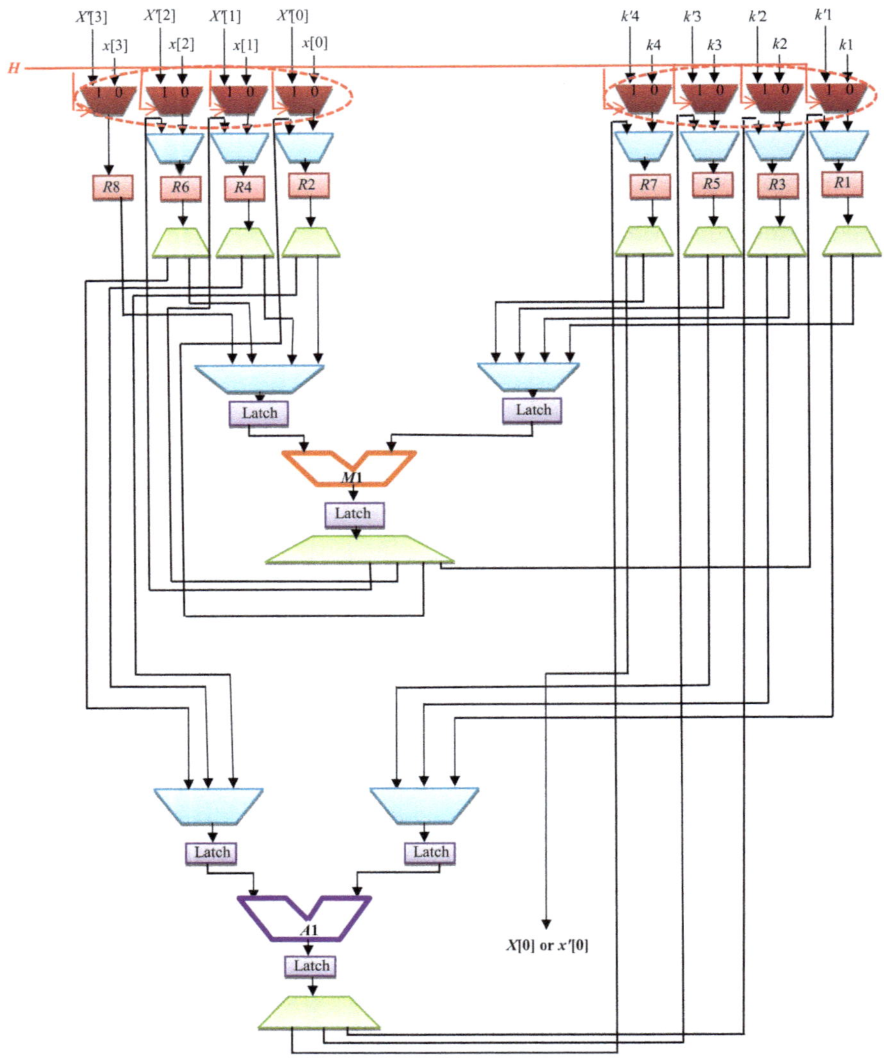

Figure 8.9 Hologram obfuscated RTL data path of 4-point DCT–IDCT integrated design (Sengupta and Rathor, 2019). Note: switching muxes of the hologram are circled in red-dotted ovals

8.6 Determination of gate count affected due to hologram-based obfuscation of DSP cores

The process of estimating the gate count of the hologram-obfuscated design (Sengupta and Rathor, 2019) and the high level transformation (HLT)-obfuscated design (Sengupta *et al.*, 2017a) can be generically represented using (8.5):

$$G_c^a = \Delta G_c^b + G_c^i \tag{8.5}$$

where

G_c^a = number of gates affected due to obfuscation.
ΔG_c^b = difference in gate count between obfuscated design and baseline.
G_c^i = number of gates modified in terms of input connectivity after obfuscation.

Using (8.5), the number of gates affected due to hologram obfuscation for IIR–FIR filter designs with respect to 15 nm NanGate technology scale is determined to be equal to 6,544 gates. On the other hand, the total gate count of both IIR and FIR filters (baseline version) combined is 5,968 + 3,280 = 9,248 gates. Therefore the difference in gate count wrt baseline (ΔG_c^b) is 9,248 − 6,544 = 2,704 gates. Since the number of gates present in the hologram-obfuscated IIR–FIR filter design is much less than the combined baseline equivalent gate count of the IIR and FIR filters, clear area savings in terms of gate count are evident. However, there is delay trade-off involved in the obfuscated-design process. More specifically, the delay overheads compared to the baseline of the IIR filter (when the switching element is H = '0' in the hologram-obfuscated design) is 77.7 ps. On the contrary, the delay overhead compared to the baseline of the FIR filter (when the switching element is H = '1' in the hologram-obfuscated design) is 233.1 ps.

Next, the gate count determination of Sengupta *et al.* (2017a) is analysed. The obfuscation process used in Sengupta and Roy (2017) is purely HLT based and includes: Redundant operation elimination (ROE), logic transformation (LT) and tree-height transformation (THT). In most cases, depending on the nature of the DSP application, not all HLTs are applicable. For example, in the case of IIR and FIR filters, only THT-based obfuscation is employable. This indicates that HLTs incur a drawback for certain applications during structural obfuscation. Let us discuss the IIR and FIR filter applications to explain this context. THT obfuscation does not affect the node count (and, thus, resources) as it only targets minimizing the critical delay of the application. This is because THT does not alter the mux–demux sizes or the functional resource count. Since it only affects the interconnectivity information of mux and demux, $\Delta G_c^b = 0$ from (8.5). This also indicates that the obfuscated design area using Sengupta and Roy (2017) is the same as the baseline version. The gate count due to the gates modified based on the change in the input–output connectivity of the mux–demux only accounts for the number of gates affected due to obfuscation (G_c^a). Referring to Figure 8.12 of the HLT obfuscated FIR data path using Sengupta *et al.* (2017a), it is clearly apparent that one input of each 2:1 muxes gets affected due to THT. Since a 2:1 mux comprises of 64 NAND gates, thus due to change in of a single input, 32 gates of each 2:1 mux gets modified. Thus the total number of gates modified in terms of input connectivity after THT obfuscation (G_c^i) is 32 + 32 = 64 gates. Similarly, G_c^i = 160 gates for THT obfuscated IIR filter using same concept. Therefore the total affected gate count due to THT obfuscated IIR and FIR filter is 64 + 160 = 224 gates.

8.7 Structural obfuscation of DSP cores (circuits) using high-level transformation (Sengupta *et al.*, 2017a)

This section demonstrates with examples an alternative structural obfuscation method for DSP circuits using high-level transformations, which is also very effective. The section gives six examples of DSP circuits in the context of its obfuscated circuit.

8.7.1 Example 1: 4-point DCT

The functions of 4-point DCT are

$$X[0] = c2 \times x[0] + c2 \times x[1] + c2 \times x[2] + c2 \times x[3]$$

$$X[1] = c1 \times x[0] + c3 \times x[1] + (-c3) \times x[2] + (-c1) \times x[3]$$

$$X[2] = c2 \times x[0] + (-c2) \times x[1] + (-c2) \times x[2] + c2 \times x[3]$$

$$X[3] = c3 \times x[0] + (-c1) \times x[1] + c1 \times x[2] + (-c3) \times x[3]$$

The generic function of 4-point DCT to compute the first sample is

$$X[0] = k1 \times x[0] + k2 \times x[1] + k2 \times x[2] + k2 \times x[3]$$

DFG of 4-point DCT after THT-based obfuscation using Sengupta *et al.* (2017a) is shown here. Based on the generic representation to compute the first sample, the respective DFG is shown in Figure 8.10.

Scheduled DFG of 4-point DCT based on resource constraint (1M, 1A) after THT-based obfuscation using Sengupta *et al.* (2017a) is shown in Figure 8.11.

The equation of 4-point DCT to compute the first sample is

$$X[0] = k1 \times x[0] + k2 \times x[1] + k3 \times x[2] + k4 \times x[3]$$

The RTL data path after THT-based obfuscation (Sengupta *et al.*, 2017a) is shown in Figure 8.12.

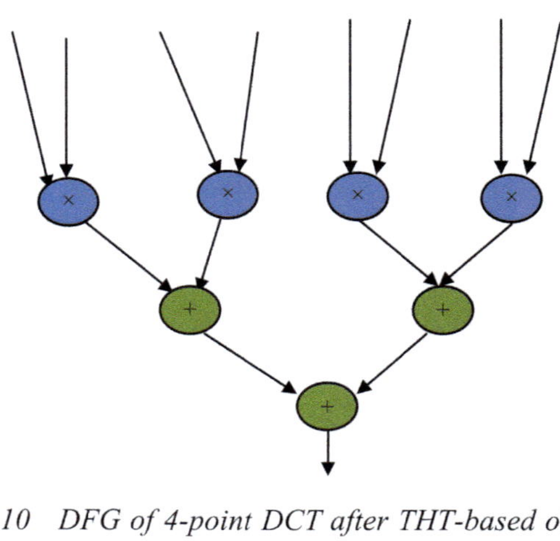

Figure 8.10 DFG of 4-point DCT after THT-based obfuscation

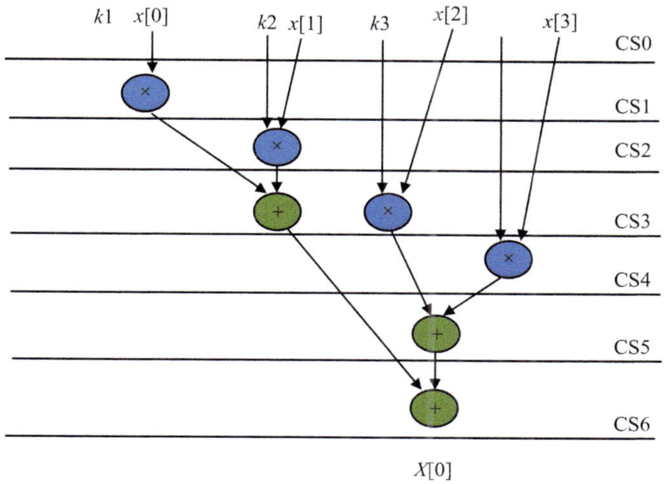

Figure 8.11 Scheduled DFG of 4-point DCT based on resource constraint (1M, 1A)

8.7.2 Example 2: 8-point DCT

The functions of 8-point DCT are (Sengupta and Mohanty, 2019):

$$
\begin{aligned}
X[0] &= c4 \times x[0] + c4 \times x[1] + c4 \times x[2] + c4 \times x[3] + c4 \times x[4] \\
&\quad + c4 \times x[5] + c4 \times x[6] + c4 \times x[7] \\
X[1] &= c1 \times x[0] + c3 \times x[1] + c5 \times x[2] + c7 \times x[3] + (-c7) \times x[4] \\
&\quad + (-c5) \times x[5] + (-c3) \times x[6] + (-c1) \times x[7] \\
X[2] &= c2 \times x[0] + c6 \times x[1] + (-c6) \times x[2] + (-c2) \times x[3] + (-c2) \times x[4] \\
&\quad + (-c6) \times x[5] + (-c6) \times x[6] + c2 \times x[7] \\
X[3] &= c3 \times x[0] + (-c7) \times x[1] + (-c1) \times x[2] + (-c5) \times x[3] + c5 \times x[4] \\
&\quad + c1 \times x[5] + c7 \times x[6] + (-c3) \times x[7] \\
X[4] &= c4 \times x[0] + (-c4) \times x[1] + (-c4) \times x[2] + c4 \times x[3] + c4 \times x[4] \\
&\quad + (-c4) \times x[5] + (-c4) \times x[6] + c4 \times x[7] \\
X[5] &= c5 \times x[0] + (-c1) \times x[1] + c7 \times x[2] + c3 \times x[3] + (-c3) \times x[4] \\
&\quad + (-c7) \times x[5] + c1 \times x[6] + (-c5) \times x[7] \\
X[6] &= c6 \times x[0] + (-c2) \times x[1] + c2 \times x[2] + (-c6) \times x[3] + (-c6) \times x[4] \\
&\quad + c2 \times x[5] + (-c2) \times x[6] + c6 \times x[7] \\
X[7] &= c7 \times x[0] + (-c5) \times x[1] + c3 \times x[2] + (-c1) \times x[3] + c1 \times x[4] \\
&\quad + (-c3) \times x[5] + c5 \times x[6] + (-c7) \times x[7]
\end{aligned}
$$

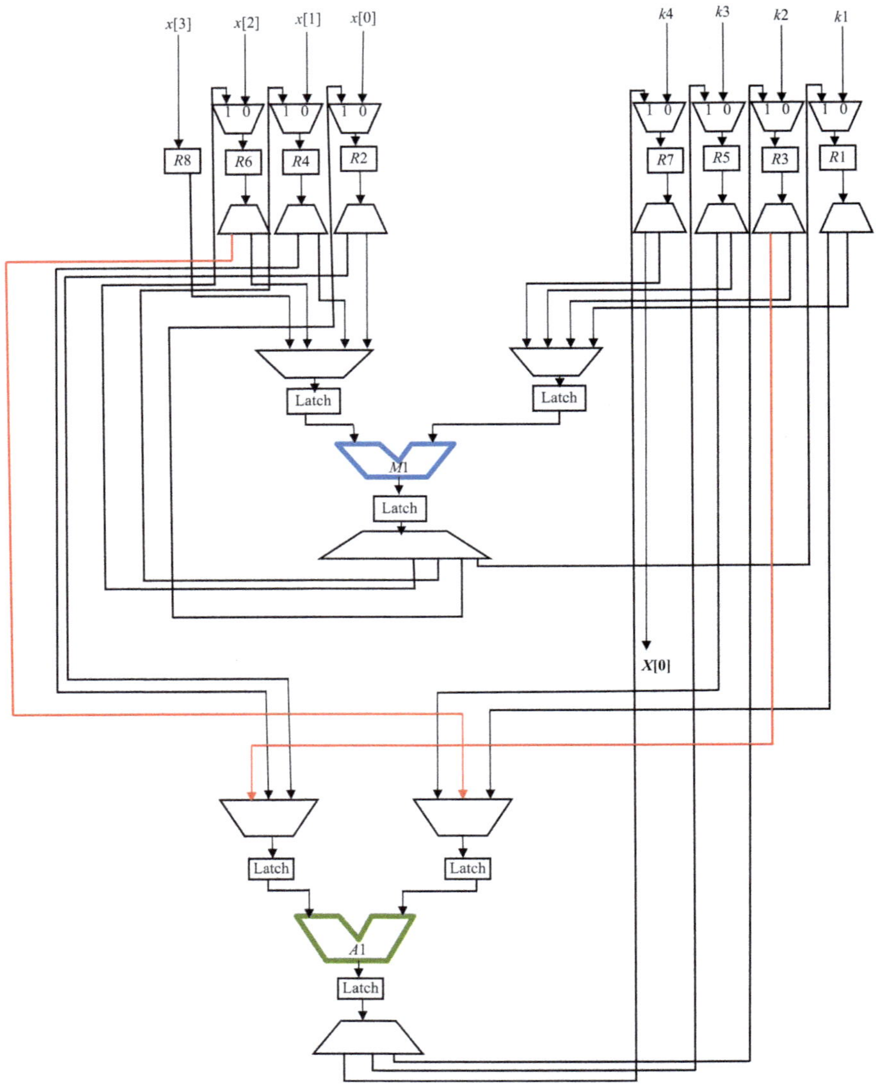

Figure 8.12 RTL data path of 4-point DCT after THT based obfuscation

The generic function of 8-point DCT to compute the first sample is

$$X[0] = k1 \times x[0] + k2 \times x[1] + k3 \times x[2] + k4 \times x[3] \\ + k5 \times x[4] + k6 \times x[5] + k7 \times x[6] + k8 \times x[7]$$

The DFG of 8-point DCT after THT-based obfuscation using Sengupta *et al.* (2017a) is shown here. Based on the generic representation to compute the first sample, the respective DFG is shown in Figure 8.13.

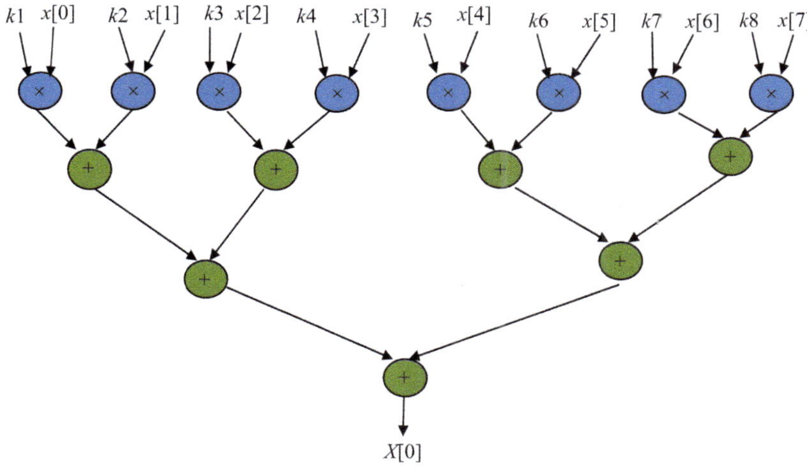

Figure 8.13 DFG of 8-point DCT after THT-based obfuscation

Scheduled and allocated DFGs of 8-point DCT based on resource constraint (1*M*, 1*A*) after THT-based obfuscation using Sengupta *et al.* (2017a) are shown in Figures 8.14 and 8.15, respectively.

The equation of 8-point DCT to compute the first sample is

$$\begin{aligned} X[0] &= k1 \times x[0] + k2 \times x[1] + k3 \times x[2] + k4 \times x[3] \\ &\quad + k5 \times x[4] + k6 \times x[5] + k7 \times x[6] + k8 \times x[7] \end{aligned}$$

The RTL data path after THT-based obfuscation (Sengupta *et al.*, 2017a) is shown in Figure 8.16.

8.7.3 Example 3: 8-point IDCT

The functions of 8-point IDCT are

$$\begin{aligned} x'[0] &= c4 \times X'[0] + c1 \times X'[1] + c2 \times X'[2] + c3 \times X'[3] \\ &\quad + c4 \times X'[4] + c5 \times X'[5] + c6 \times X'[6] + c7 \times X'[7] \\ x'[1] &= c4 \times X'[0] + c3 \times X'[1] + c6 \times X'[2] + (-c7) \times X'[3] \\ &\quad + (-c4) \times X'[4] + (-c1) \times X'[5] + (-c2) \times X'[6] + (-c5) \times X'[7] \\ x'[2] &= c4 \times X'[0] + c5 \times X'[1] + (-c6) \times X'[2] + (-c1) \times X'[3] \\ &\quad + (-c4) \times X'[4] + c7 \times X'[5] + c2 \times X'[6] + c3 \times X'[7] \\ x'[3] &= c4 \times X'[0] + c7 \times X'[1] + (-c2) \times X'[2] + (-c5) \times X'[3] \\ &\quad + c4 \times X'[4] + c3 \times X'[5] + (-c6) \times X'[6] + (-c1) \times X'[7] \\ x'[4] &= c4 \times X'[0] + (-c7) \times X'[1] + (-c2) \times X'[2] + c5 \times X'[3] \\ &\quad + c4 \times X'[4] + (-c3) \times X'[5] + (-c6) \times X'[6] + c1 \times X'[7] \\ x'[5] &= c4 \times X'[0] + (-c5) \times X'[1] + (-c6) \times X'[2] + c1 \times X'[3] \\ &\quad + (-c4) \times X'[4] + (-c7) \times X'[5] + c2 \times X'[6] + (-c3) \times X'[7] \\ x'[6] &= c4 \times X'[0] + (-c3) \times X'[1] + (-c6) \times X'[2] + c7 \times X'[3] \\ &\quad + (-c4) \times X'[4] + c1 \times X'[5] + (-c2) \times X'[6] \times c5 \times X'[7] \\ x'[7] &= c4 \times X'[0] + (-c1) \times X'[1] + c2 \times X'[2] + (-c3) \times X'[3] + c4 \times X'[4] \\ &\quad + (-c5) \times X'[5] + c6 \times X'[6] + (-c7) \times X'[7] \end{aligned}$$

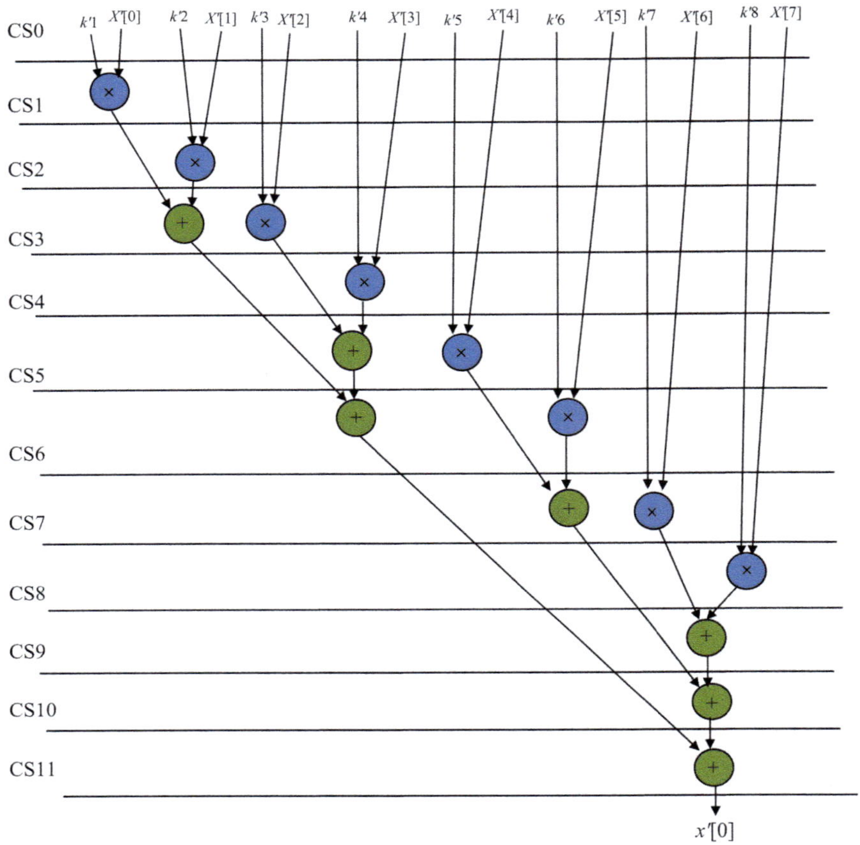

Figure 8.14 Scheduled DFG of 8-point DCT based on resource constraint (1M, 1A)

The generic function of 8-point DCT to compute the first sample is

$$x'[0] = k'1 \times X'[0] + k'2 \times X'[1] + k'3 \times X'[2] + k'4 \times X'[3] \\ + k'5 \times X'[4] + k'6 \times X'[5] + k'7 \times X'[6] + k'8 \times X'[7]$$

The DFG of 8-point IDCT after THT-based obfuscation using Sengupta *et al.* (2017a) is shown here. Based on the generic representation to compute the first sample, the respective DFG is shown in Figure 8.17.

The scheduled DFG of 8-point IDCT based on resource constraint (1*M*, 1*A*) after THT-based obfuscation using Sengupta *et al.* (2017a) is shown in Figure 8.18.

The equation of 8-point IDCT to compute the first sample is

$$x'[0] = k'1 \times X'[0] + k'2 \times X'[1] + k'3 \times X'[2] + k'4 \times X'[3] \\ + k'5 \times X'[4] + k'6 \times X'[5] + k'7 \times X'[6] + k'8 \times X'[7]$$

The RTL data path after THT-based obfuscation (Sengupta *et al.*, 2017a) is shown in Figure 8.19.

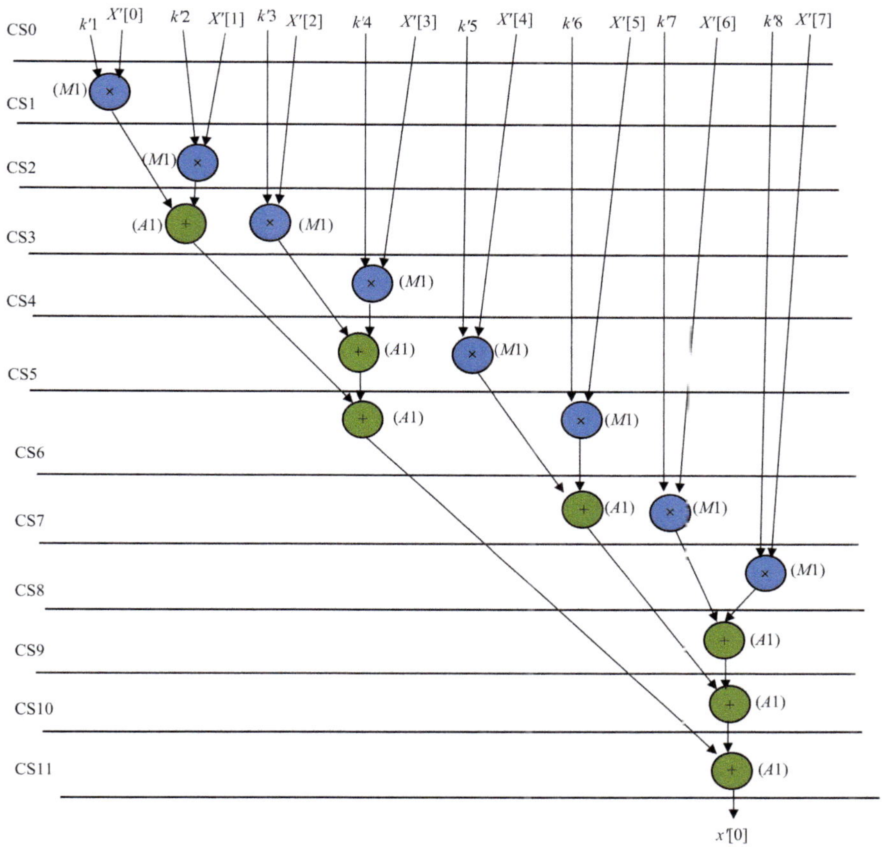

Figure 8.15 Scheduled and allocated DFG of 8-point DCT based on resource constraint (1M, 1A)

8.7.4 Example 4: FIR filter

The FIR filter function is

$$Y'[n] = h0 \times X'[n] + h1 \times X'[n-1] + h2 \times X'[n-2] + h3 \times X'[n-3].$$

The DFG of FIR filter after THT-based obfuscation using Sengupta *et al.* (2017a) is shown in Figure 8.20.

The scheduled DFG of FIR filter based on resource constraint ($1M$, $1A$) after THT-based obfuscation using Sengupta *et al.* (2017a) is shown in Figure 8.21.

$$\text{FIR: } Y'[n] = h0 \times X'[n] + h1 \times X'[n-1] + h2 \times X'[n-2] + h3 \times X'[n-3].$$

The data path after HLT-based obfuscation (Sengupta *et al.*, 2017a) is shown in Figure 8.22.

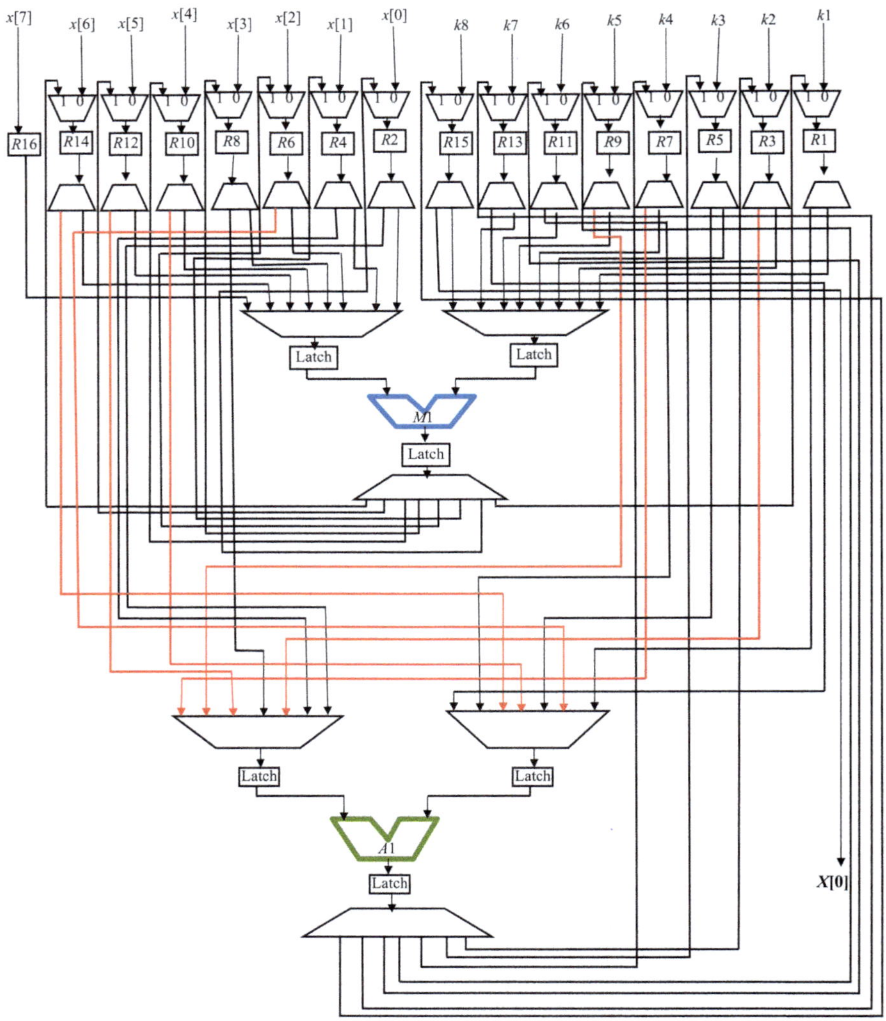

Figure 8.16 RTL datapath of 8-point DCT after THT-based obfuscation

8.7.5 Example 5: IIR filter

The function of IIR filter is

$$Y[n] = b0 \times X[n] + b1 \times X[n-1] + b2 \times X[n-2] + b3 \times X[n-3] \\ + (-a1) \times Y[n-1] + (-a2) \times Y[n-2] + (-a3) \times Y[n-3].$$

The DFG of IIR filter after THT-based obfuscation using Sengupta *et al.* (2017a) is shown in Figure 8.23.

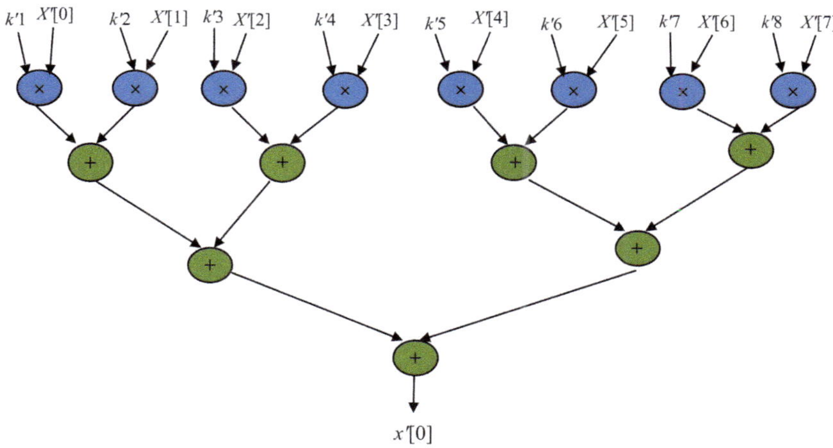

Figure 8.17 The DFG of 8-point IDCT after THT-based obfuscation

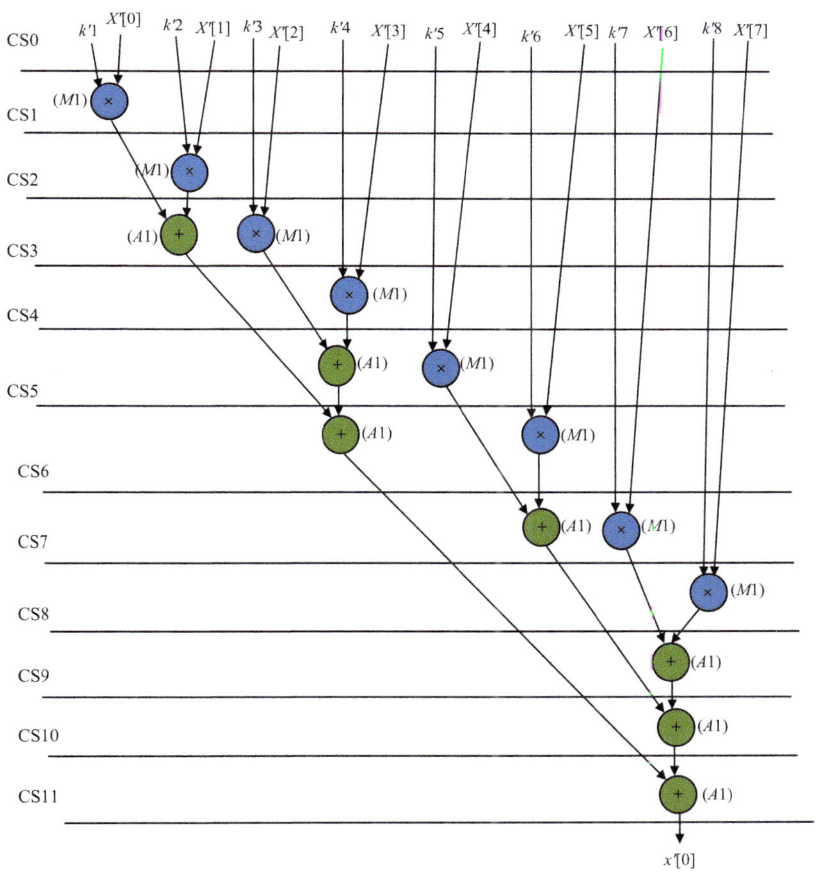

Figure 8.18 The scheduled DFG of 8-point IDCT based on resource constraint (1M, 1A)

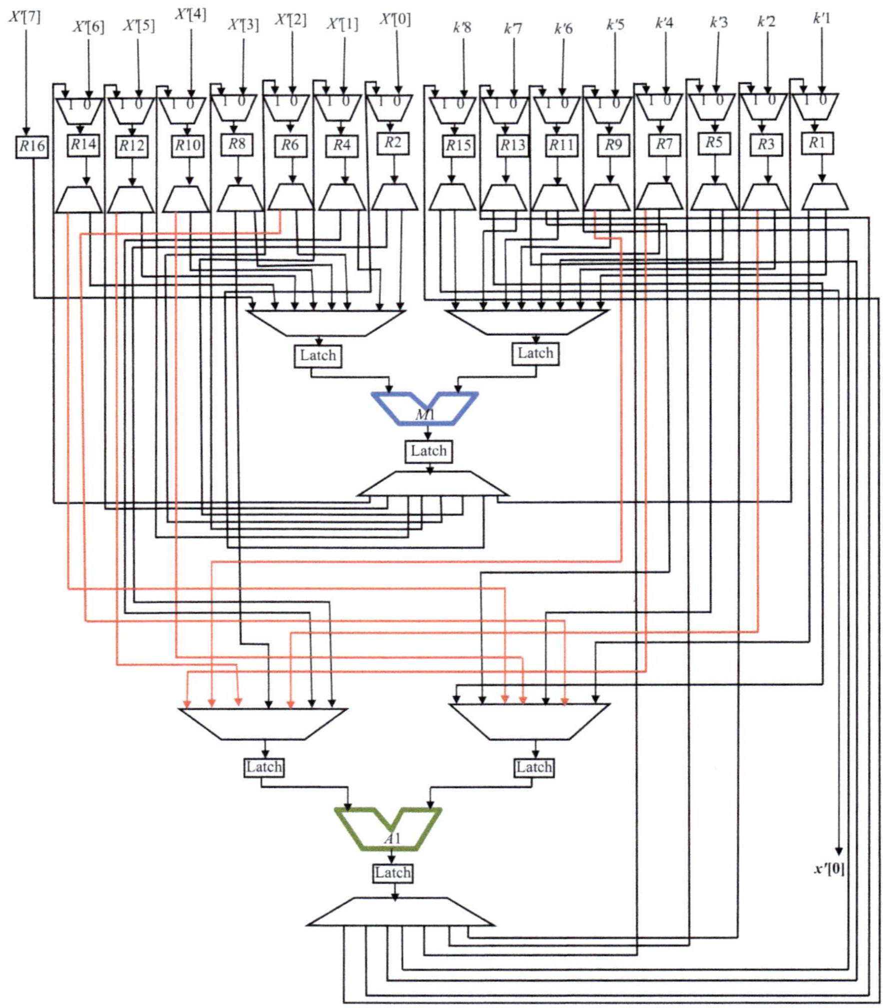

Figure 8.19 RTL datapath of 8-point IDCT after HLT-based obfuscation

The scheduled DFG of IIR filter based on resource constraint (1*M*, 1*A*) after THT-based obfuscation using Sengupta *et al.* (2017a) is shown in Figure 8.24.

$$\text{IIR: } Y[n] = b0 \times X[n] + b1 \times X[n-1] + b2 \times X[n-2] + b3 \times X[n-3] + (-a1) \times Y[n-1] + (-a2) \times Y[n-2] + (-a3) \times Y[n-3].$$

The data path after HLT-based obfuscation (Sengupta *et al.*, 2017a) is shown in Figure 8.25.

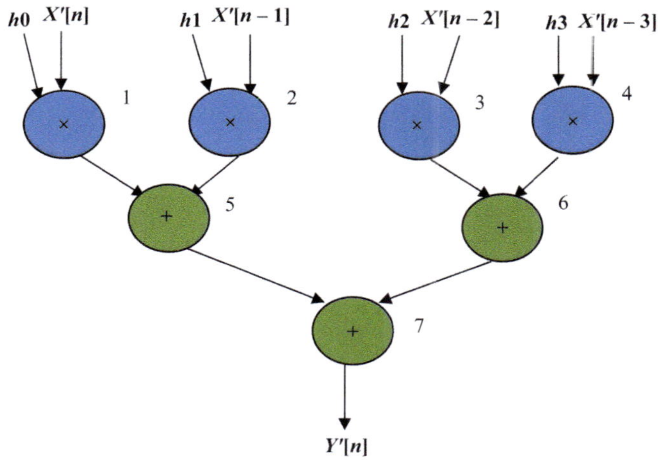

Figure 8.20 The DFG of FIR filter after THT-based obfuscation

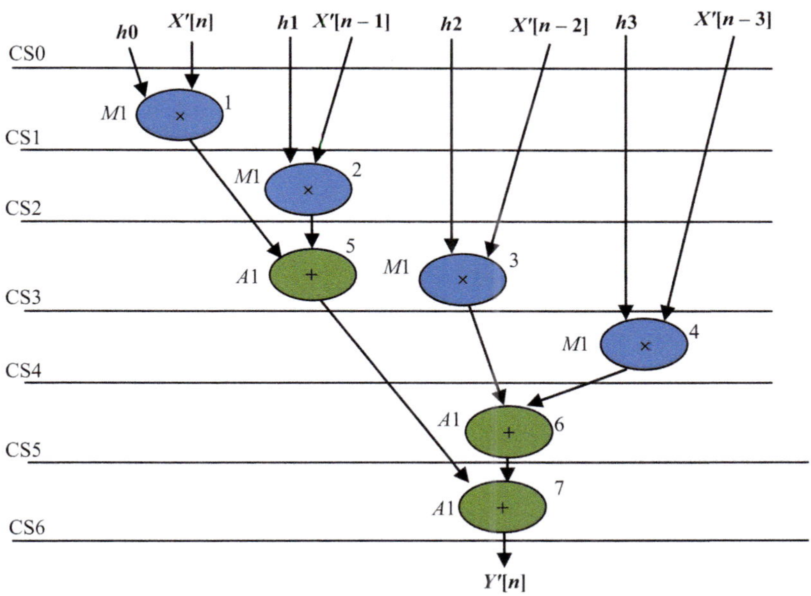

Figure 8.21 The scheduled DFG of FIR filter based on resource constraint (1M, 1A)

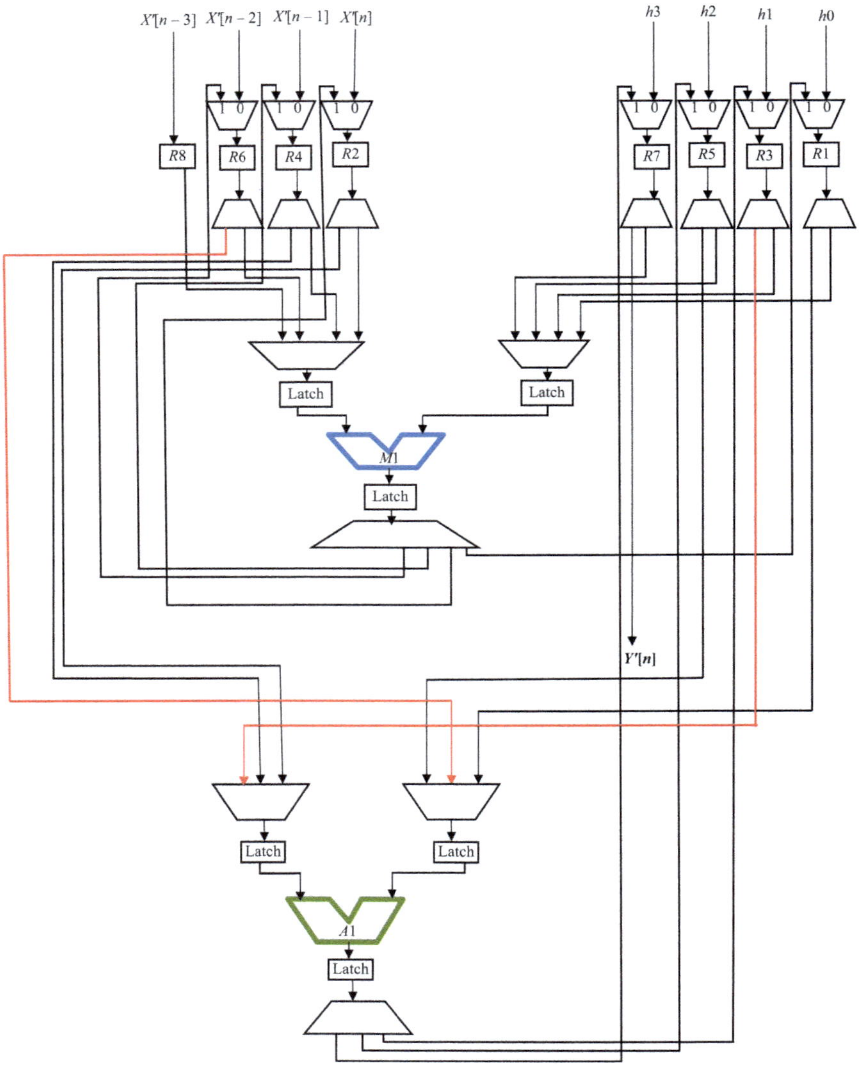

Figure 8.22 RTL datapath of FIR filter after HLT-based obfuscation

8.7.6 Example 6: 4-point IDCT

The functions of 4-point IDCT are

$$
\begin{aligned}
x'[0] &= c2 \times X'[0] + c1 \times X'[1] + c2 \times X'[2] + c3 \times X'[3] \\
x'[1] &= c2 \times X'[0] + c3 \times X'[1] + (-c2) \times X'[2] + (-c1) \times X'[3] \\
x'[2] &= c2 \times X'[0] + (-c3) \times X'[1] + (-c2) \times X'[2] + c1 \times X'[3] \\
x'[3] &= c2 \times X'[0] + (-c1) \times X'[1] + c2 \times X'[2] + (-c3) \times X'[3]
\end{aligned}
$$

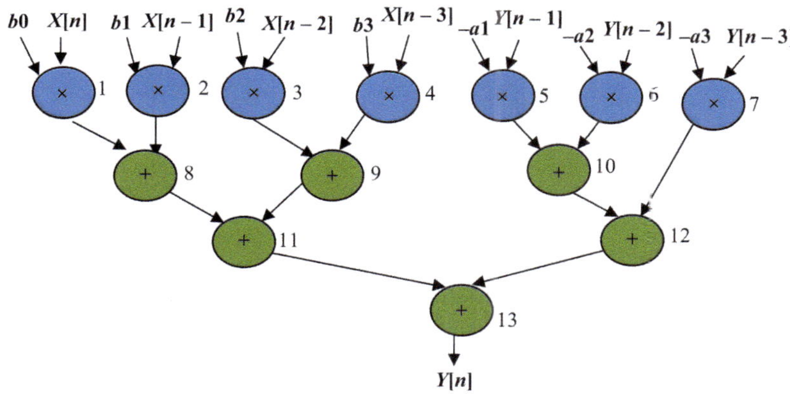

Figure 8.23 The DFG of IIR filter after THT-based obfuscation

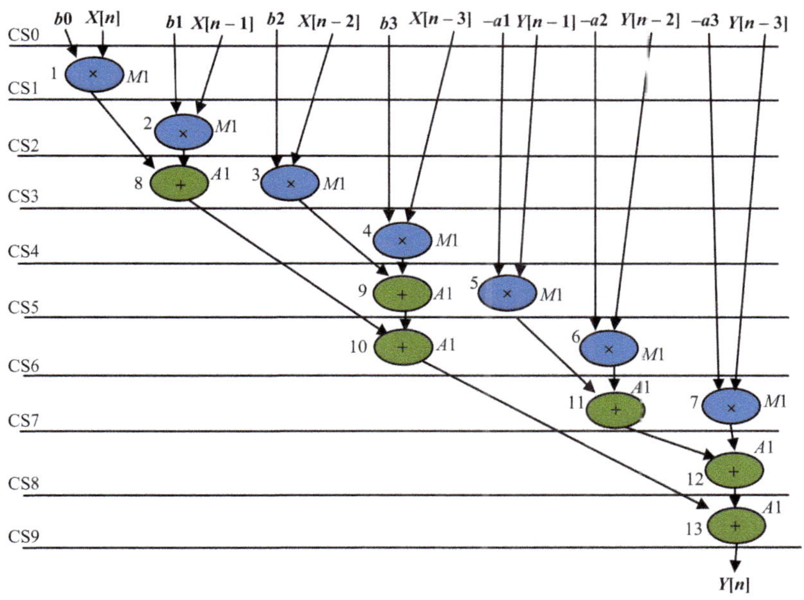

Figure 8.24 The scheduled DFG of IIR filter based on resource constraint (1M, 1A)

The generic function of 4-point DCT to compute the first sample is

$$x'[0] = k'1 \times X'[0] + k'2 \times X'[1] + k'3 \times X'[2] + k'4 \times X'[3]$$

The DFG of 4-point IDCT after THT-based obfuscation using Sengupta *et al.* (2017a) is shown here. Based on the generic representation to compute the first sample, the respective DFG is shown in Figure 8.26.

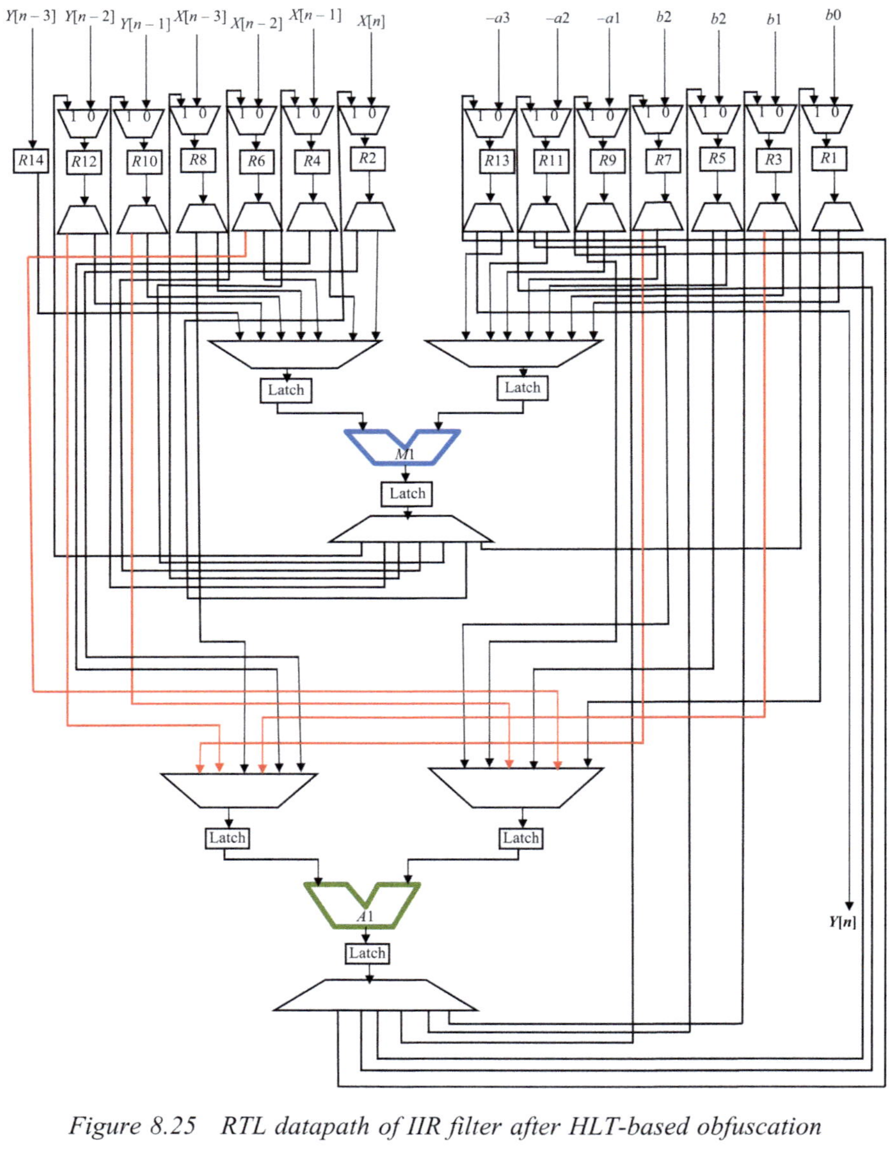

Figure 8.25 RTL datapath of IIR filter after HLT-based obfuscation

The scheduled DFG of 4-point IDCT based on resource constraint (1*M*, 1*A*) after THT-based obfuscation using Sengupta *et al.* (2017a) is shown in Figure 8.27.

The equation of 4-point IDCT to compute the first sample is

$$x'[0] = k'1 \times X'[0] + k'2 \times X'[1] + k'3 \times X'[2] + k'4 \times X'[3]$$

The RTL data path after THT-based obfuscation (Sengupta *et al.*, 2017a) is shown in Figure 8.28.

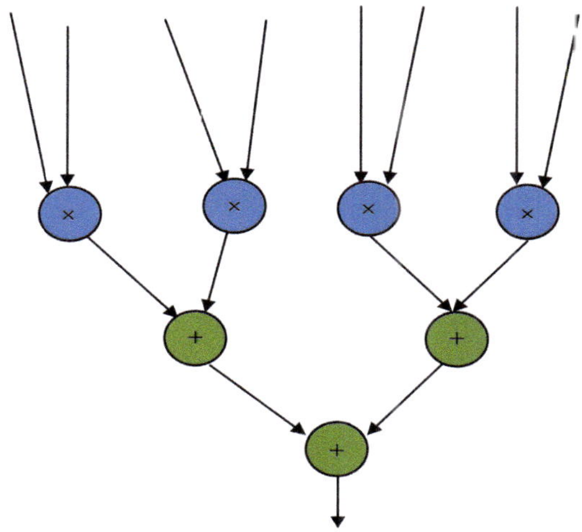

Figure 8.26 The DFG of 4-point IDCT after THT-based obfuscation

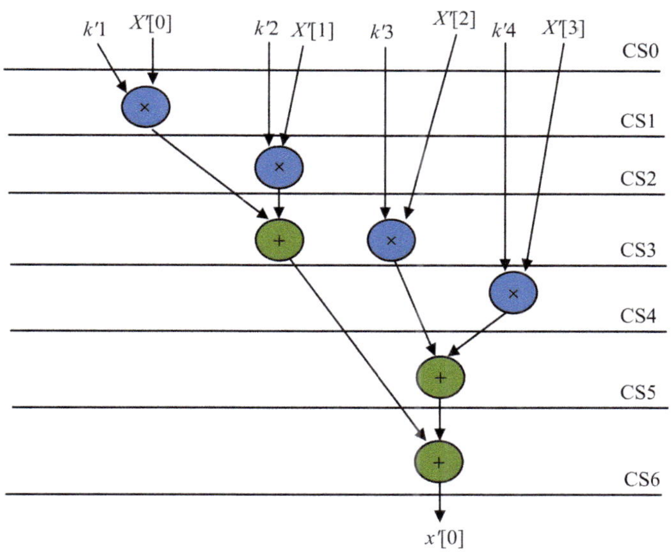

Figure 8.27 The scheduled DFG of 4-point IDCT based on resource constraint (1M, 1A)

8.8 Design of non-obfuscated (baseline versions) DSP circuits

This section demonstrates the design process with examples of non-obfuscated DSP circuits.

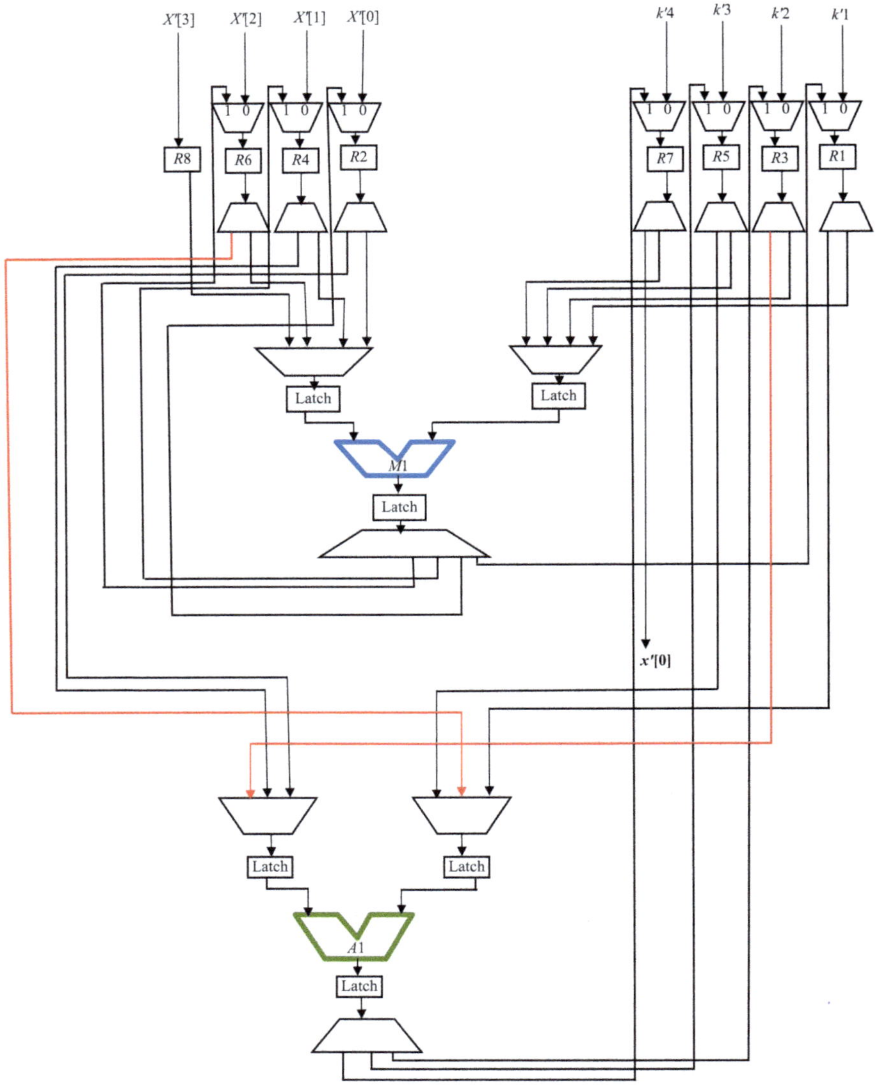

Figure 8.28 RTL datapath of 4-point IDCT after HLT-based obfuscation

8.8.1 Example 1: 4-point DCT (Sengupta and Mohanty, 2019)

The generic equation of 4-point DCT to compute the first sample is

$$X[0] = k1 \times x[0] + k2 \times x[1] + k3 \times x[2] + k4 \times x[3]$$

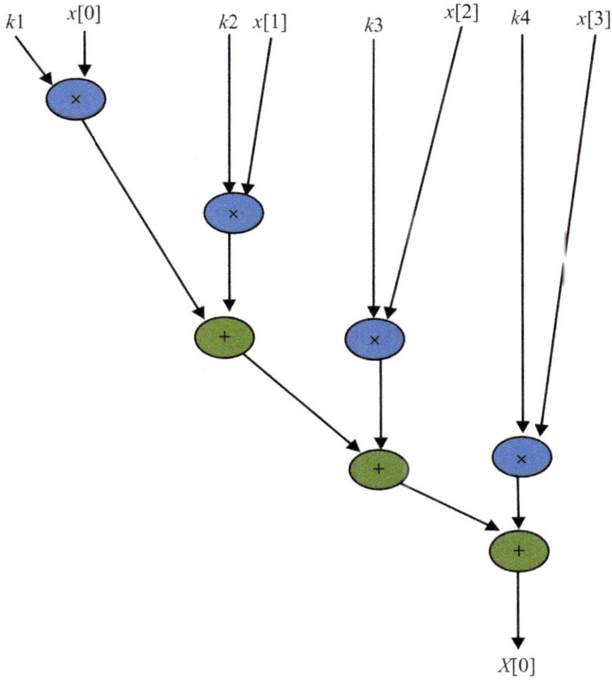

Figure 8.29 The DFG of 4-point DCT

The DFG of 4-point DCT based on the generic representation to compute the first sample is shown in Figure 8.29.

The scheduled DFG of 4-point DCT based on resource constraints (1*A*, 1*M*) is shown in Figure 8.30.

The generic equation of 4-point DCT to compute the first sample is

$$X[0] = k1 \times x[0] + k2 \times x[1] + k3 \times x[2] + k4 \times x[3].$$

The respective data path is shown in Figure 8.31.

8.8.2 Example 2: 4-point IDCT

The generic equation of 4-point IDCT to compute the first sample is

$$x'[0] = k'1 \times X'[0] + k'2 \times X'[1] + k'3 \times X'[2] + k'4 \times X'[3]$$

The DFG of 4-point IDCT based on generic representation to compute the first sample is shown in Figure 8.32.

The scheduled DFG of 4-point IDCT based on resource constraints (1*A*, 1*M*) is shown in Figure 8.33.

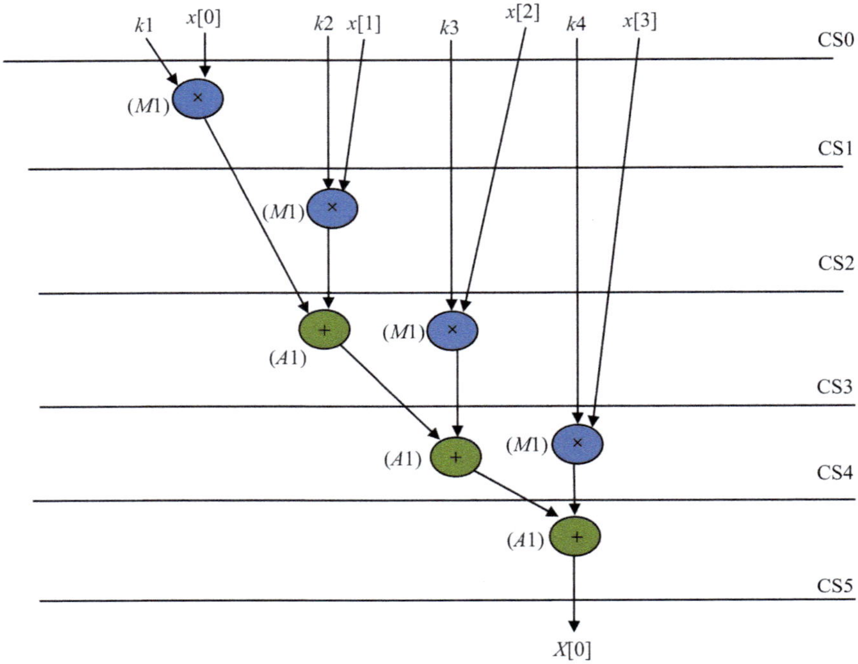

Figure 8.30 The scheduled DFG of 4-point DCT based on resource constraints (1A, 1M)

The generic equation of 4-point IDCT to compute the first sample is

$$x'[0] = k'1 \times X'[0] + k'2 \times X'[1] + k'3 \times X'[2] + k'4 \times X'[3]$$

The respective data path is shown in Figure 8.34.

8.8.3 Example 3: 4-point DIT–FFT

The equations of 4-point decimation in time (DIT)–FFT are

$$\begin{aligned} X'[0] &= H'[0] \times W_4^0 + G'[0] \\ X'[1] &= H'[1] \times W_4^1 + G'[1] \\ X'[2] &= H'[0] \times W_4^2 + G'[0] \\ X'[3] &= H'[1] \times W_4^3 + G'[1] \end{aligned}$$

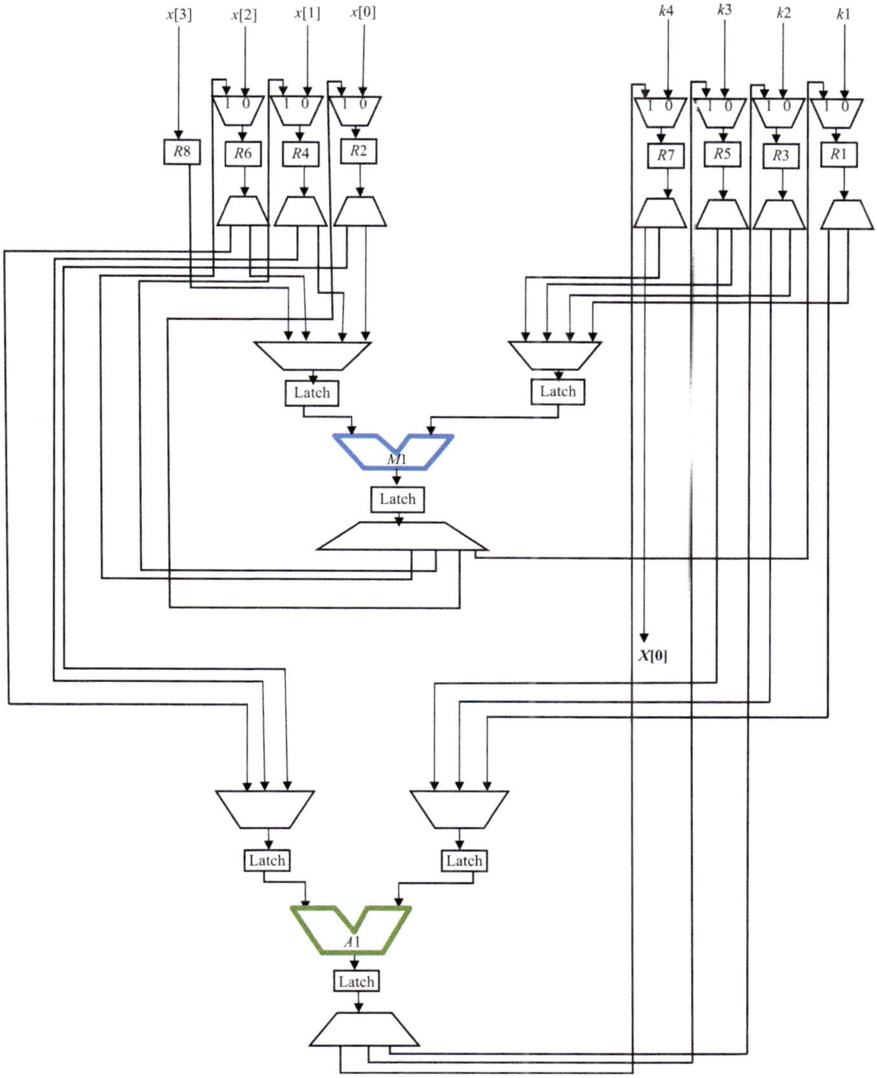

Figure 8.31 Datapath of 4-point DCT

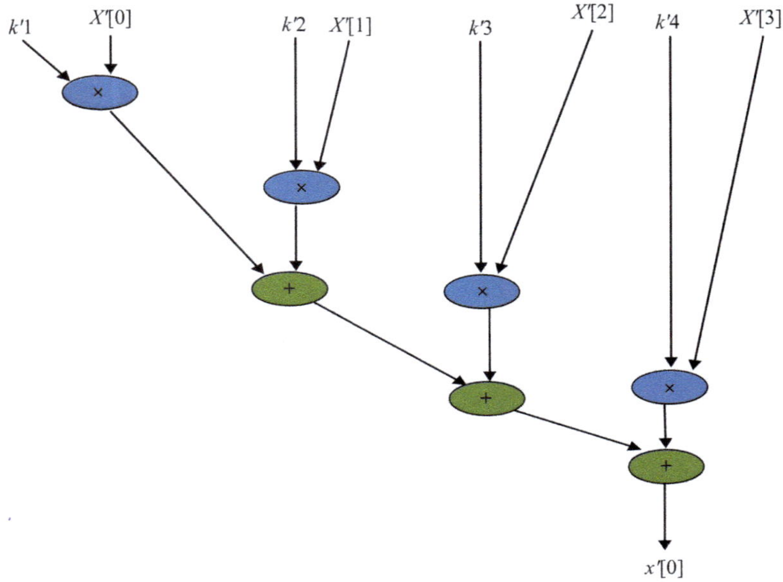

Figure 8.32 DFG of 4-point IDCT

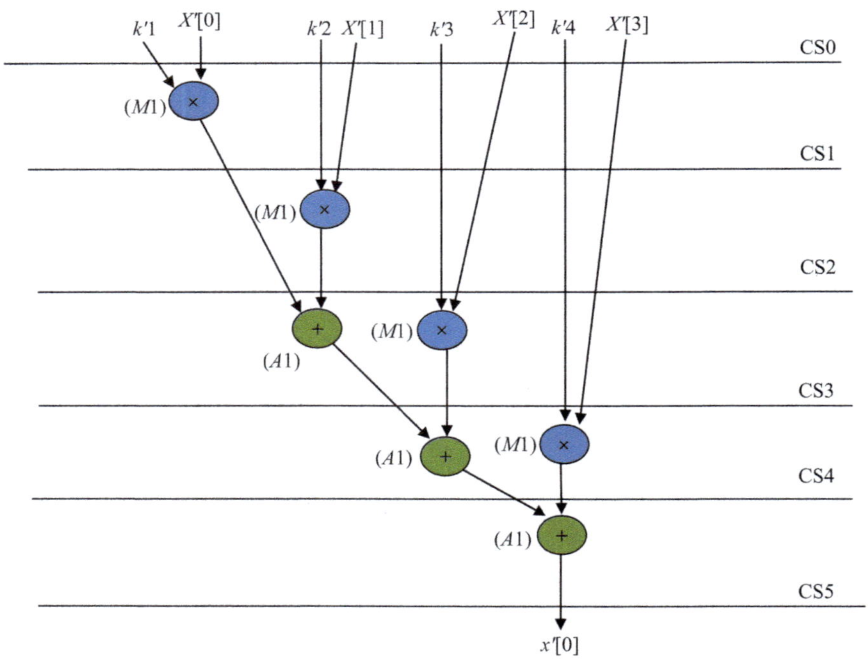

Figure 8.33 The scheduled DFG of 4-point IDCT based on resource constraints (1A, 1M)

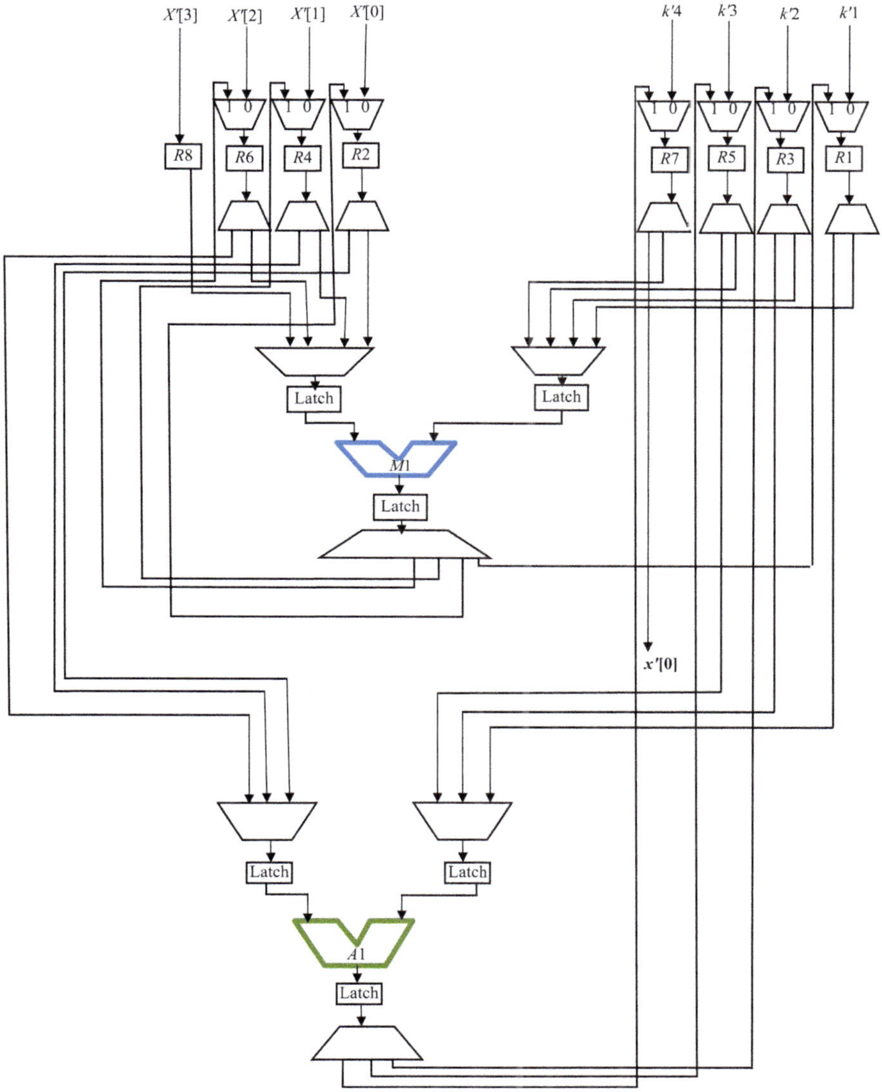

Figure 8.34 Datapath of 4-point IDCT

The DFG of 4-point DIT–FFT is shown in Figure 8.35.

The scheduling of 4-point DIT–FFT is shown in Figure 8.36.

The first and third equations of 4-point DIT–FFT and their corresponding data paths, shown in Figure 8.37.

$$X'[0] = H'[0] \times W_4^0 + G'[0]$$
$$X'[2] = H'[0] \times W_4^2 + G'[0]$$

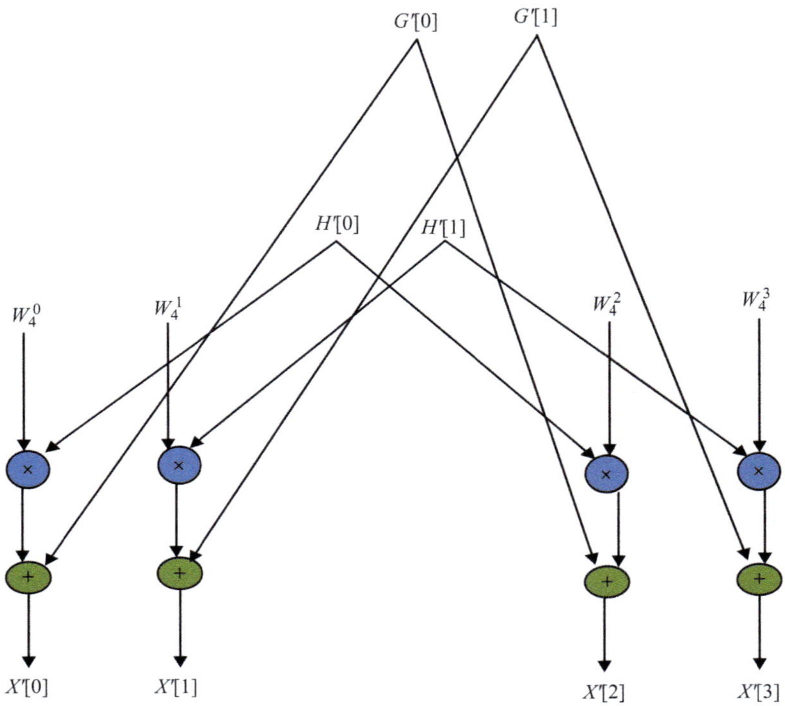

Figure 8.35 DFG of 4-point DIT–FFT

The second and fourth equations of 4-point DIT–FFT and their corresponding data paths, shown in Figure 8.38.

$$X'[1] = H'[1] \times W_4^1 + G'[1]$$
$$X'[3] = H'[1] \times W_4^3 + G'[1]$$

8.8.4 Example 4: 8-point DIT–FFT

The equations of 8-point DIT–FFT are

$$X[0] = H[0] \times W_8^0 + G[0]$$
$$X[1] = H[1] \times W_8^1 + G[1]$$
$$X[2] = H[2] \times W_8^2 + G[2]$$
$$X[3] = H[3] \times W_8^3 + G[3]$$
$$X[4] = H[0] \times W_8^4 + G[0]$$
$$X[5] = H[1] \times W_8^5 + G[1]$$
$$X[6] = H[2] \times W_8^6 + G[2]$$
$$X[7] = H[3] \times W_8^7 + G[3]$$

The DFG of 8-point DIT–FFT is shown in Figure 8.39.

The scheduled DFG of 8-point DIT–FFT is shown in Figure 8.40.

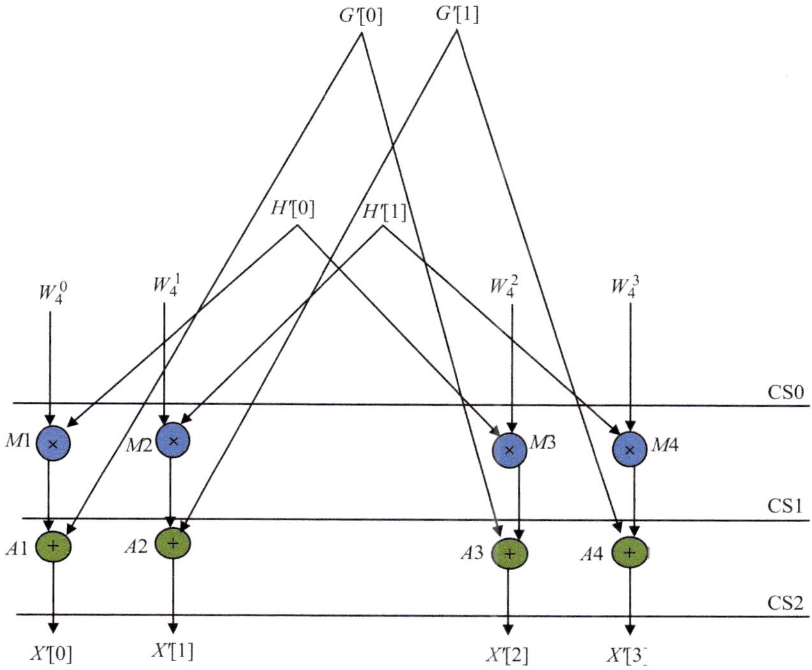

Figure 8.36 The scheduled DFG of 4-point DIT–FFT

The first and fifth equations of 8-point DIT–FFT and their corresponding data paths, shown in Figure 8.41.

$$X[0] = H[0] \times W_8^0 + G[0]$$
$$X[4] = H[0] \times W_8^4 + G[0]$$

The second and sixth equations of 8-point DIT–FFT and their corresponding data paths, shown in Figure 8.42.

$$X[1] = H[1] \times W_8^1 + G[1]$$
$$X[5] = H[1] \times W_8^5 + G[1]$$

The third and seventh equations of 8-point DIT–FFT and their corresponding data paths, shown in Figure 8.43.

$$X[2] = H[2] \times W_8^2 + G[2]$$
$$X[6] = H[2] \times W_8^6 + G[2]$$

The fourth and eighth equations of 8-point DIT–FFT and their corresponding data paths, shown in Figure 8.44.

$$X[3] = H[3] \times W_8^3 + G[3]$$
$$X[7] = H[3] \times W_8^7 + G[3]$$

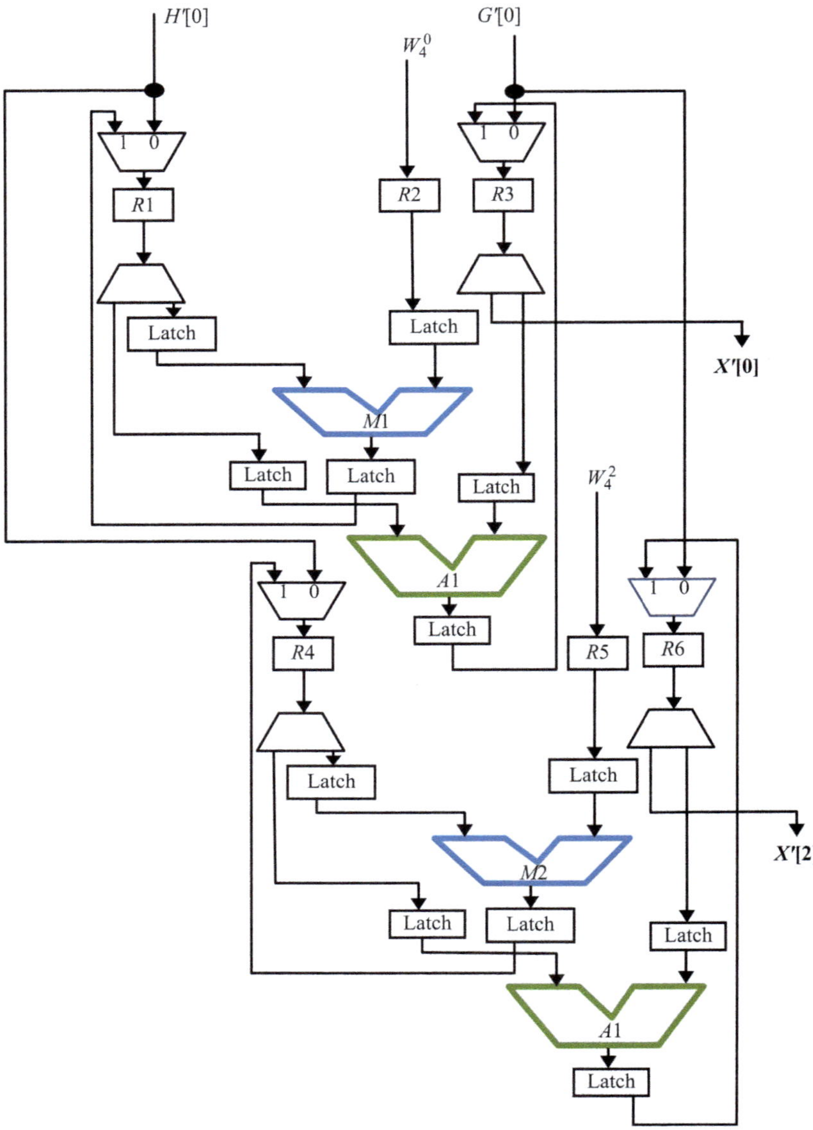

Figure 8.37 RTL datapath corresponding to the first and third equations of 4-point DIT–FFT

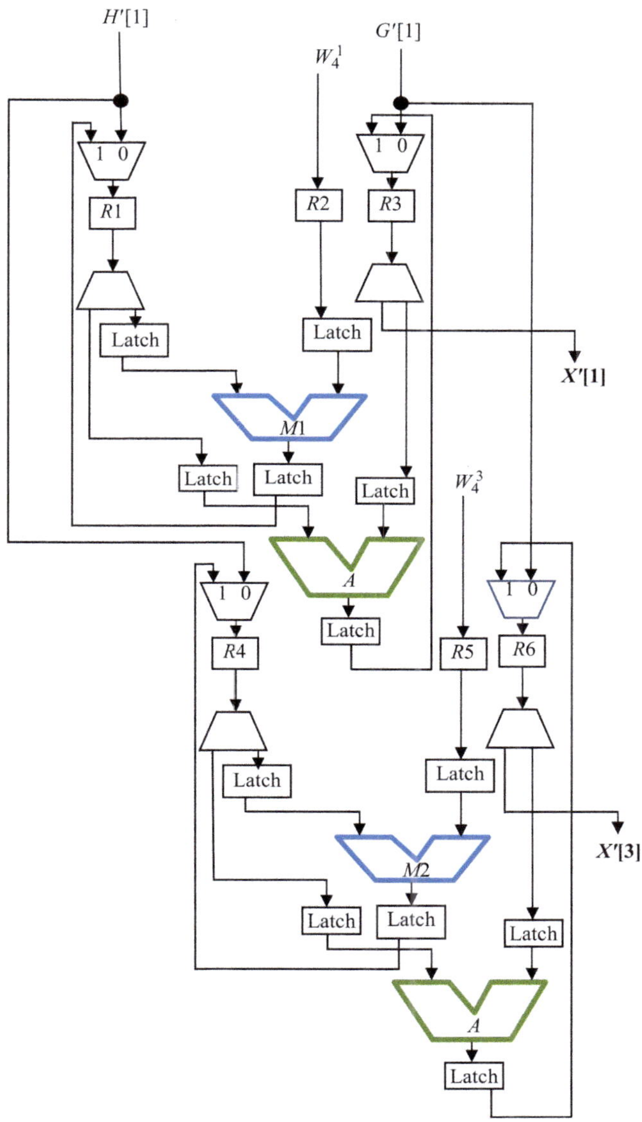

Figure 8.38 RTL datapath corresponding to the second and fourth equations of 4-point DIT–FFT

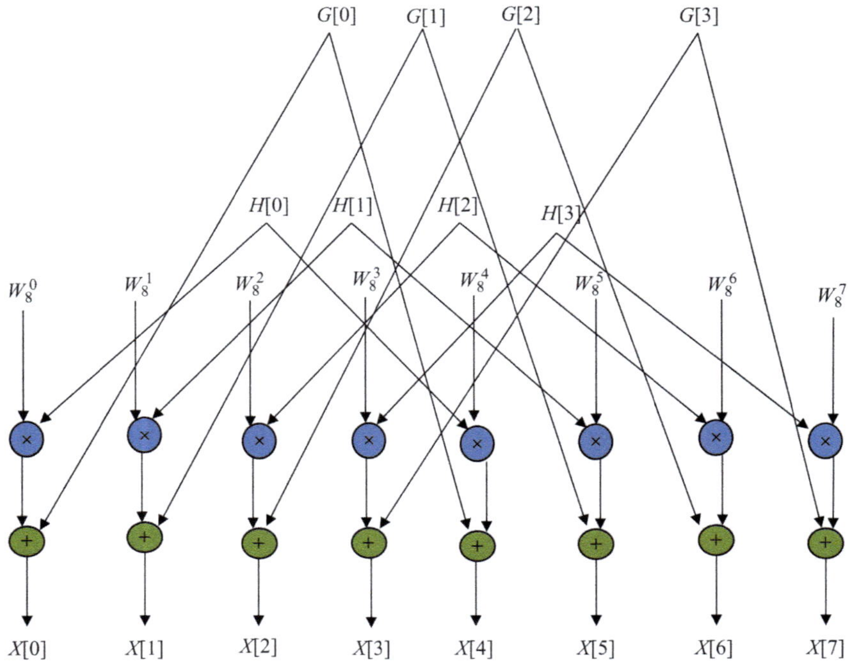

Figure 8.39 DFG of 8-point DIT–FFT

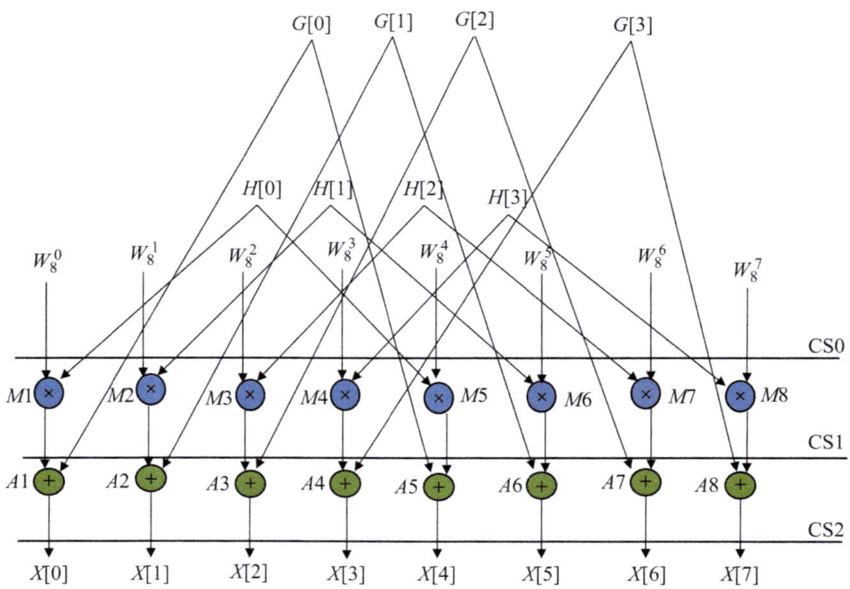

Figure 8.40 Scheduled DFG of 8-point DIT–FFT

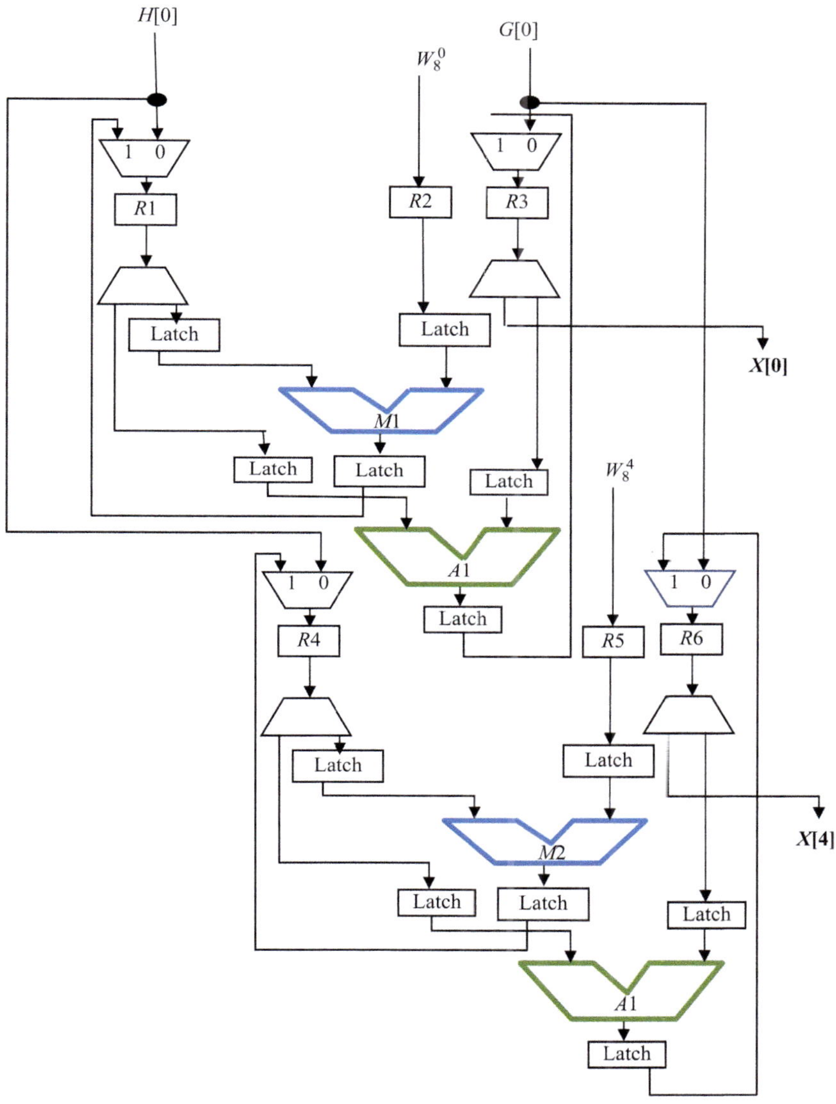

Figure 8.41 RTL datapath corresponding to the first and fifth equations of 8-point DIT–FFT

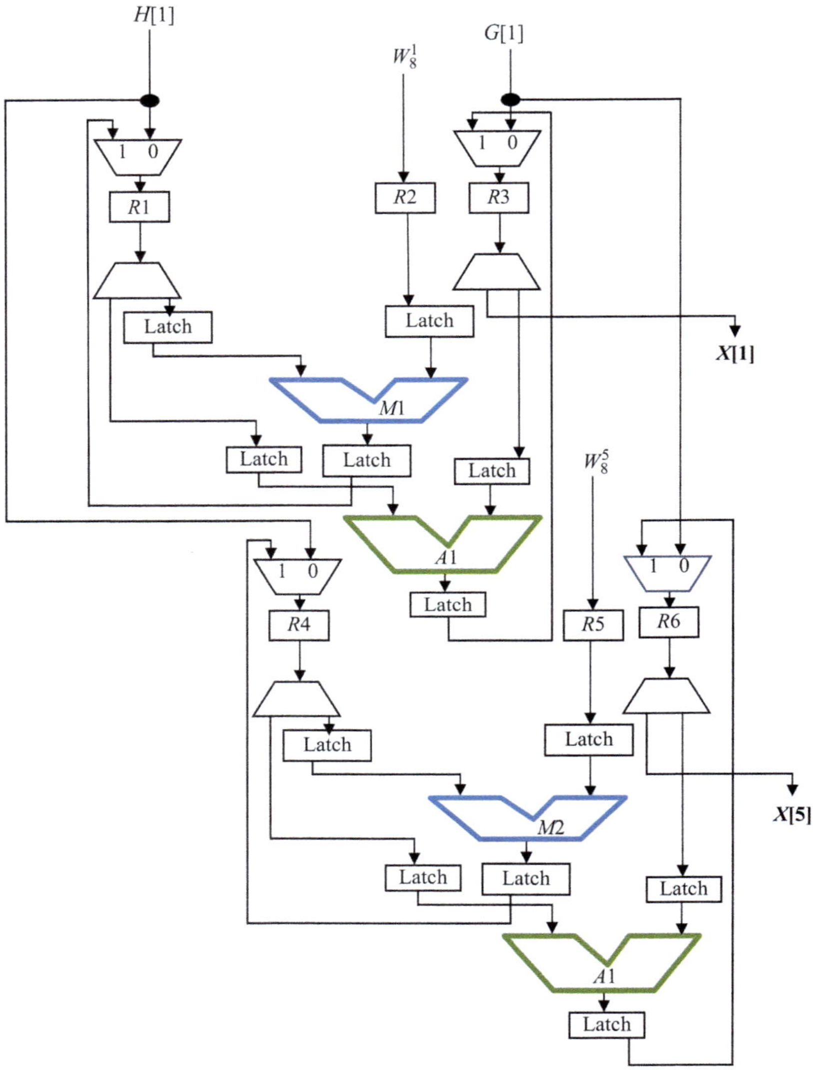

Figure 8.42 RTL datapath corresponding to the second and sixth equations of 8-point DIT–FFT

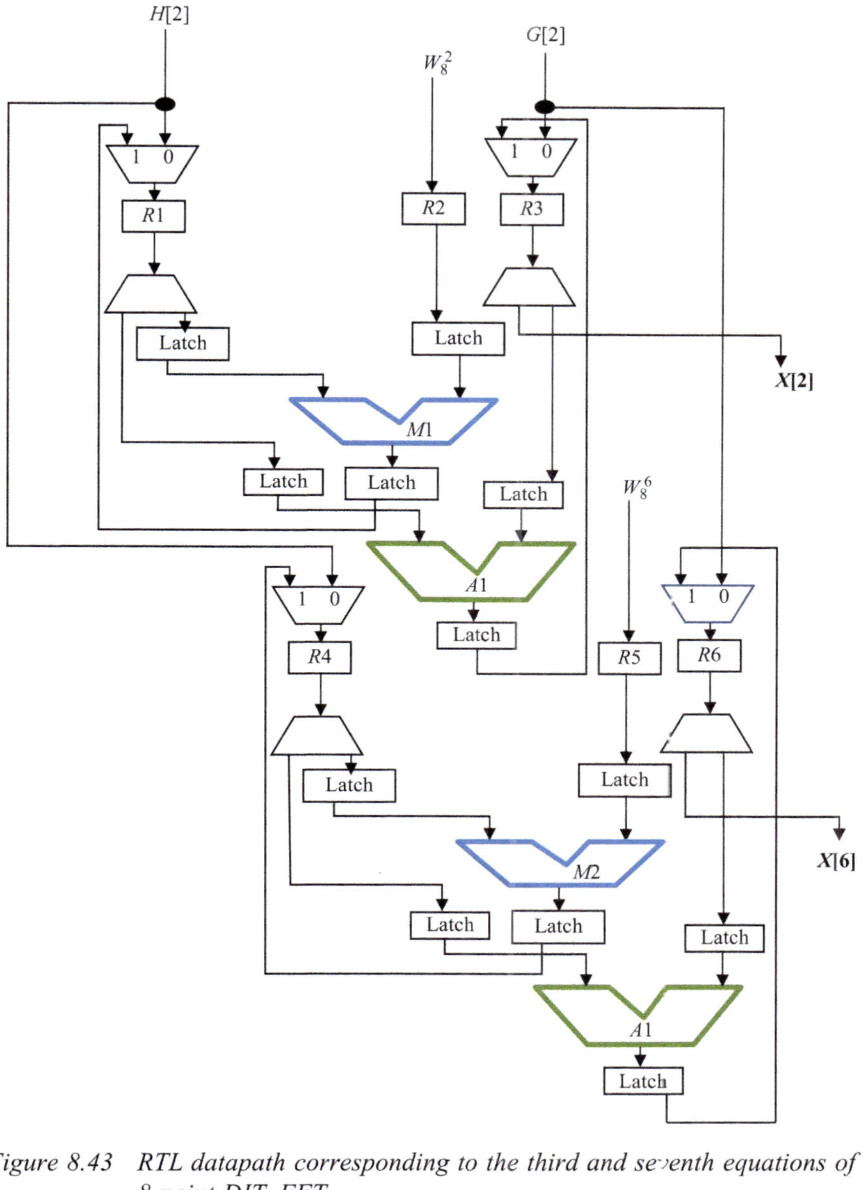

Figure 8.43 RTL datapath corresponding to the third and seventh equations of 8-point DIT–FFT

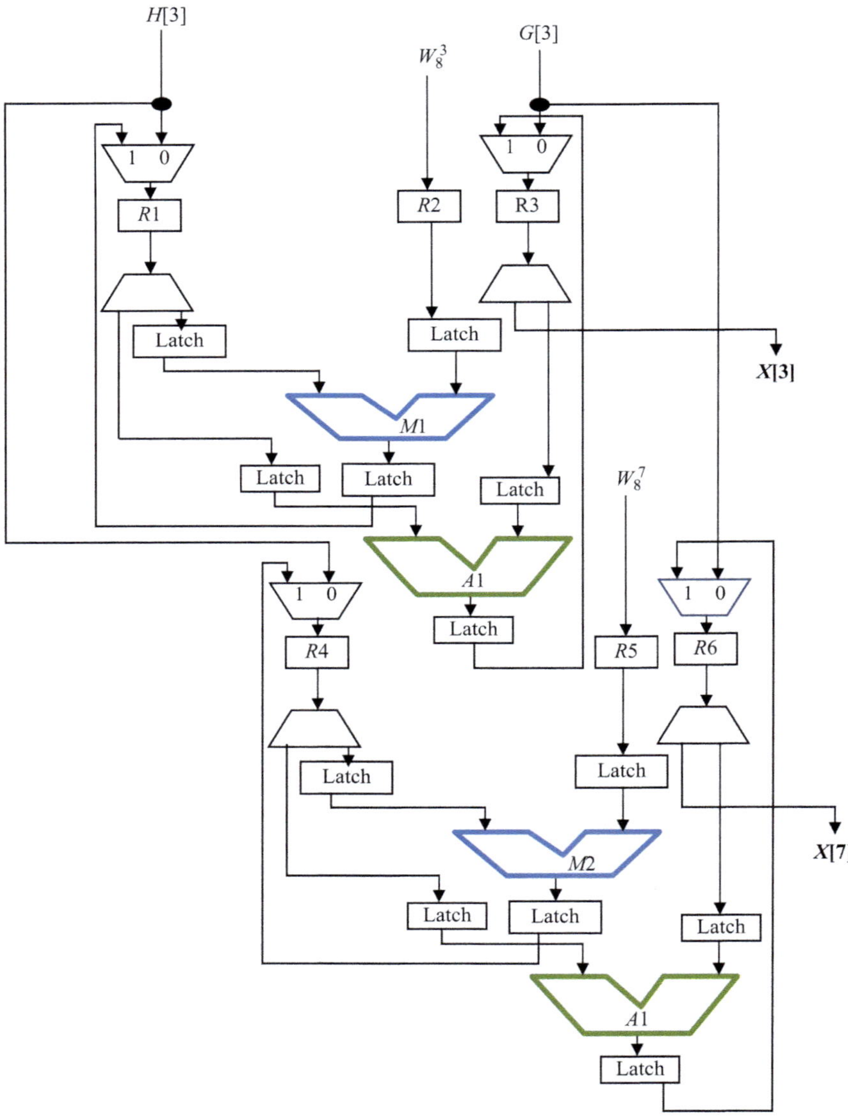

Figure 8.44 RTL datapath corresponding to the fourth and eight equations of 8-point DIT–FFT

8.8.5 Example 5: 8-point DCT

The functions of 8-point DCT are

$$X[0] = c4 \times x[0] + c4 \times x[1] + c4 \times x[2] + c4 \times x[3] + c4 \times x[4] + c4 \times x[5] + c4 \times x[6] + c4 \times x[7]$$

$$X[1] = c1 \times x[0] + c3 \times x[1] + c5 \times x[2] + c7 \times x[3] + (-c7) \times x[4] + (-c5) \times x[5] + (-c3) \times x[6] + (-c1) \times x[7]$$

$$X[2] = c2 \times x[0] + c6 \times x[1] + (-c6) \times x[2] + (-c2) \times x[3] + (-c2) \times x[4] + (-c6) \times x[5] + (-c6) \times x[6] + c2 \times x[7]$$

$$X[3] = c3 \times x[0] + (-c7) \times x[1] + (-c1) \times x[2] + (-c5) \times x[3] + c5 \times x[4] + c1 \times x[5] + c7 \times x[6] + (-c3) \times x[7]$$

$$X[4] = c4 \times x[0] + (-c4) \times x[1] + (-c4) \times x[2] + c4 \times x[3] + c4 \times x[4] + (-c4) \times x[5] + (-c4) \times x[6] + c4 \times x[7]$$

$$X[5] = c5 \times x[0] + (-c1) \times x[1] + c7 \times x[2] + c3 \times x[3] + (-c3) \times x[4] + (-c7) \times x[5] + c1 \times x[6] + (-c5) \times x[7]$$

$$X[6] = c6 \times x[0] + (-c2) \times x[1] + c2 \times x[2] + (-c6) \times x[3] + (-c6) \times x[4] + c2 \times x[5] + (-c2) \times x[6] + c6 \times x[7]$$

$$X[7] = c7 \times x[0] + (-c5) \times x[1] + c3 \times x[2] + (-c1) \times x[3] + c1 \times x[4] + (-c3) \times x[5] + c5 \times x[6] + (-c7) \times x[7]$$

The generic representation of 8-point DCT to compute the first sample is

$$X[0] = k1 \times x[0] + k2 \times x[1] + k3 \times x[2] + k4 \times x[3] + k5 \times x[4] + k6 \times x[5] + k7 \times x[6] + k8 \times x[7]$$

The DFG of generic representation of 8-point DCT to compute the first sample is shown in Figure 8.45.

The scheduled DFG of 8-point DCT based on resource constraints (1*A*, 1*M*) is shown in Figure 8.46.

The respective data path is shown in Figure 8.47.

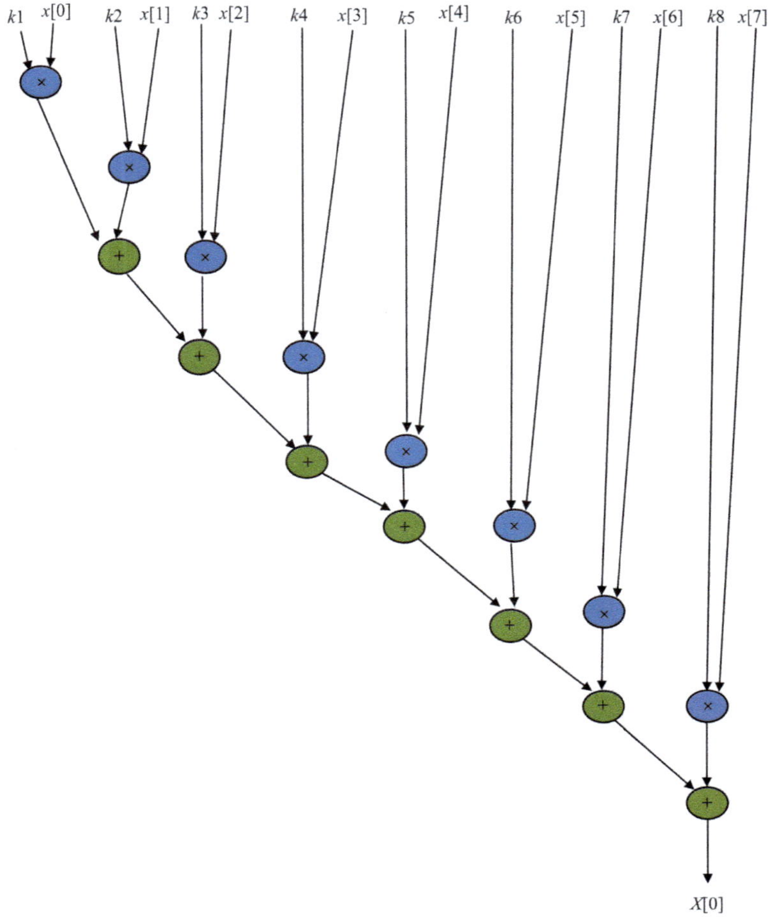

Figure 8.45 DFG of 8-point DCT

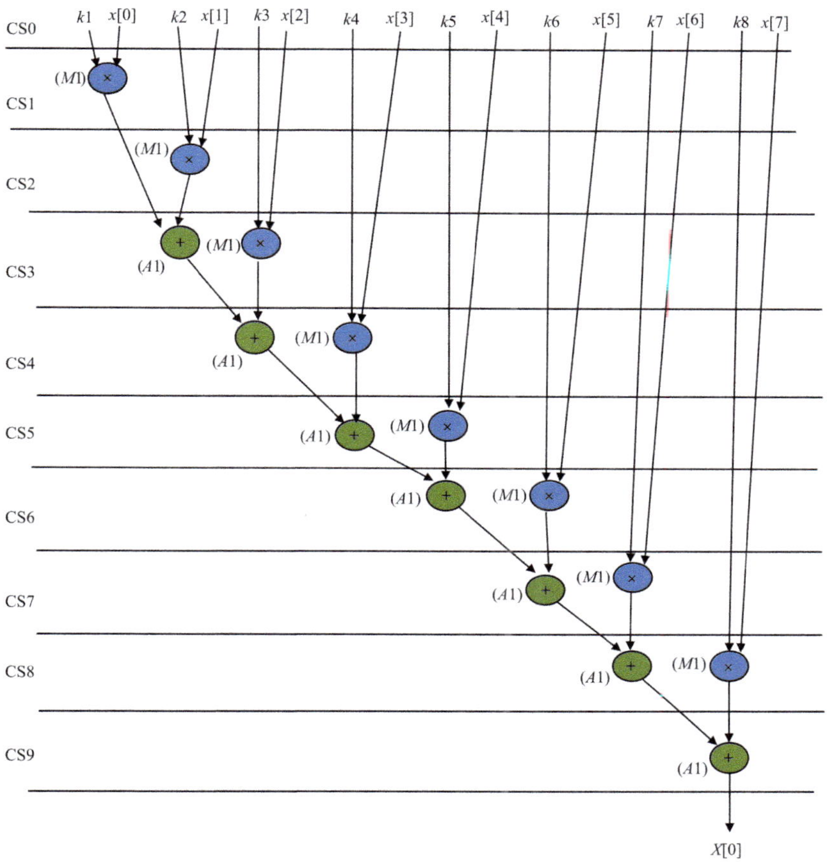

Figure 8.46 Scheduled DFG of 8-point DCT based on resource constraints (1A, 1M)

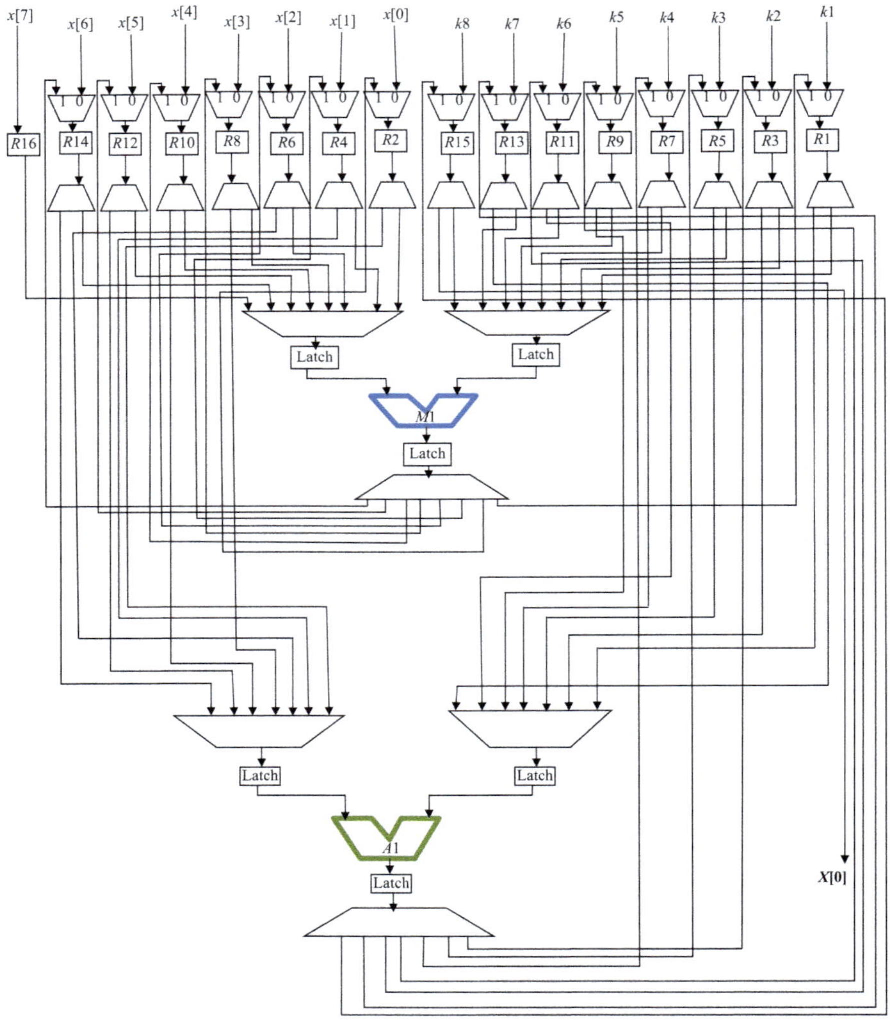

Figure 8.47 RTL datapath of 8-point DCT

8.8.6 Example 6: 8-point IDCT

The functions of 8-point IDCT are

$$
\begin{aligned}
x'[0] &= c4 \times X'[0] + c1 \times X'[1] + c2 \times X'[2] + c3 \times X'[3] \\
&\quad + c4 \times X'[4] + c5 \times X'[5] + c6 \times X'[6] + c7 \times X'[7] \\
x'[1] &= c4 \times X'[0] + c3 \times X'[1] + c6 \times X'[2] + (-c7) \times X'[3] \\
&\quad + (-c4) \times X'[4] + (-c1) \times X'[5] + (-c2) \times X'[6] + (-c5) \times X'[7] \\
x'[2] &= c4 \times X'[0] + c5 \times X'[1] + (-c6) \times X'[2] + (-c1) \times X'[3] \\
&\quad + (-c4) \times X'[4] + c7 \times X'[5] + c2 \times X'[6] + c3 \times X'[7] \\
x'[3] &= c4 \times X'[0] + c7 \times X'[1] + (-c2) \times X'[2] + (-c5) \times X'[3] \\
&\quad + c4 \times X'[4] + c3 \times X'[5] + (-c6) \times X'[6] + (-c1) \times X'[7] \\
x'[4] &= c4 \times X'[0] + (-c7) \times X'[1] + (-c2) \times X'[2] + c5 \times X'[3] \\
&\quad + c4 \times X'[4] + (-c3) \times X'[5] + (-c6) \times X'[6] + c1 \times X'[7] \\
x'[5] &= c4 \times X'[0] + (-c5) \times X'[1] + (-c6) \times X'[2] + c1 \times X'[3] \\
&\quad + (-c4) \times X'[4] + (-c7) \times X'[5] + c2 \times X'[6] + (-c3) \times X'[7] \\
x'[6] &= c4 \times X'[0] + (-c3) \times X'[1] + (-c6) \times X'[2] + c7 \times X'[3] \\
&\quad + (-c4) \times X'[4] + c1 \times X'[5] + (-c2) \times X'[6] + c5 \times X'[7] \\
x'[7] &= c4 \times X'[0] + (-c1) \times X'[1] + c2 \times X'[2] + (-c3) \times X'[3] \\
&\quad + c4 \times X'[4] + (-c5) \times X'[5] + c6 \times X'[6] + (-c7) \times X'[7]
\end{aligned}
$$

The generic representation of 8-point IDCT to compute the first sample is

$$
\begin{aligned}
x'[0] = {} & k'1 \times X'[0] + k'2 \times X'[1] + k'3 \times X'[2] + k'4 \times X'[3] \\
& + k'5 \times X'[4] + k'6 \times X'[5] + k'7 \times X'[6] + k'8 \times X'[7]
\end{aligned}
$$

The DFG of the generic representation of 8-point IDCT to compute the first sample is shown in Figure 8.48.

The scheduled DFG of 8-point IDCT based on resource constraints ($1A$, $1M$) is shown in Figure 8.49.

The respective data path is shown in Figure 8.50.

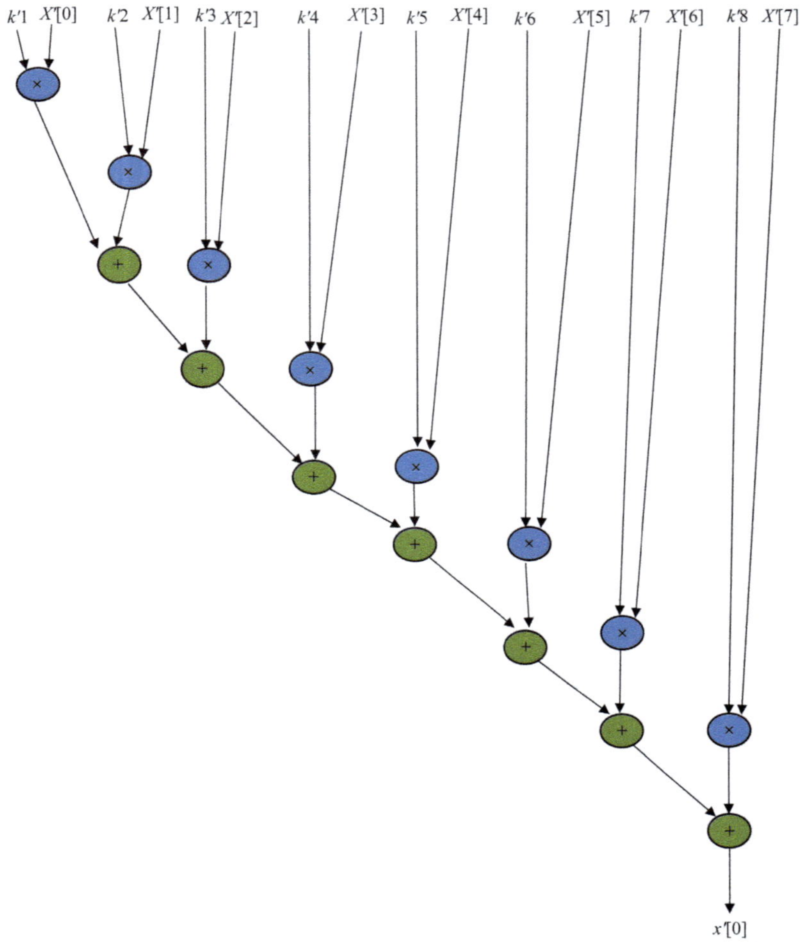

Figure 8.48 DFG of 8-point IDCT

8.8.7 Example 7: FIR filter

The equation of FIR filter is

$$Y'[n] = h0 \times X'[n] + h1 \times X'[n-1] + h2 \times X'[n-2] + h3 \times X'[n-3]$$

The DFG of FIR filter is shown in Figure 8.51.

The scheduled DFG of FIR based on resource constraints (1*A*, 1*M*) is shown in Figure 8.52.

The respective data path is shown in Figure 8.53.

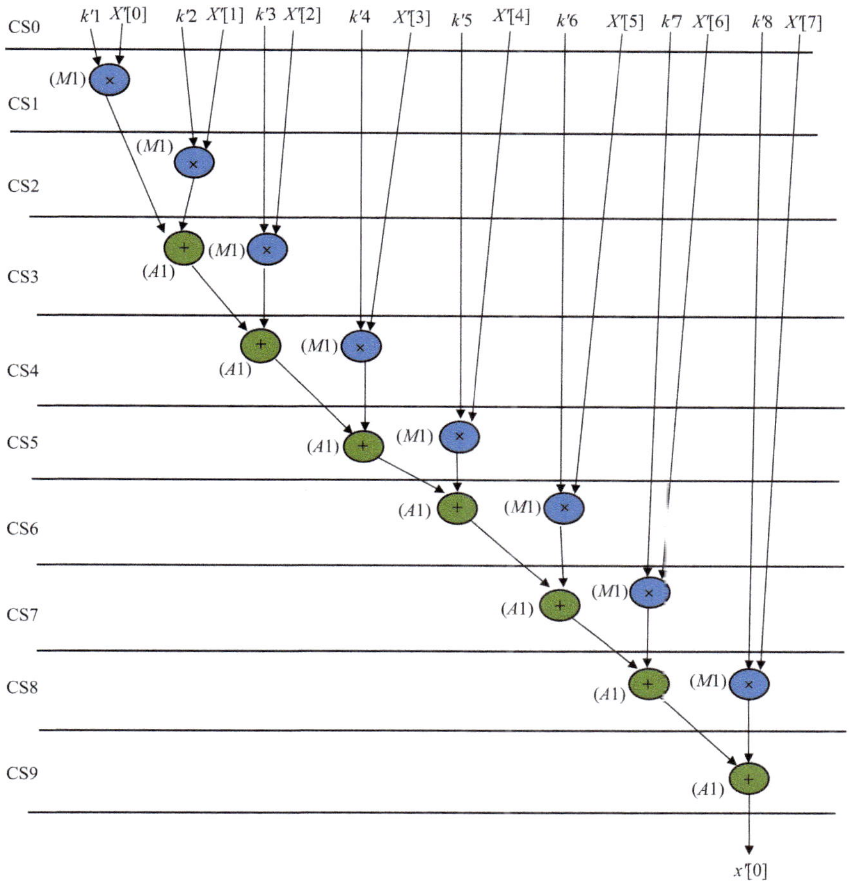

Figure 8.49 Scheduled DFG of 8-point IDCT based on resource constraints (1A, 1M)

8.8.8 Example 8: IIR filter

The equation of IIR filter is

$$\begin{aligned} Y[n] = {} & b0 \times X[n] + b1 \times X[n-1] + b2 \times X[n-2] \\ & + b3 \times X[n-3] + (-a1) \times Y[n-1] \\ & + (-a2) \times Y[n-2] + (-a3) \times Y[n-3] \end{aligned}$$

The DFG of IIR filter is shown in Figure 8.54.

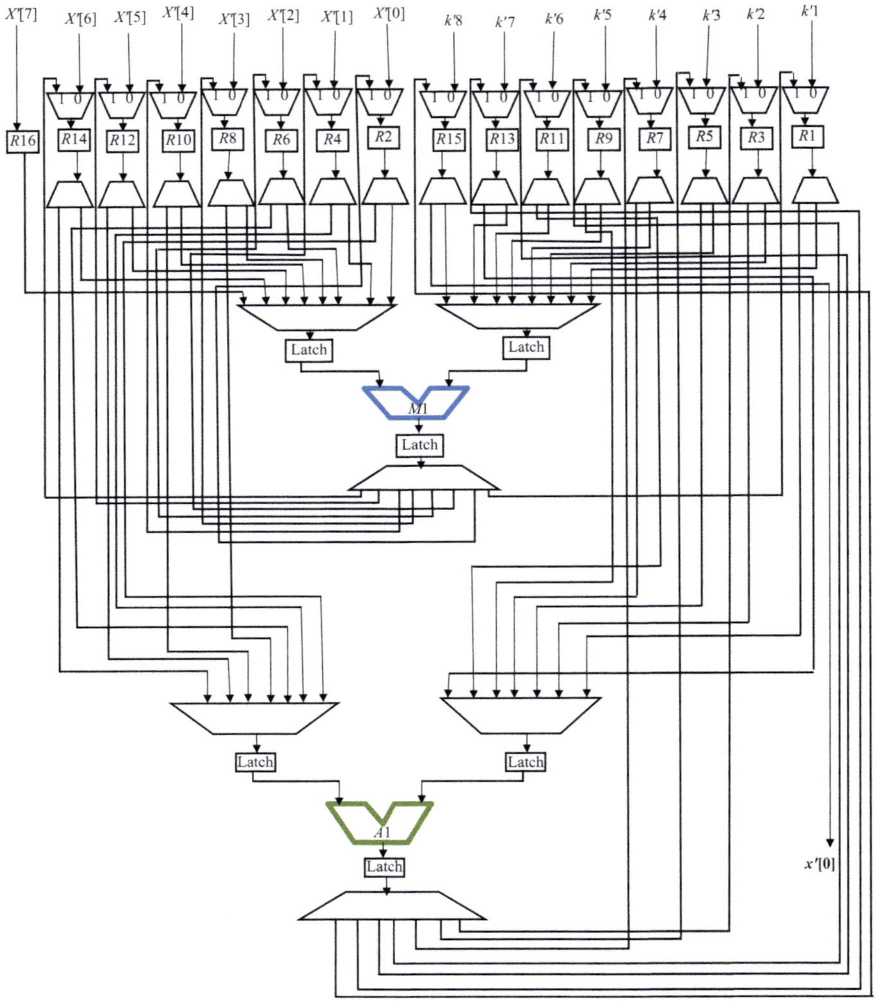

Figure 8.50 RTL datapath of 8-point IDCT

The scheduled DFG of IIR based on resource constraints (1*A*, 1*M*) is shown in Figure 8.55.

$$IIR: Y[n] = b0 \times X[n] + b1 \times X[n-1] + b2 \times X[n-2] + b3 \times X[n-3] + (-a1) \times Y[n-1] + (-a2) \times Y[n-2] + (-a3) \times Y[n-3].$$

The respective data path is shown in Figure 8.56.

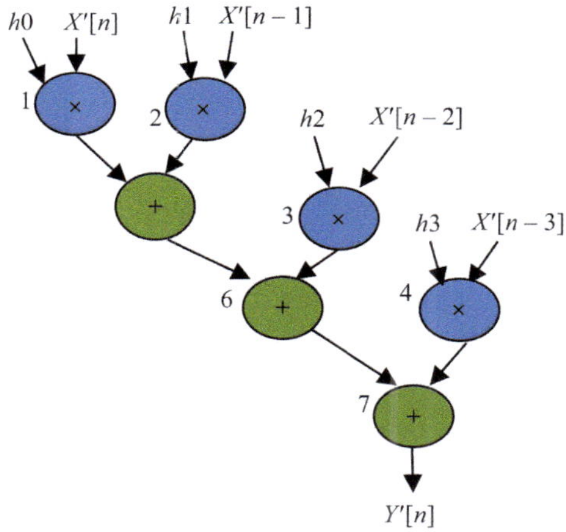

Figure 8.51 DFG of FIR filter

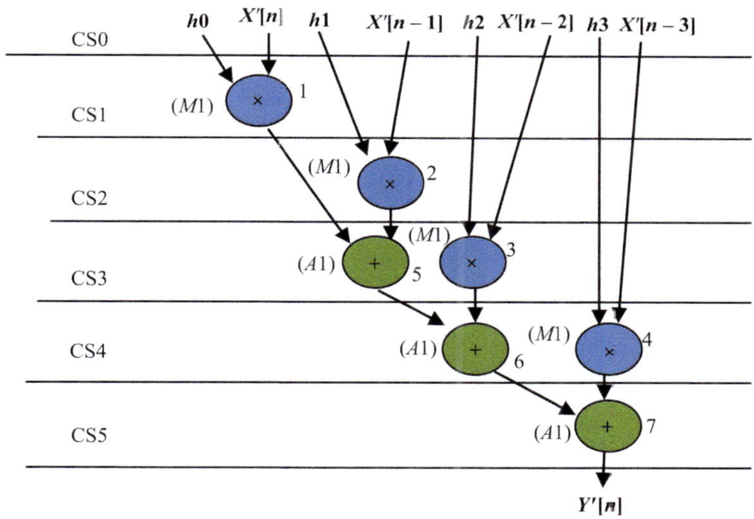

Figure 8.52 Scheduled DFG of FIR filter based on resource constraints (1A, 1M)

8.9 Analyses of case studies

The following five hologram-obfuscated designs produced have been analysed in terms of the security and estimated area/delay overheads (Sengupta and Rathor, 2019):

1. 8-point DCT–4-point IDCT hologram-based obfuscated design
2. IIR–FIR filter hologram-based obfuscated design

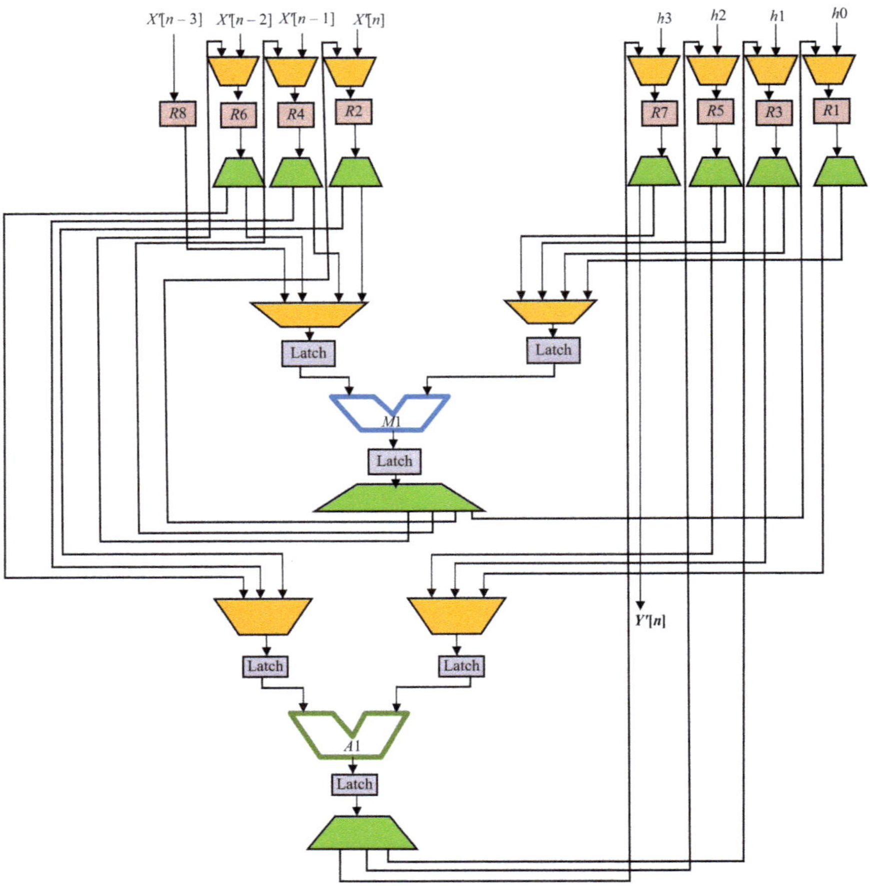

Figure 8.53 RTL datapath of FIR filter

3. 8-point DIT–FFT–4-point DIT–FFT hologram-based obfuscated design
4. 4-point DCT–4-point IDCT hologram-based obfuscated design
5. 8-point DCT–8-point IDCT hologram-based obfuscated design

To assess the effectiveness of the hologram-based structural obfuscation approach (Sengupta and Rathor, 2019) it has been compared with another structural-obfuscation approach (for DSP cores) based on HLTs (Sengupta *et al.*, 2017a). The same DSP applications (FIR, IIR, 4-point DCT, 4-point IDCT, 8-point DCT, 8-point IDCT, 4-point DIT-FFT and 8-point DIT-FFT) have also been obfuscated based on the HLT obfuscation approach (Sengupta *et al.*, 2017a) in order to compare the approaches statistically. The strength of obfuscation

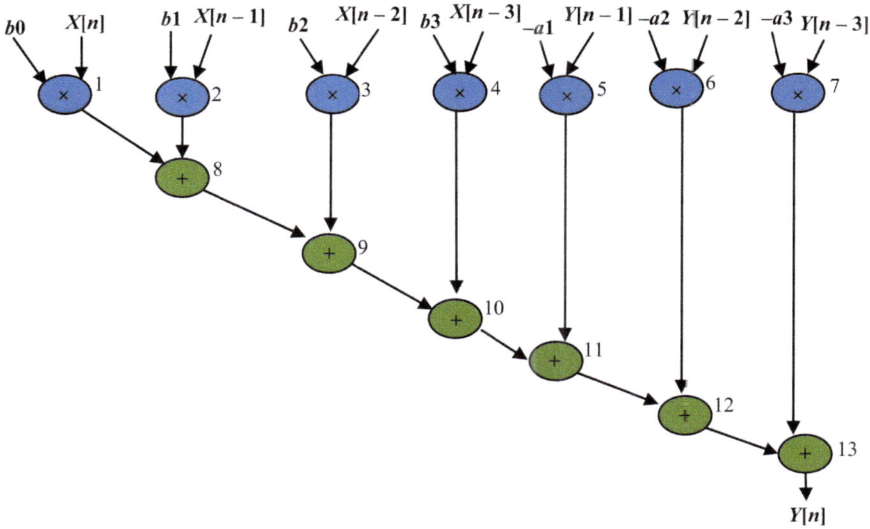

Figure 8.54 DFG of IIR filter

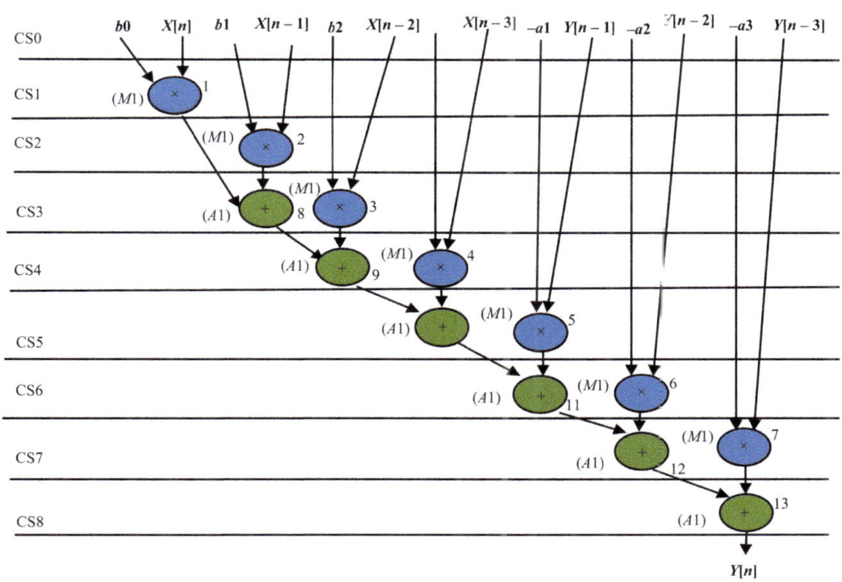

Figure 8.55 Scheduled DFG of IIR filter based on resource constraints (1A, 1M)

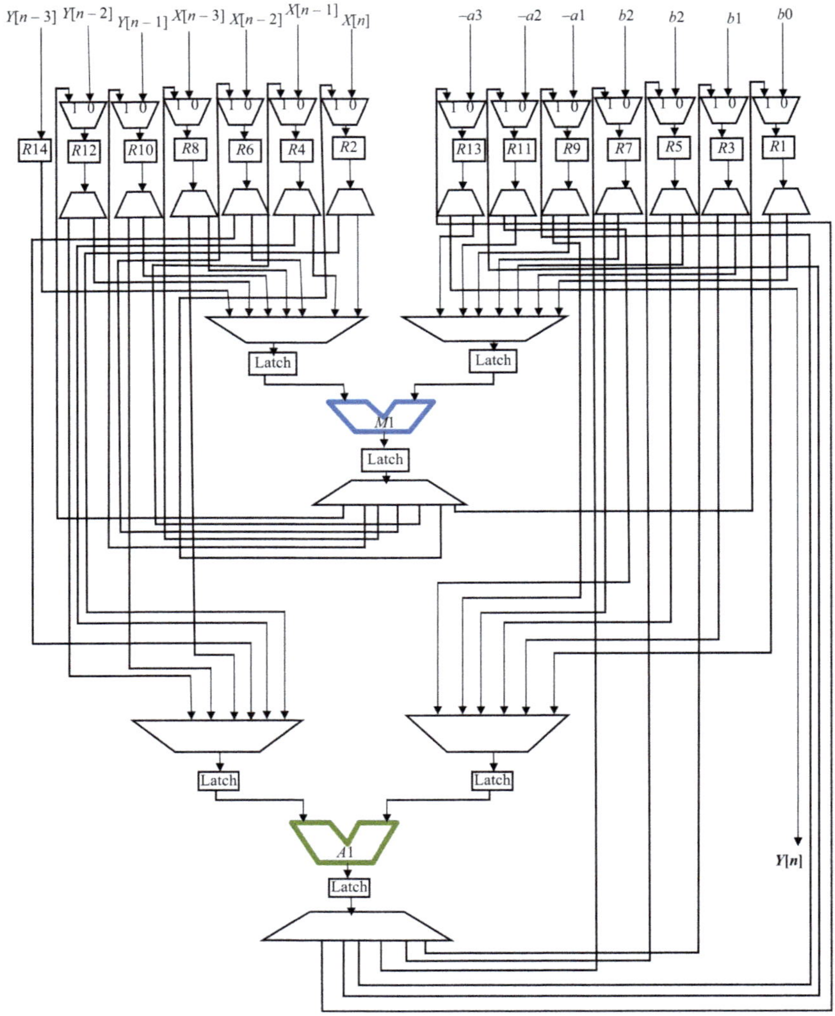

Figure 8.56 RTL datapath of IIR filter

(security), design delay and area analysis of hologram-based structural obfuscation approach from Sengupta and Rathor (2019) have been discussed in the following sub-sections.

8.9.1 Strength of obfuscation analysis

The strength of obfuscation of an obfuscated design is defined as

$$S_{ob} = \frac{G_c^a}{G_c^T} \tag{8.6}$$

Table 8.1 The study of the effect of hologram obfuscation on gate count (Sengupta and Rathor, 2019)

DSP applications	**Gate count of baseline (un-obfuscated design)**	**Gate count of hologram obfuscated design**	**Gate count reduction (G_c^a) (through Sengupta and Rathor (2019))**	**Affected gate count in % using Sengupta and Rathor (2019)**
8-pt DCT + 4-pt DCT	63,936	52,352	11,584	18.1
IIR + FIR	9,248	6,544	2,704	29.2
8-pt DIT–FFT + 4-pt DIT–FFT	14,016	9,856	4,160	29.7
4-pt DCT + 4-pt IDCT	26,240	14,400	11,840	45.1
8-pt DCT + 8-pt IDCT	101,632	55,424	46,208	45.5

where

S_{ob} = strength of obfuscation.
G_c^a = number of gates affected due to obfuscation.
G_c^T = total number of gates in a baseline (un-obfuscated) design.

Here the strength of obfuscation indicates the security ability of an obfuscated design against RE attacks. According to (8.6), the security strength becomes higher on increasing the value of number of gates affected due to obfuscation (G_c^a). The number of gates affected due to hologram-based obfuscation has been presented in Table 8.1. This table also reflects the total gate count of both the baseline design $\left(G_c^T\right)$ and hologram-obfuscated design along with the gate percentage affected due to hologram-based obfuscation using Sengupta and Rathor (2019). Similarly, Table 8.2 presents the gate count of HLT-based obfuscated design and the number of gates affected due to HLT-based obfuscation using Sengupta *et al.* (2017a). The strength of obfuscation comparison between hologram-based obfuscation and HLT-based obfuscation has been shown in Table 8.3. It is evident from the table that the hologram-based obfuscation approach is capable to affect a larger number of gates compared to the HLT-based obfuscation approach. Thus it is evident that the hologram-based obfuscation approach achieves higher strength of obfuscation than the HLT-based obfuscation approach (Sengupta *et al.*, 2017a). Additionally, the HLT-based obfuscation approach is not capable to obscure the DIT–FFT DSP application. This is because the HLTs (Sengupta *et al.*, 2017a) are not applicable to this application, thus rendering no effect on the gates of the design. However, the hologram-based obfuscation approach is capable of affecting the gate count significantly as shown in Table 8.3. Hence for such DSP applications, the hologram-based obfuscation approach additionally plays a crucial role in providing security through structural obfuscation. (*Note: the HLT obfuscated*

Table 8.2 The study of the effect of HLT obfuscation on gate count (Sengupta et al., 2017a)

DSP applications	Gate count of baseline (un-obfuscated)	Gate count of HLT obfuscation (Sengupta *et al.*, 2017a)	Number of gates modified due to change in input connectivity	Number of gates affected (using (8.5))
8-pt DCT	50,816	50,816	1,792	1,792
4-pt DCT	13,120	13,120	256	256
IIR	5,968	5,968	160	160
FIR	3,280	3,280	64	64
8-pt DIT–FFT	9,344	9,344	0	0
4-pt DIT–FFT	4,672	4,672	0	0
8-pt IDCT	50,816	50,816	1,792	1,792
4-pt IDCT	13,120	13,120	256	256

Table 8.3 Comparative study of the strength of obfuscation between the hologram obfuscation approach; Sengupta and Rathor (2019) and Sengupta et al. (2017a)

DSP applications	Number of gates affected G_c^a		Affected gate count in % (strength of obfuscation)	
	Sengupta and Rathor (2019)	Sengupta *et al.* (2017a)	Sengupta and Rathor (2019)	Sengupta *et al.* (2017a)
8-pt DCT + 4-pt DCT	11,584	2,048	18.1	3.2
IIR + FIR	2,704	224	29.2	2.4
8-pt DIT–FFT + 4-pt DIT–FFT	4,160	0	29.7	0
4-pt DCT + 4-pt IDCT	11,840	512	45.1	2.0
8-pt DCT + 8-pt IDCT	46,208	3,584	45.5	3.5

designs obtained using Sengupta et al. (2017a) for different DSP applications are provided in Section 8.7.)

8.9.2 Design area analysis

The hologram-based obfuscation approach achieves huge area savings as it obfuscates two DSP designs simultaneously. The reduction in gate count of a (DSP1 + DSP2) obfuscated design with respect to total gate count of two baseline (DSP1 and DSP2) designs is shown in Table 8.1 (Sengupta and Rathor, 2019). However, HLT-based obfuscation (Sengupta *et al.*, 2017a) does not affect the total number of gates. More explicitly, the gate count of a baseline design does not change (increase or decrease) after performing HLT-based obfuscation, as shown in Table 8.2 (the underlying reason is discussed in Section 8.8). Hence, the hologram-based obfuscation approach is more area-efficient than the HLT-based obfuscation approach.

8.9.3 Design delay analysis

The additional switching muxes of a hologram-obfuscated design are accountable for extra propagation delays (Sengupta and Rathor, 2019). For some hologram-obfuscated DSP designs (e.g., IIR-FIR filter and 8-point DCT–4-point DCT) the switching muxes of size 2:1 are used at the input and output of the designs. Since all the input muxes are executed in parallel, delay overhead is estimated as a sum of two 2:1 muxes (input mux + output mux). For some hologram-obfuscated DSP designs (e.g., 4-point DCT–IDCT, 8-point DIT–FFT–4-point DIT–FFT and 8-point DCT–IDCT) the switching muxes of size 2:1 are used only at the input of the designs. Therefore only the propagation delay of a 2:1 mux incurs delay overhead for such hologram-obfuscated DSP designs (Sengupta and Rathor, 2019). On the contrary, the delay overhead due to HLT-based obfuscation (Sengupta *et al.*, 2017a) depends on the type of DSP application to be obfuscated, resource constraints applied and types of HLTs applicable for the design. Therefore the delay overhead comparison of both approaches is not straightforward. However, for DSP applications tested with hologram-based obfuscation, the delay overhead of a hologram-obfuscated design compared to a HLT-based obfuscation (with resource constraints of 1*A* and 1*M*) is either zero or nominal. In the case of 4-point DCT–IDCT and 8-point DCT–IDCT, the delay overhead of hologram-based obfuscation compared to HLT-based obfuscation is zero, whereas in the case of an IIR–FIR filter and 8-point DCT–4-point DCT, the delay overhead is nominal.

8.10 Conclusion

In this chapter we discussed a structural-obfuscation approach based on security hologram feature (Sengupta and Rathor, 2019). Like a hologram, two (or more) DSP applications can be camouflaged into a single design in which each application manifests or evolves on application of the correct switching-element value. This integration of two applications into a single design leads to significant structural modification in the architecture of both applications and, thus, it is capable to obfuscate both designs simultaneously. As the hologram-based structural obfuscation approach obfuscates two designs simultaneously by camouflaging them into a single RTL data path, huge area savings are possible.

This chapter provides a comprehensive perspective on the hologram-based obfuscation of DSP circuits. At the end of this chapter, the reader is able to understand the following concepts:

- Necessity of protecting IP cores against piracy and RE threats.
- Hologram-based structural obfuscation.
- Security features of hologram-based obfuscation.

- Rules of applying hologram-based obfuscation for DSP applications.
- Design of a hologram-obfuscated architecture.
- Resemblance of a hologram-obfuscated design with a hologram.
- Process of gate count determination of baseline and obfuscated design architectures.
- Comparative analyses of several structural-obfuscation approaches.

8.11 Questions and exercises

1. What is a hologram?
2. How can an security hologram concept be applied to structural obfuscation?
3. What is a switching element in a hologram?
4. What is rule-1 used in hologram-based structural obfuscation?
5. What is rule-2 used in hologram-based structural obfuscation?
6. What is rule-3 used in hologram-based structural obfuscation?
7. State the transfer function of an FIR filter.
8. State the transfer function of an IIR filter.
9. State the transfer function of DCT.
10. State the transfer function of IDCT.
11. Schedule DCT using (2*A*, 2*M*) resource constraint.
12. In the hologram obfuscated IIR–FIR filter design, what is the switching element?
13. How is gate count determined in a baseline DCT data path?
14. How is the affected gate count of a hologram-obfuscated design evaluated?
15. How is the strength of obfuscation evaluated in a hologram-obfuscated design?
16. State the area/delay trade-off involved during the design of a hologram-obfuscated design.
17. In which cases are ROE and LT not applicable for structural obfuscation?
18. How is THT-based obfuscation performed?
19. What is the outcome of performing THT obfuscation?
20. Compare the advantages of hologram-based obfuscation over HLT-based obfuscation.

References

Y. Lao and K. K. Parhi, "Obfuscating DSP circuits via high-level transformations," IEEE Transactions on Very Large Scale Integration (VLSI) Systems, vol. 23, no. 5, Pages: 819–830, 2015.

M. C. McFarland, A. C. Parker, and R. Camposano, "Tutorial on high-level synthesis," in DAC '88 Proceedings of the 25th ACM/IEEE Design Automation, vol. 27, no. 1, Pages: 330–336, 1988.

R. Schneiderman, "DSPs evolving in consumer electronics applications," IEEE Signal Processing Magazine, vol. 27, no. 3, Pages: 6–10, 2010.

A. Sengupta, "Intellectual property cores: Protection designs for CE products," IEEE Consumer Electronics Magazine, vol. 5, no. 1, Pages: 83–88, 2016.

A. Sengupta, "Hardware security of CE devices [hardware matters]," IEEE Consumer Electronics Magazine, vol. 6, no. 1, Pages: 130–133, 2017.

A. Sengupta and S. P. Mohanty, "Trojan security aware DSP IP core and integrated circuits," IP Core Protection and Hardware-Assisted Security for Consumer Electronics, Pages: 79–122, 2019, e-ISBN: 9781785618000.

A. Sengupta and M. Rathor, "Protecting DSP kernels using robust hologram-based obfuscation," IEEE Transactions on Consumer Electronics, vol. 65, no. 1, Pages: 99–108, 2019.

A. Sengupta and D. Roy, "Protecting an intellectual property core during architectural synthesis using high-level transformation based obfuscation," IET Electronics Letters, vol. 53, no. 13, Pages: 849–851, 2017.

A. Sengupta, R. Sedaghat, and Z. Zeng, "A high level synthesis design flow with a novel approach for efficient design space exploration in case of multiparametric optimization objective," Microelectronics Reliability, vol. 50, no. 3, Pages: 424–437, 2010.

A. Sengupta, D. Roy, S. P. Mohanty, and P. Corcoran, "DSP design protection in CE through algorithmic transformation based structural obfuscation," IEEE Transactions on Consumer Electronics, vol. 63, no. 4, Pages: 467–476, 2017a.

A. Sengupta, S. Bhadauria, and S. P. Mohanty, "Low-cost security aware HLS methodology," IET Computers & Digital Techniques, vol. 11, no. 2, Pages: 68–79, 2017b.

A. Sengupta, S. Bhadauria, and S. P. Mohanty, "TL-HLS: Methodology for low cost hardware Trojan security aware scheduling with optimal loop unrolling factor during high level synthesis," IEEE Transactions on CAD of Integrated Circuits and Systems, vol. 36, no. 4, Pages: 655–668, 2017c.

R. Torrance and D. James, "The state-of-the-art in IC reverse engineering," Canadian Healthcare Engineering Society, vol. 5747, Pages: 363–381, 2009.

Matrix Technologies, Hologram Features, 2016. [Online]. Available: http://www.matrixtechnologies.in/hologram-features.html, last accessed on May 2019.

Chapter 9
Security of functionally obfuscated DSP cores

Anirban Sengupta[1] *and Mahendra Rathor*[1]

This chapter discusses the security of functionally obfuscated digital signal processing (DSP) cores. Some of the most prominent aspects of this chapter include descriptions of the attack models, points of vulnerability for launching reverse engineering (RE), security properties of the functional-obfuscation process, security against removal attack in the context of DSP cores, hardware-design methods of various state-of-the-art security solutions and a comparative perspective of different approaches used in the context of DSP cores.

The chapter is organized as follows: Section 9.1 is the introduction to the chapter; Section 9.2 discusses attack scenarios and threat models of functional obfuscation; Section 9.3 describes the functional obfuscation of DSP cores; Section 9.4 discusses the selected approaches for security of functionally obfuscated DSP cores against removal attacks; Section 9.5 presents analyses of case studies using selected approaches; Section 9.6 concludes the chapter.

9.1 Introduction

As we know, consumer electronics systems encompass a wide range of smart devices such as multi-spectral cameras, smart meters, smart tablets, etc. These smart devices comprise reusable intellectual property (IP) cores either as a standalone application-specific integrated circuits or included as a component on a system on chip platform (Schneiderman, 2010; Sengupta, 2017). Regardless of the form of realization, IP cores are an integral part of data-intensive computation engines. IP, due to its intellectual value representing several years of research and development, design and validation needs to be secured. It is also widely acknowledged that these sophisticated IPs can be infringed/abused by attackers through different forms of threat models including RE (Sengupta, 2016, 2017; Torrance and James, 2009). Further, once designed, an IP core once designed (till netlist) can be easily used in different process technologies to convert it into a hard IP format. This also entices an attacker to steal the IP netlist and use it without

[1]Computer Science and Engineering, Indian Institute of Technology Indore, Indore, India

authorization (Castillo *et al.*, 2007; Plaza and Markov, 2015; Roy *et al.*, 2008; Sengupta *et al.*, 2017a; Lao and Parhi, 2015). Thus there needs to be active form of protection to secure IP cores against standard threats. The most common form of active protection is called functional obfuscation (Yasin *et al.*, 2016; Zhang, 2015; Sengupta *et al.*, 2019). Section 9.2 discusses, in detail, the different attack scenarios and their countermeasures using functional obfuscation. This chapter also highlights the functional-obfuscation process for DSP cores.

9.2 Attack scenarios and threat models of functional obfuscation

An IP core realized as an integrated circuit (IC) is susceptible to various hardware threats during different stages of its design flow. The hardware threats for an IP core include (1) piracy, (2) counterfeiting or cloning, (3) hardware Trojan insertion and (4) overbuilding. The hardware threats arise because an adversary in an untrustworthy design house may launch a RE attack on design representations through (1) register transfer level netlist, (2) GDS-II (layout/physical design), (3) masks (during fabrication), (4) non-activated IC (during packaging) and (5) activated IC (obtained from the open market). In an untrustworthy regime an IP core can be attacked or misused (Yasin *et al.*, 2016; Chakraborty and Bhunia, 2009; Sengupta *et al.*, 2017b, 2018). Two typical attack scenarios of an IP core are depicted in Figure 9.1.

In the current era of hardware security IP protection, functional obfuscation has emerged as a capable security mechanism that can be used to protect an IP core against RE threats. Functional obfuscation is a mechanism of obscuring the functionality of a design by locking it logically through some key gates (Sengupta and

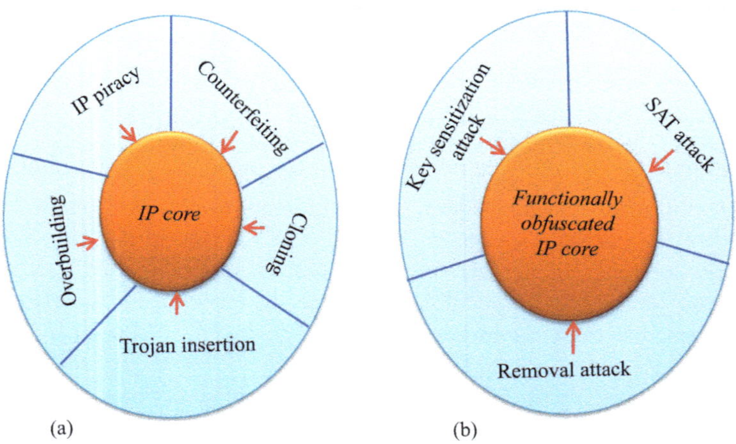

Figure 9.1 Attack scenarios for (a) an un-obfuscated and (b) a functionally obfuscated IP core

Mohanty, 2018; Sengupta *et al.*, 2019). As the secret key is not known to the attacker in advance, he/she is unable to use IP cores illegally through RE. More explicitly, an attacker cannot pirate, counterfeit, clone or overbuild an IP core, thus being incapable of inserting hardware Trojans.

However, employing functional obfuscation invites a different class of attacks. Since the idea of functional obfuscation is based on locking IP cores using key gates, key-based attacks therefore become automatically applicable. Additionally, an attacker may also attempt to remove or deactivate the key gates. Figure 9.1(b) depicts the attack scenario for a functionally obfuscated IP core. The vulnerability of functional obfuscation towards different attacks is discussed as follows (Sengupta *et al.*, 2019):

1. *Key sensitization attack:* An attacker can extract the keys of a locked (functionally obfuscated) netlist through a key sensitization attack. To mount this attack, the attacker also requires a functional IC (which can be obtained from open market) along with the obfuscated netlist (which may be obtained through RE). Once the attacker has the obfuscated netlist and a functional IC, he/she can sensitize the key through the following possibilities (Yasin *et al.*, 2016):
    (i) The attacker can sensitize a key-bit at primary output by controlling the primary inputs of the design. To do so, the attacker needs to identify the input pattern that can sensitize the correct key-bits to the primary output.
    (ii) The attacker tries to find isolated key-gates in the obfuscated netlist. This is because the key-bits are easy to sensitize through isolated key-gates. If a key-gate is not connected to other key-gates (through any path) in the obfuscated design, then it is termed as an isolated key-gate.
    (iii) If the attacker finds a run (or sequence) of key-gates in the obfuscated design, then he/she can substitute it by a single gate. This leads to reduction in key-bits. Thus the run of key-gates renders obtaining key-bits easier for the attacker.
    (iv) In the path of sensitization of a key-bit, the effect of other key-bits can be nullified by muting the relevant key-gates.
2. *Boolean Satisfiability (SAT) attack:* To mount this attack, an attacker needs a locked netlist and an activated or functional IC. The SAT attack algorithm first generates distinguished input–output (IO) pairs using a SAT formula. These distinguished IO pairs are exploited to eliminate wrong key combinations. A subset of wrong key combinations can be eliminated using each distinguished IO pair. The SAT attack algorithm generates the distinguished IO pair iteratively as long as the elimination of all the wrong keys is accomplished (Subramanyan *et al.*, 2015; Xie and Srivastava, 2019).
3. *Removal attack:* In this attack, an attacker is assumed to have access to the obfuscated netlist. The attacker attempts to eliminate the additionally inserted key-gates (IP core locking logic) by detecting them using sophisticated algorithms or tools (Yasin *et al.*, 2017).

9.3 Functional obfuscation of DSP cores

DSP cores used in contemporary and smart electronic systems are also vulnerable to IP piracy, cloning, counterfeiting, overbuilding and hardware Trojan insertion threats. The security of DSP cores against these threats can be provided using functional obfuscation. The functional obfuscation of DSP cores is performed by inserting a certain number of IP core-locking blocks (ILBs) at appropriate locations into the design (Sengupta *et al.*, 2019). The number of ILBs inserted is determined using a rule based on some parameters. The structure of an ILB is constituted using a combination of basic XOR, XNOR, AND, NAND, OR and NOT gates. A sample ILB structure is shown in Figure 9.2. As shown in the figure, the internal gates in an ILB are interconnected in such a fashion that higher security through functional obfuscation could be achieved. Further, these ILBs are structurally reconfigurable, i.e., the gate structures of the ILB architecture can be configured according to the values at the key-bits. In other words, the ILB gate structure used (in the obfuscated netlist) is never fixed to make it undetectable to an attacker. Different ILB gate structures are used based on the reconfiguration to confuse the attacker and thwart removal attacks. This is possible because various structures of ILBs can be generated using different combinations of same basic gates (i.e., XOR, XNOR, AND, NAND, OR, NOT gates). Yet, each ILB structure is capable of providing a similar security strength (Sengupta *et al.*, 2019).

The various security features of ILBs used in functionally obfuscated DSP cores are discussed as follows (Sengupta *et al.*, 2019):

1. *Multi-pair wise security:* The key-bits of an ILB are multi-pair wise secured because a single key-bit cannot be sensitized to output without controlling (knowing) the other key-bits of the same ILB. This is because the path of sensitization of a key-bit is interfered by the other multiple key-bits. Hence the ILBs are capable to secure a functionally obfuscated DSP core against the key sensitization–based attack.
2. *Prohibiting key-gates isolation:* As shown in the ILB structure (Figure 9.2), a path essentially exists between any two key-gates. Thus the isolated key-gates are not present in the ILB structure. Hence, obtaining key-bits through isolated key-gates becomes infeasible.

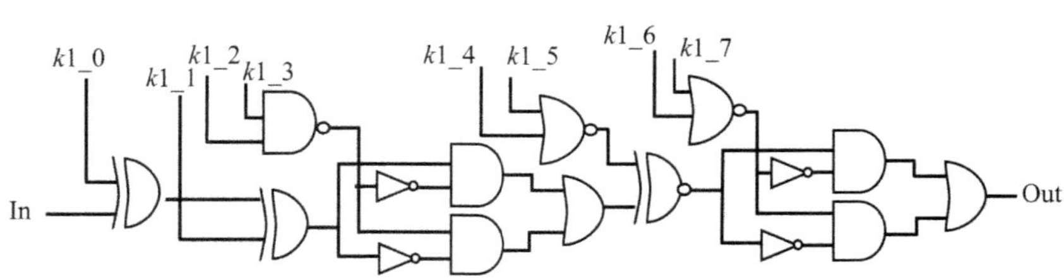

Figure 9.2 A sample ILB structure (requires an 8-bit key k1 to activate)

3. *Resilient to muting of key-gates:* The key-gates in the ILB structure are intertwined in such a fashion that the effect of any of key-gate cannot be muted. Hence the attacker's attempt of sensitizing key-bits to output by muting the effect of remaining key-gates is thwarted.
4. *Resilient to run of key-gates:* As shown in the ILB structure (Figure 9.2), there is no such sequence of key-gates that can be substituted by a single key-gate. Hence the attacker's attempt to sensitize a key-bit by increasing the valid key-space is thwarted.

The total number of ILBs (T_{ilb}) inserted into a DSP core is determined using the equation:

$$T_{\text{ilb}} = (\text{total number of functional units}) \times (\text{number of output bits of a functional unit}) \tag{9.1}$$

For example, if there are three functional units (FUs) in a design and the size of each FU is 32-bit, then the total output bits of FUs in the design is $3 \times 32 = 96$. ILBs are inserted at each output bit of all instances of each FU of the DSP core. Thus a total of 96 ILBs are inserted in the DSP core.

9.4 Selected approaches for securing functionally obfuscated DSP cores against removal attacks

The security against removal attack using three different approaches, viz. the custom AES-128–based approach (Sengupta *et al.*, 2019), custom SHA-512–based approach (Sengupta and Rathor, 2019) and custom anti-removal logic (ARL) unit–based approach (Rathor and Sengupta, 2019), are primarily discussed in this chapter. To learn more about the handling of other attack scenarios such as key sensitization and SAT attacks, Sengupta and Mohanty (2019) are suggested for reading.

9.4.1 Threat model

Although functional obfuscation is employed for DSP cores to secure them against RE threats, a functionally obfuscated DSP core is, however, also vulnerable to a different class of attacks such as key-sensitization, SAT and removal attacks (Sengupta *et al.*, 2019). More details on these attacks are explained in (Sengupta and Mohanty, 2019). Interested readers may refer to the relevant chapters for lucid elaboration. This chapter discusses only those techniques that can be used to secure a functionally obfuscated DSP core against removal attack. The vulnerability of a functionally obfuscated DSP core towards removal attack nullifies the purpose of functional obfuscation. In this attack, an attacker is assumed to have access to the obfuscated netlist. The attacker attempts to eliminate the ILBs (locking logic) by detecting those using automation algorithms or tools. The security techniques against removal attack are discussed in Section 9.4.2.

9.4.2 Discussion on selected approaches

This subsection discusses following three approaches that are used to secure a functionally obfuscated DSP core against removal attacks.

9.4.2.1 Security of functionally obfuscated DSP core against removal attacks using custom AES-128–based ILBs–key generation hardware (Sengupta *et al.*, 2019)

Features of the approach

- In this approach (Sengupta *et al.*, 2019), functional obfuscation of DSP cores is performed using ILBs (intertwined combinational logic of basic gates). To safeguard ILBs against removal attacks, no fixed architecture of ILBs have been used. Instead, the ILB structure is derived based on the output of custom AES-128 unit. This produces new ILB architecture depending on the 128-bit input and 128-bit fixed secret key. Since the attacker is unaware of any fixed ILB architecture, detection of ILB in the complex DSP netlist is thus highly challenging. Further, post-synthesis, the ILB gate structure changes and gets camouflaged (unrecognizable) in the form of basic gates in the entire design netlist. Additionally, since the authors have designed a custom AES hardware structure (not publicly available), detection of AES itself post-synthesis in the design netlist (comprising of DSP core and ILBs) is also highly challenging.
- The authors have selected AES-128 because it is a one-way, random, function that provides strong security where decoding the input from the output is not possible without the knowledge of the fixed secret key. Additionally, AES is highly secure as it comprises non-linear logic responsible for bit manipulation using S-box. It also comprises a diffusion layer, key addition layer and key scheduler that contribute to the robustness of the module. As mentioned, AES-128 uses 128-bit input and 128-bit secret key (a total of 256-bit input) to generate 128-bit encrypted keys of ILBs. This indicates that one instance of AES-128 can reconfigure 16 ILBs only. Therefore to reconfigure 64 ILBs, 512 bits of ILB keys should be known. Thus for an attacker to know the architecture of 64 reconfigured ILBs (in case of larger designs such as JPEG), 1,024 bits of input (4 times 128-bit input + 4 times 128-bit secret key) is required to decode the 512 bits of ILB keys.
- The security technique using custom AES-128 is a relatively heavyweight solution against removal attack. This is because in order to reconfigure up to 64 ILBs, four instances of custom AES-128 units are needed to be integrated with the functionally obfuscated DSP design. This incurs substantial area or power overheads.

Overview

As discussed earlier, this approach uses a custom AES-128 unit to generate encrypted keys of ILBs. Based on these encrypted keys, ILBs of a functionally obfuscated design are structurally reconfigured. Figure 9.3 shows the overview of the protection provided using a custom AES-128 unit against removal attacks (Sengupta *et al.*, 2019).

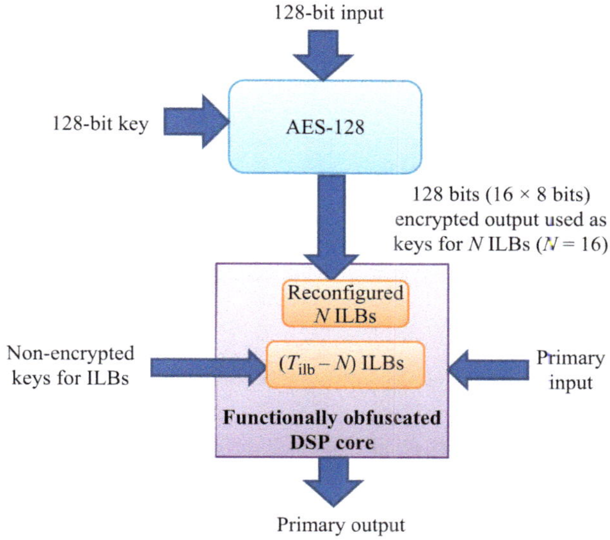

Figure 9.3 Overview of protection scenario of a functionally obfuscated DSP core against removal attacks using custom AES-128 (Sengupta et al., 2019)

Hardware design and integration with obfuscated DSP core

The AES-128 hardware is executed through four basic steps: (1) byte substitution, (2) shift row, (3) column mixing, (4) key mixing. The execution flow of these steps for round *i* of AES-128 is shown in Figure 9.4. The corresponding column diffusion matrix equations are also indicated in the figure. *Note: only the first four outputs (C0–C3) are shown.* The custom hardware design of the AES-128 unit is shown in Figure 9.5. As highlighted in the figure, the byte substitution step is performed using S-box. This step adds non-linearity in the algorithm and ensures that each output bit is a function of all input bits. Further, the shift row and column mixing steps add the diffusion property in the cipher. In the column diffusion step, the state values stored in the register *Bn* (*B*0–*B*15) are processed with multiplication by 2 in the Galois field (GF (2^8)). The hardware design of the 2 × *Bn* operation is shown in Figure 9.6. The handling of modular reduction portion and two possible outputs of *Bn* using the hardware design are highlighted in the figure. Furthermore, the key-mixing step is based on the key-expansion process. The hardware design used to expand the keys through the key scheduler and the *G*-function is shown in Figure 9.7. The top-level implementation of the custom hardware of AES-128 in a logic synthesis tool (Quartus II v13.0) is shown in Figure 9.8. As a solution to the removal attack on ILBs, custom AES-128 hardware is integrated with a functionally obfuscated DSP core by connecting the output of AES-128 to the input key-bits of ILBs. According to the value of key-bits obtained, the ILBs of functionally obfuscated DSP core are structurally reconfigured to thwart removal attacks. A top-level diagram of obfuscated FIR filter design integrated with custom AES-128 is

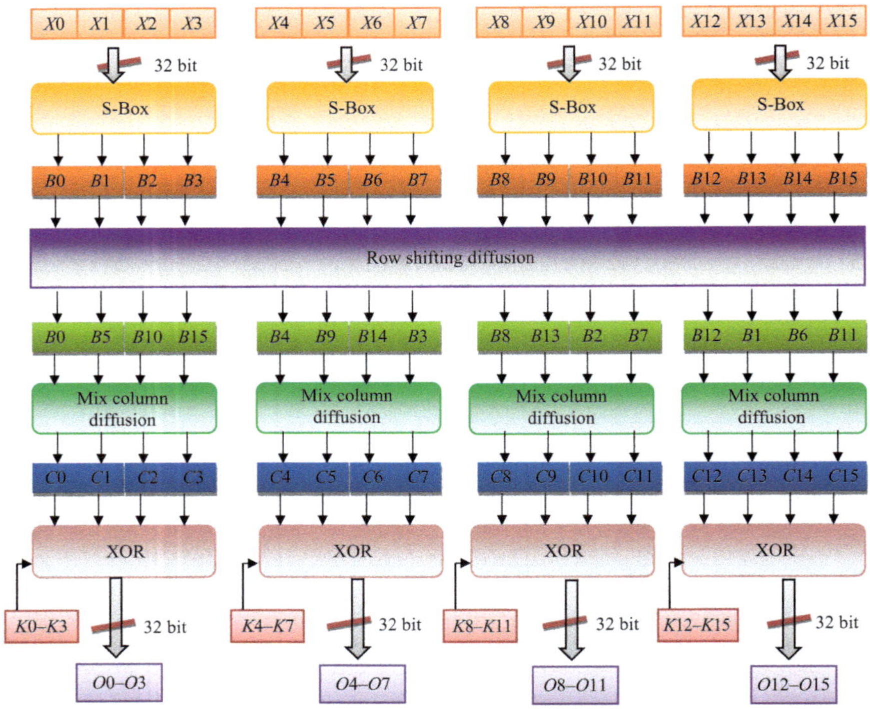

Based on the matrix multiplication performed in the mix-column diffusion layer, the following equations are obtained on which Galois Field (GF) arithmetic is performed.

$$C0 = (02 \cdot B0) + (03 \cdot B5) + (01 \cdot B10) + (01 \cdot B15)$$

$$C1 = (01 \cdot B0) + (02 \cdot B5) + (03 \cdot B10) + (01 \cdot B15)$$

$$C2 = (01 \cdot B0) + (01 \cdot B5) + (02 \cdot B10) + (03 \cdot B15)$$

$$C3 = (03 \cdot B0) + (01 \cdot B5) + (01 \cdot B10) + (02 \cdot B15)$$

Figure 9.4 Round 'i' (1 to 10) of AES-128

shown in Figure 9.9. The custom AES-128 integrated with the gate-level structure of FIR filter design is shown in Figure 9.10.

9.4.2.2 Security of functionally obfuscated DSP cores against removal attacks using custom SHA-512–based ILBs–key generation hardware (Sengupta and Rathor, 2019)

Features of the approach

- This approach (Sengupta and Rathor, 2019) secures the functionally obfuscated DSP cores against removal attack on ILBs. To secure ILBs, their

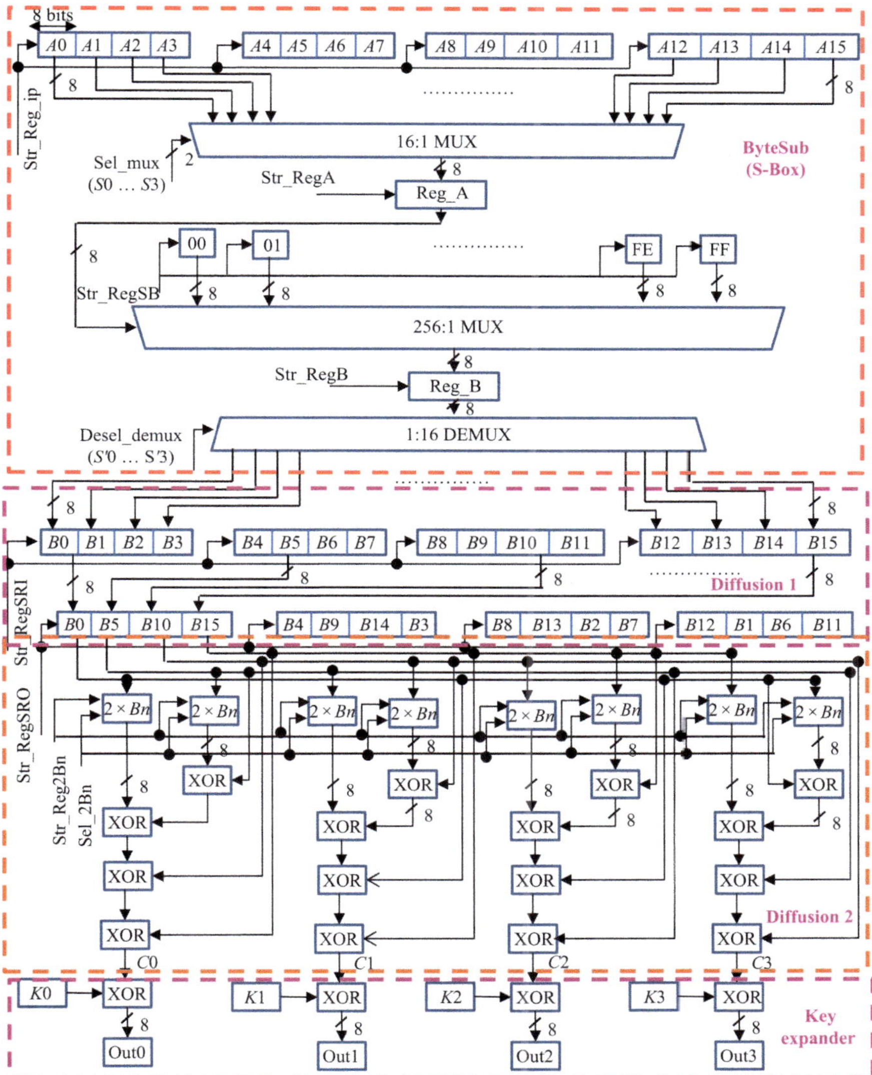

Figure 9.5 Custom hardware design of AES-128

structures are reconfigured according to the output of custom SHA-512–based ILB–key generation logic. Thus the detection of ILBs in the complex DSP netlist becomes difficult because its architecture is not known to the attacker in advance. Further, post-synthesis, the ILBs gate structure change and get camouflaged (unrecognizable) the form of basic gates in the entire design netlist. Thus a removal attack on ILBs becomes extremely difficult. Additionally, since

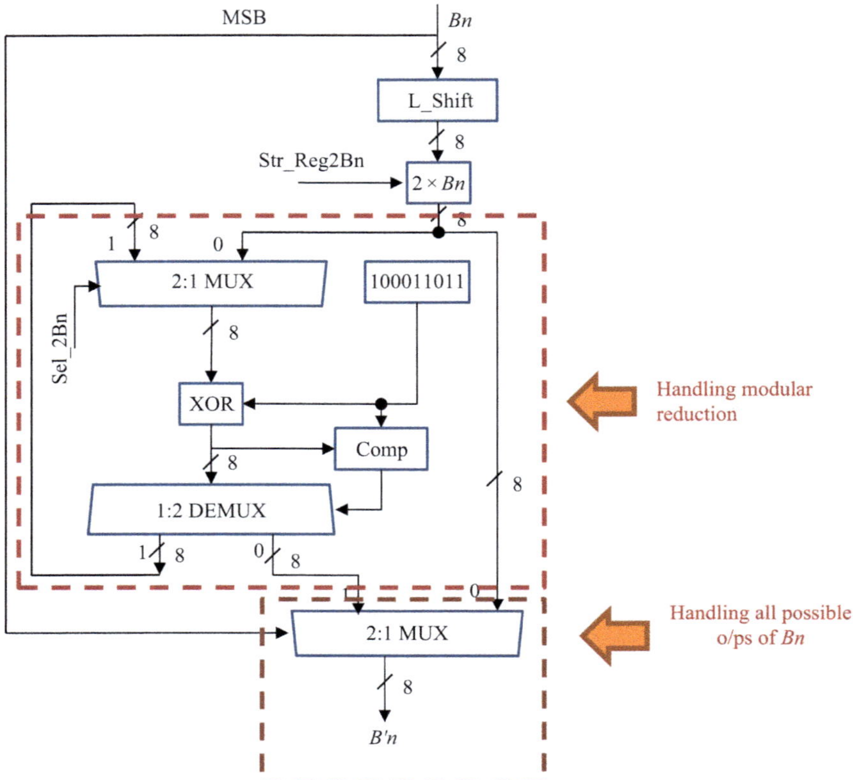

Figure 9.6 Hardware design of 2 × Bn operation

the authors have designed a custom SHA-512–based ILBs–key generation hardware (not publicly available), therefore, post-synthesis, its detection in the design netlist (comprising of DSP core and ILBs) is highly challenging (Sengupta and Rathor, 2019).

- The cryptohash function (SHA-512) provides some strong security features (Sengupta and Rathor, 2019) such as the generated hash digest always being unique corresponding to the initial information bits. It is very difficult to find two distinct inputs that produce the same hash value through the same hash function of SHA-512 (collision resistance feature). Furthermore, the generated hash digest is always deterministic, i.e., it always produces same hash value for a given input. Additionally, the probability of generating every hash value in the output is almost same (uniformity feature). Since the custom SHA-512–based ILBs–key generation hardware is capable to generate the 512-bit secure hash digest (used as ILB keys), hence up to 64 ILBs (each requiring an 8-bit key) can be structurally reconfigured according to their key-bits. To know the 512-bit hash-digest, an attacker needs to find 1,024 bits of plain text input in

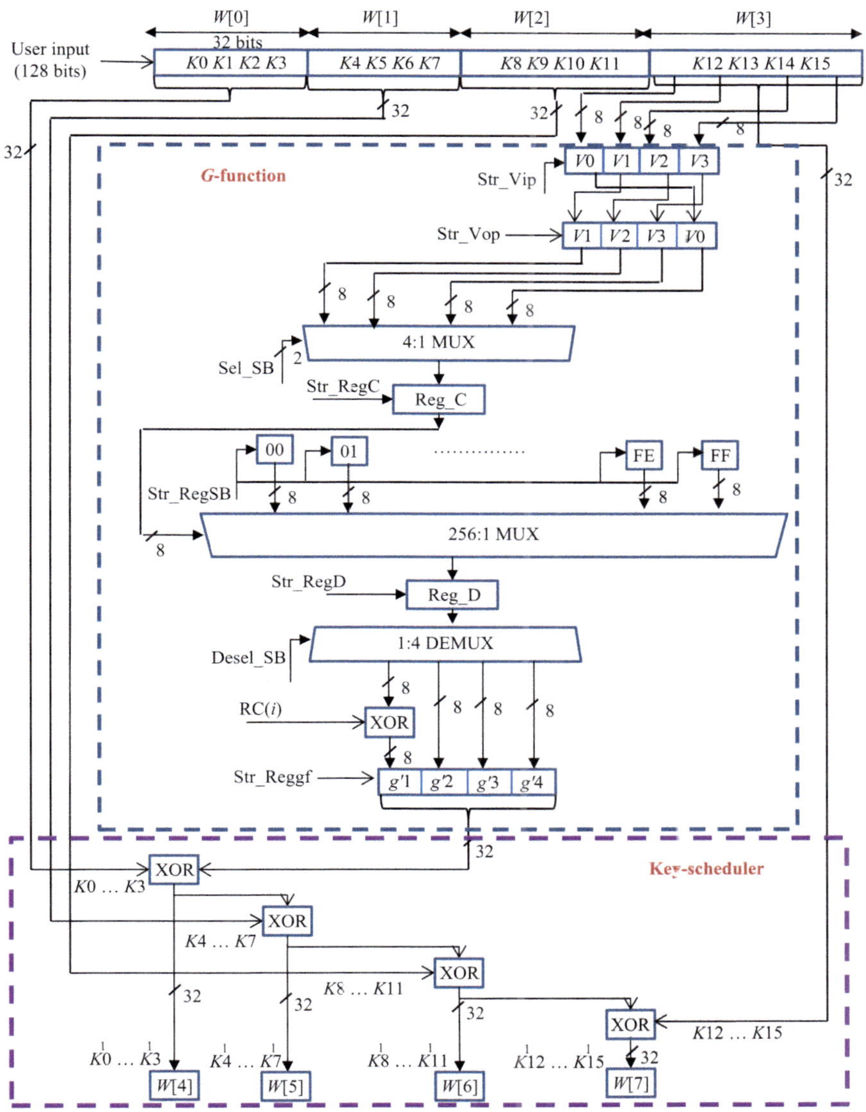

Figure 9.7 Hardware design of key-expansion process of AES-128

the SHA-512 algorithm. Therefore for an attacker to know the architecture of 64 reconfigured ILBs, 512 bits of ILBs keys are required to be decoded from 1,024 bits of plain text input of SHA-512 (Sengupta and Rathor, 2019).

- In order to reconfigure up to 64 ILBs, one SHA-512–based key generation block needs to be integrated with a functionally obfuscated DSP design. It is important to know that it incurs a considerably lower area overhead than four

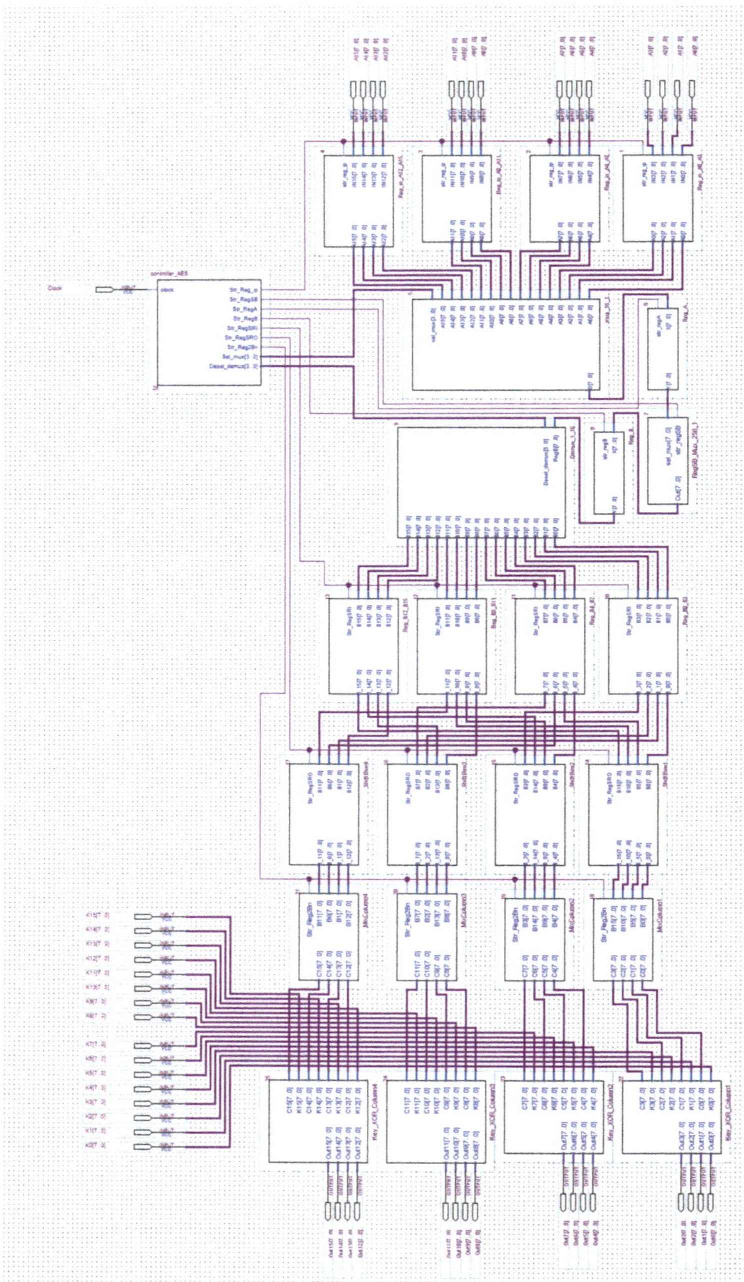

Figure 9.8 Custom AES-128 hardware implemented in Quartus II v13.0

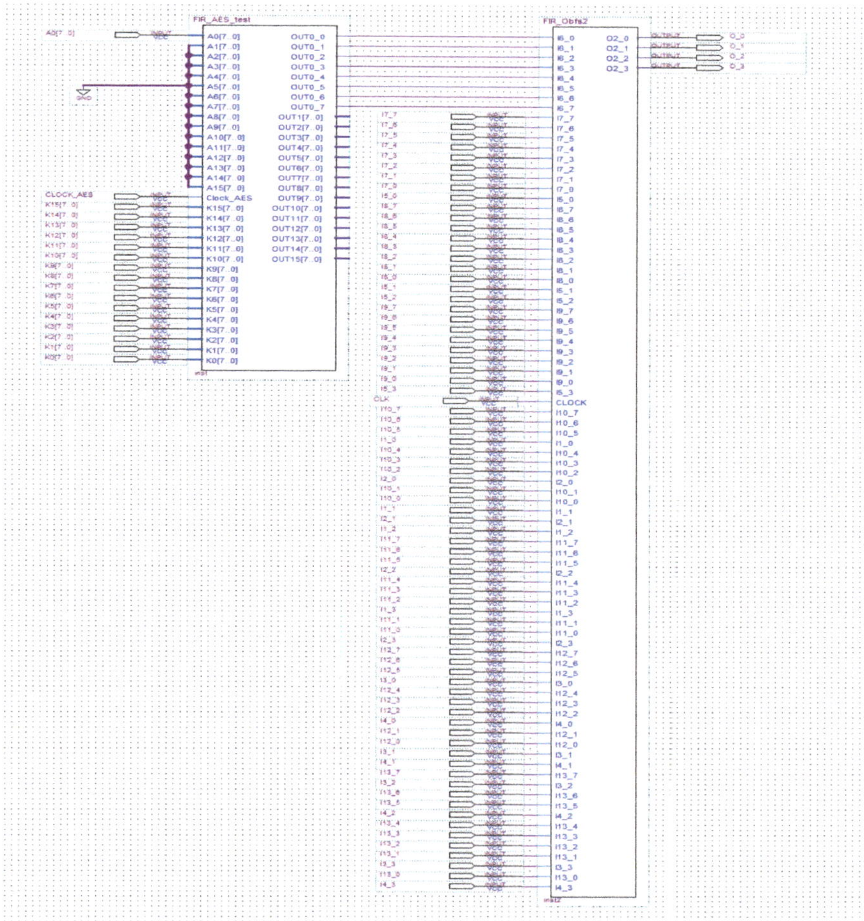

Figure 9.9 Top-level FIR obfuscated with AES

instances of AES-128 required to structurally reconfigure the same number of ILBs. Therefore the security technique using SHA-512–based key generation hardware is considered more lightweight than AES hardware in terms of design overheads (Sengupta and Rathor, 2019).

Advantage of using SHA-512–based key generation hardware over AES-128

- The potential features of SHA-512–based key generation logic, such as collision resistance, uniformity and being deterministic in nature (defined earlier in Section 9.4.2.2), renders it advantageous over AES-128 (Sengupta and Rathor, 2019).
- The AES-128 generates encrypted output depending on the fixed secret key, therefore it is susceptible to side-channel attack (which is used to extract the

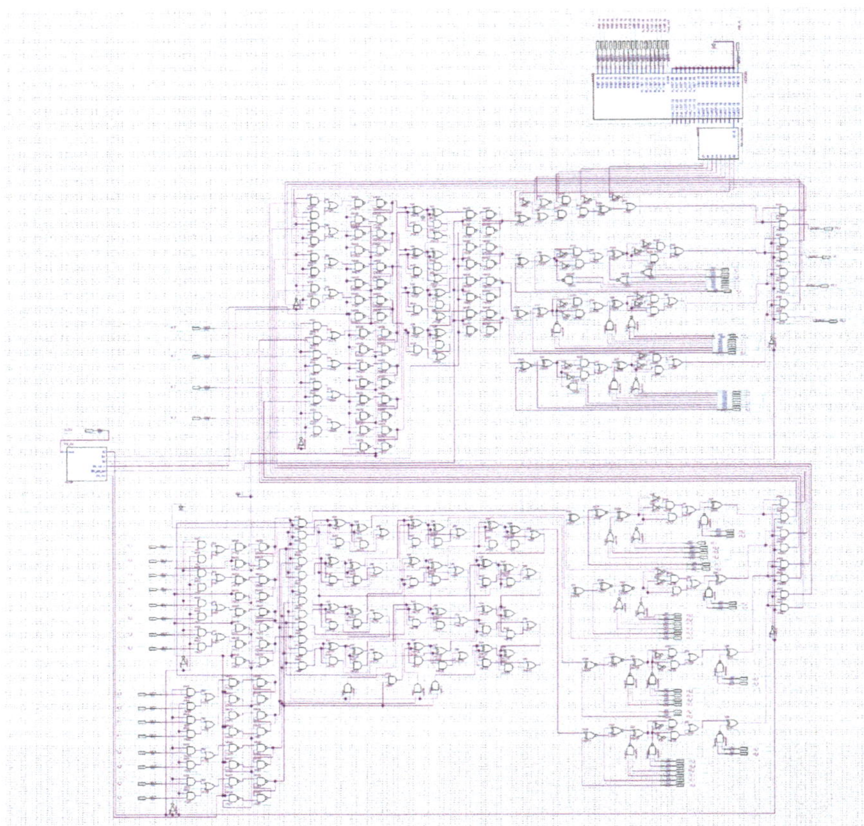

Figure 9.10 FIR obfuscated with AES

keys). On the contrary, the generated hash output of SHA-512 does not depend on the fixed secret key. Hence, the possibility of side-channel attack is eliminated on using SHA-512–based key generation logic (Sengupta and Rathor, 2019).

- Using single AES-128 hardware, a maximum 16 ILBs of an obfuscated DSP core can be reconfigured. On the contrary, a single SHA-512–based key generation hardware is capable to reconfigure up to 64 ILBs. Hence, SHA-512–based key encryption hardware provides more security than AES-128 (Sengupta and Rathor, 2019).
- Since an AES-128 block is capable to produce keys for a maximum 16 ILBs only, more than one AES needs to be used for larger designs where the number of ILBs is high (around 50 or more). Integration of more than one AES with a functionally obfuscated design leads to significant increase in area or power overheads. On the contrary, single SHA-512–based key generation hardware is

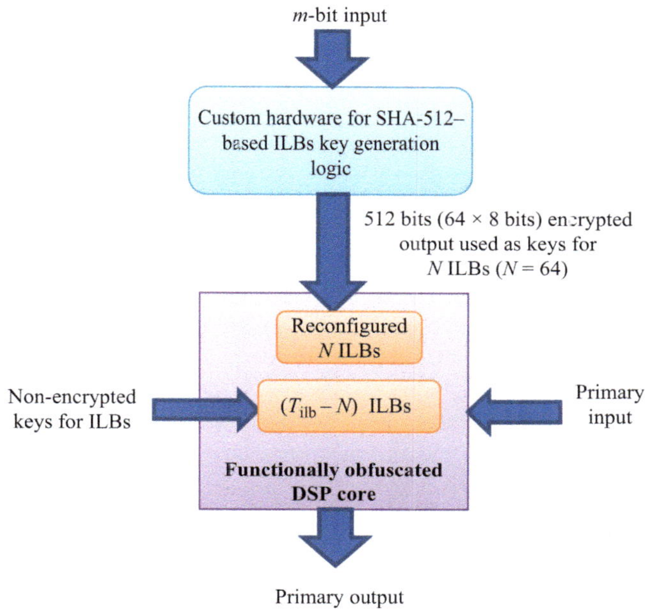

Figure 9.11 Overview of protection scenario of a functionally obfuscated DSP core against the removal attack using SHA-512–based ILBs-key generation hardware (Sengupta and Rathor, 2019)

capable to produce keys for up to 64 ILBs of a functionally obfuscated design. Hence the integration of single SHA-512–based key generation hardware with a functionally obfuscated design (of large size) incurs lower design area or power overheads compared to AES-128 (Sengupta and Rathor, 2019).

Overview

Figure 9.11 provides an overview of securing a functionally obfuscated design using SHA-512–based ILBs-key generation hardware against removal attack (Sengupta and Rathor, 2019). As shown in the figure, the 512-bit hash digest of the SHA-512–based key generation hardware is fed as keys to the ILBs of the functionally obfuscated DSP design. Based on the encrypted key-bits (512-bit hash digest), up to 64 ILBs of the functionally obfuscated design can be structurally reconfigured. This hinders the detection of any specific ILB architecture by an attacker, thus thwarting the removal attack (Sengupta and Rathor, 2019).

As shown in Figure 9.11, the input to the custom hardware of SHA-512–based ILBs–key generation logic (Sengupta and Rathor, 2019) is of arbitrary length (m bits) and output is $N \times 8$ bits, where N is the number of ILBs to be reconfigured and ‘8’ is the number of key bits per ILB. The maximum value of N can be 64 because the total length of the hash digest is of 512 bits and one ILB needs an 8-bit key to be activated. Further, the keys of the remaining ILBs of the functionally obfuscated design are kept non-encrypted as shown in Figure 9.11. The secure encryption of ILB keys using SHA-512–based key generation logic leads to robust structural reconfiguration of a number of ILBs simultaneously. Further, post-synthesis, the reconfigured ILBs structures are camouflaged with the DSP circuit, thus resulting in enhanced security against removal attacks (Sengupta and Rathor, 2019).

The protection against the removal attack using this approach is ensured due to four reasons (Sengupta and Rathor, 2019). (1) The synthesis of SHA-512–based key generation hardware is performed for an arbitrary length input (*m* bits). Further, the SHA-512–based key generation hardware is integrated with a functionally obfuscated design. This integration is performed by connecting output bits of SHA-512–based key generation hardware to the keys of the subset of ILBs of the functionally obfuscated design. Based on the number of ILBs to be structurally reconfigured, the outputs bits of the SHA-512–based key generation hardware are used as encrypted keys for ILBs. *The encrypted 8-bit key (for an ILB) determines the organization of the internal gate structure, thus reconfiguring the ILB structurally.* Since the SHA-512–based key generation hardware is capable to reconfigure a larger number of ILBs, hence rendering more ILBs camouflaged from an attacker's perspective. (2) Complete synthesis of the functionally obfuscated DSP core (having reconfigured ILBs) along with SHA-512–based key generation hardware further enhances the camouflaging of ILB structures post-synthesis. This is because post-synthesis the individual components of the entire design becomes indistinguishable. Thus the removal attack on ILBs and on SHA-512–based key generation hardware itself is impeded. (3) The SHA-512–based key generation hardware used is not publicly available as it is custom designed. Therefore the complete internal architecture of the SHA-512–based key generation hardware is not known to the attacker in advance. (4) The structure of ILBs are not fixed; rather, these are decided by the key-generation hardware and its arbitrary *m*-bit input. Since the attacker is not aware of the architecture of custom SHA-512–based key generation hardware, its *m*-bit input and hash output, the identification of ILBs becomes extremely difficult, thus thwarting the removal attack (Sengupta and Rathor, 2019) in a functionally obfuscated DSP netlist.

The custom SHA-512–based key generation hardware consists of two logics (Sengupta and Rathor, 2019), namely (1) SHA-512 custom logic and (2) ILB keys-extraction logic as shown in Figure 9.12. An arbitrary length (*m*-bit) input and initial hash buffer value are fed to the SHA-512 custom logic and the 512-bit hash digest is generated at the output. The initial standard values of eight hash buffers (**a**, **b**, **c**, **d**, **e**, **f**, **g**, **h**) are given in Table 9.1. Each hash buffer is 64-bit in size. The custom SHA-512 logic first converts the *m*-bit input into a block of size 1,024-bit. This conversion process is discussed in the following text. Further, the custom SHA-512 logic processes the 1,024-bit block and hash-buffer values and updates the hash buffer in each round. The number of rounds of execution of SHA-512 logic is customized (based on designer). After the final round, the eight hash buffers carry the 512-bit digest. Further, the ILB keys-extraction logic extracts the

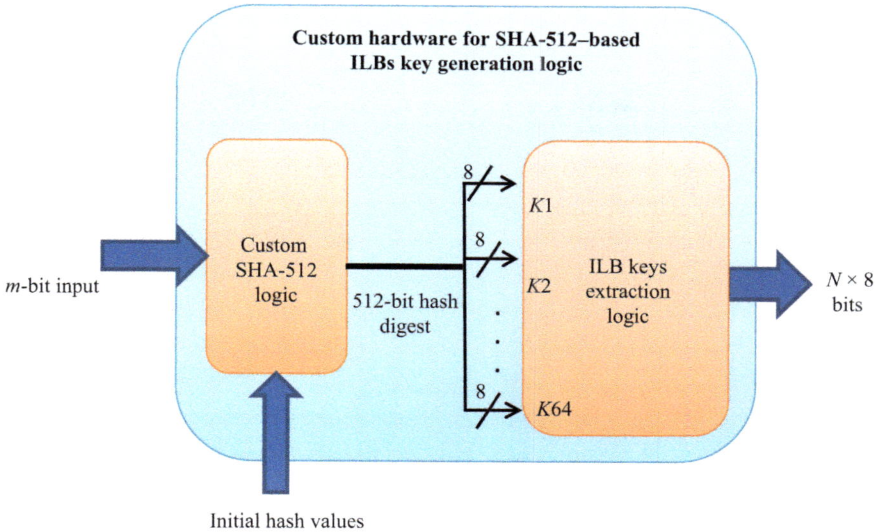

Figure 9.12 Block diagram of custom hardware for SHA-512–based ILBs-key generation logic (Sengupta and Rathor, 2019)

Table 9.1 Eight hash-buffer values

Hash buffer	Value	Hash buffer	Value
a	6A09E667F3BCC908	**e**	510E527FADE682D1
b	BB67AE8584CAA73B	**f**	9B05688C2B3E6C1F
c	3C6EF372FE94F82B	**g**	1F83D9ABFB41BD6B
d	A54FF53A5F1D36F1	**h**	5BE0CD19137E2179

chunk of 8-bits (from 512-bit hash digest) that act as encrypted keys for ILBs. How many chunks of 8-bits are extracted depends on the number of ILBs to be reconfigured. To reconfigure N number of ILBs of a functionally obfuscated DSP core, the output from the keys-extraction logic is extracted in the form of $N \times 8$ bits. The hardware of ILB keys-extraction logic is also custom designed. Since the custom keys-extraction logic is responsible for providing the encrypted key-bits for the ILBs (resulting in to ILB reconfiguration), security against removal attack is enhanced (Sengupta and Rathor, 2019).

Hardware design and integration with obfuscated DSP core

The hardware design of custom SHA-512–based key generation logic is shown in Figure 9.13 (Sengupta and Rathor, 2019). The input to the design is a bit-string of arbitrary length (m-bit) and the output is the encrypted keys for the ILBs of a

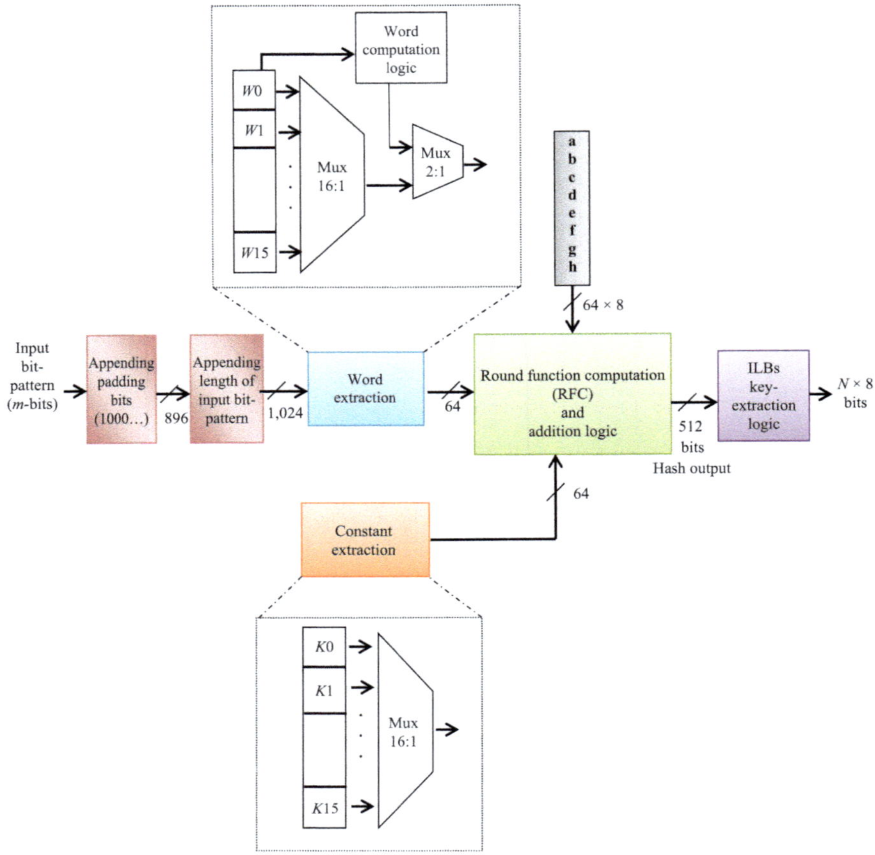

Figure 9.13 Block diagram of custom SHA-512–based key generation hardware (Sengupta and Rathor, 2019)

functionally obfuscated design. The process of obtaining encrypted keys is described below through the following steps (Sengupta and Rathor, 2019):

- *Appending padding bits:* In this step, the input bit-string of arbitrary length is transformed into a block of 896-bit by padding the following bit-sequence: '10000 . . . .' (assuming the length of input bit-string is less than 896 bits).
- *Appending input bit-string length:* The block of 896-bit is further transformed in to a block of 1,024-bit by padding the 128-bit representation of the length of input bit-string.
- *Word extraction:* In this step, a 64-bit word is extracted to be used in the round function computation (RFC) step. As shown in Figure 9.10, the word is either extracted from the 1,024-bit block (obtained in previous step) or computed using word computation logic hardware. The word computation logic is the

same as that is used in the standard SHA algorithm. For the initial 16 rounds, 64-bit size words ($W0$–$W15$) are extracted from the 1,024-bit block (i.e., rounds $R0, R1, \ldots, R15$ use words $W0, W1, \ldots, W15$, respectively). For the 17th round onwards, a 64-bit word (W_r) is computed from the word computation logic.

- *Constant extraction:* The 16 constants ($K0, K1, \ldots, K15$) used in the custom SHA-512 logic are all standard values (same as standard SHA). The size of each constant is 64-bit. The extraction of one constant (K_r) from the set of 16 constants is performed for each round of RFC.
- *Hash-buffer processing:* In the first round, this step provides the initial hash-buffer values to the RFC step. For subsequent rounds of RFC, the updated hash-buffer values are obtained from this step.
- *RFC:* The RFC depends on the previous three steps, viz. word extraction, constant extraction and hash buffer processing. A 64-bit word (W_r), a constant (K_r) and eight hash-buffer values (**a**, **b**, **c**, **d**, **e**, **f**, **g**, **h**) obtained from previous steps are processed in RFC hardware to compute the updated hash values in each round. The RFC hardware performs the following six basic functions: (1) MAJORITY function, (2) SUMMATION/ROTATION 'a', (3) CHOOSE function, (4) SUMMATION/ROTATION 'e', (5) MIX function T1, (6) MIX function T2. The hardware of RFC based on these functions is shown in Figure 9.14. This figure also shows the hardware structure of each of the six functions. The rotation function (ROTR) shown in the figure performs the circular right shift by 28 bits on the hash buffer **a**/**e**. The new or updated hash buffer values generated after the Rth round have been represented as: $\mathbf{a}_r$, $\mathbf{b}_r$, $\mathbf{c}_r$, $\mathbf{d}_r$, $\mathbf{e}_r$, $\mathbf{f}_r$, $\mathbf{g}_r$, $\mathbf{h}_r$. The hash-buffer values after the final round represent the 512-bit hash output. The designer decides the number of rounds for which the RFC hardware runs.
- *ILBs keys extraction:* The 512-bit hash digest (output of final round of RFC) is further fed to the ILBs key-extraction logic. This logic performs the extraction of $N \times 8$ bits (from 512-bit hash output) that act as keys of N number of ILBs of a functionally obfuscated design, where the maximum possible value of N can be 64. Thus up to 64 ILBs of a functionally obfuscated design can be structurally reconfigured using SHA-512–based key generation hardware. Suppose N number of ILBs of a functionally obfuscated design is to be structurally reconfigured. Therefore $N \times 8$ bits are extracted (from the 512-bit hash output) by the key-extraction logic. Further, according to these $N \times 8$ bits, N ILBs structure are reconfigured by reorganizing their internal gates such that the functionality of the ILBs does not change (output of ILBs is valid only on applying correct keys).

The custom SHA-512–based ILBs-key generation hardware is integrated with a functionally obfuscated DSP core by connecting its output with the keys of ILBs to be reconfigured (Sengupta and Rathor, 2019). The integration of SHA-512–based key generation hardware eliminates the vulnerability of the functionally obfuscated DSP design against removal attack. Figure 9.15 shows the implemented custom SHA-512–based key generation hardware and Figure 9.16 shows the

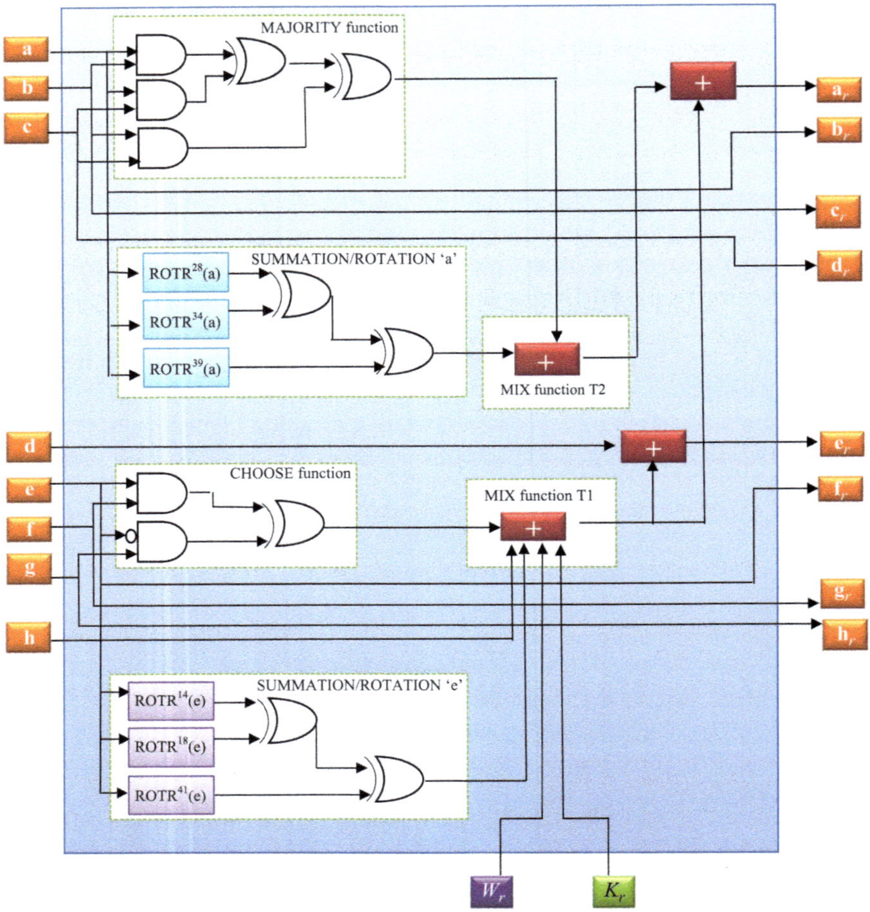

Figure 9.14 Round function computation hardware of custom SHA-512 (Sengupta and Rathor, 2019)

implemented custom RFC hardware in CAD synthesis tool. The implemented custom SHA-512–based key generation hardware integrated with the gate-level structure of obfuscated FIR filter design is shown in Figure 9.17.

9.4.2.3 Security of a functionally obfuscated DSP core against removal attack using custom lightweight anti-removal logic unit (Rathor and Sengupta, 2019)

Features of the approach

- In this approach (Rathor and Sengupta, 2019), to secure ILBs of a functionally obfuscated DSP core against the removal attack, ILB (locking logic) is structurally reconfigured according to the output of a lightweight ARL unit. The detection of ILBs in the complex DSP netlist is difficult because the ILB

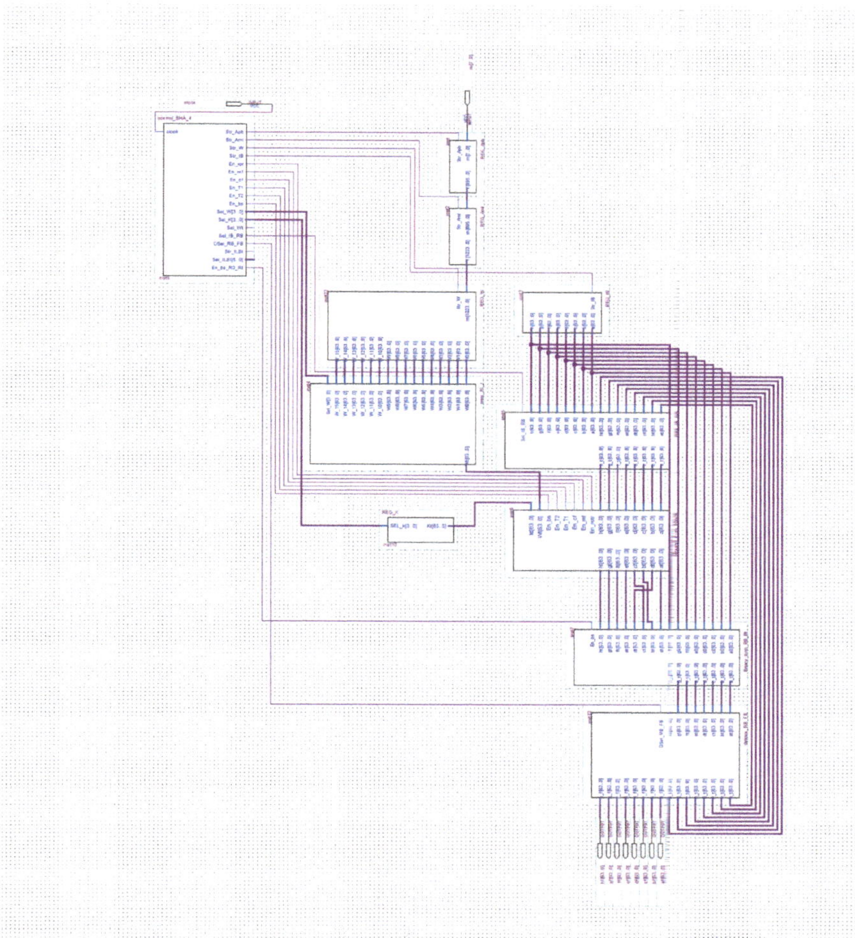

Figure 9.15 Custom SHA-512–based key generation hardware (Sengupta and Rathor, 2019)

structure is not known to the attacker in advance. This is because the ILB structure depends on the architecture and complex challenge inputs of the ARL unit (Rathor and Sengupta, 2019), which is only known to the designer. Further, post-synthesis, the ILB gate structure gets camouflaged in the entire design netlist and becomes undetectable. Thus a removal attack on ILBs becomes extremely difficult. Additionally, since the designed lightweight ARL unit can be architecturally customized (i.e., complexity of AND–OR network can be tuned as per designers choice), post-synthesis detection of the custom ARL unit in the design netlist (comprising of DSP core and ILBs) is highly challenging (Rathor and Sengupta, 2019).

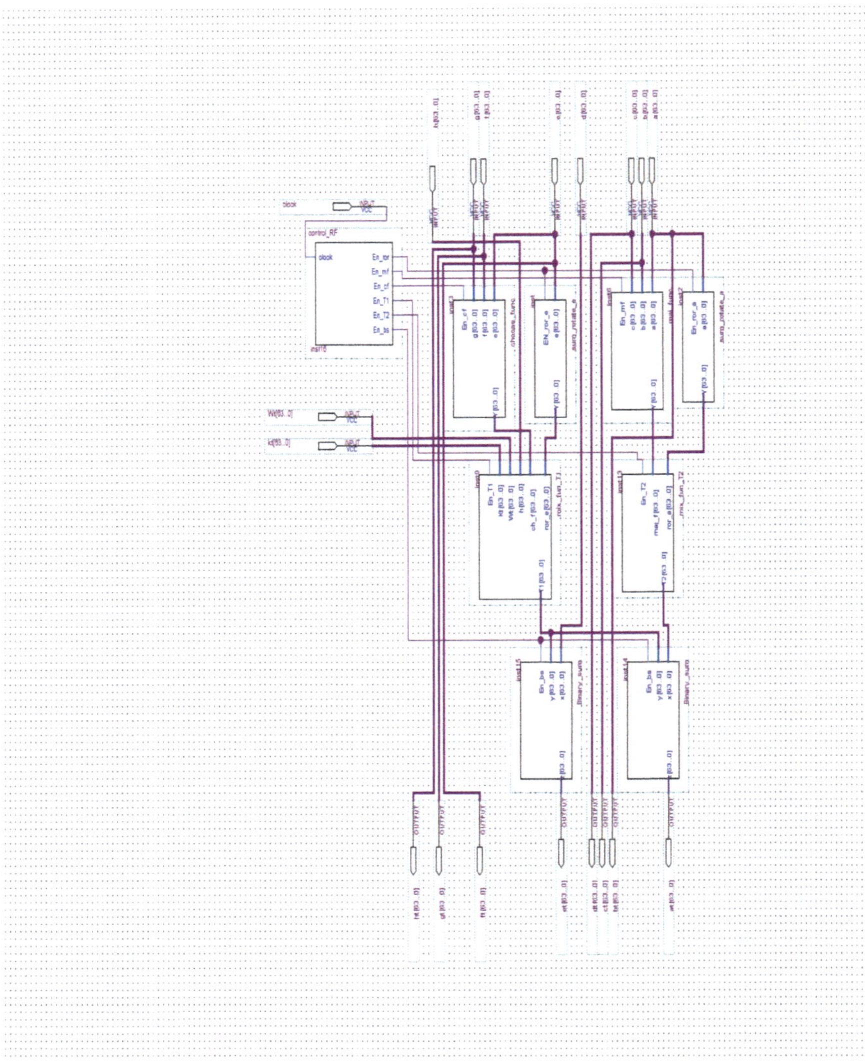

Figure 9.16 Custom RFC hardware used in SHA-512

- The custom lightweight ARL unit provides some strong security features, for example, the architecture of the ARL unit is hidden (and not fixed). The number of ARL blocks used in an ARL unit is not fixed, but rather depends on the number of ILBs to be configured (Rathor and Sengupta, 2019). Only the designer has the knowledge of the architecture of ARL unit, thus it is flexible for custom design (rendering it unknown to the attacker). Further, the length of the two challenge–bits sets S and C (n and m) is also decided by the designer.

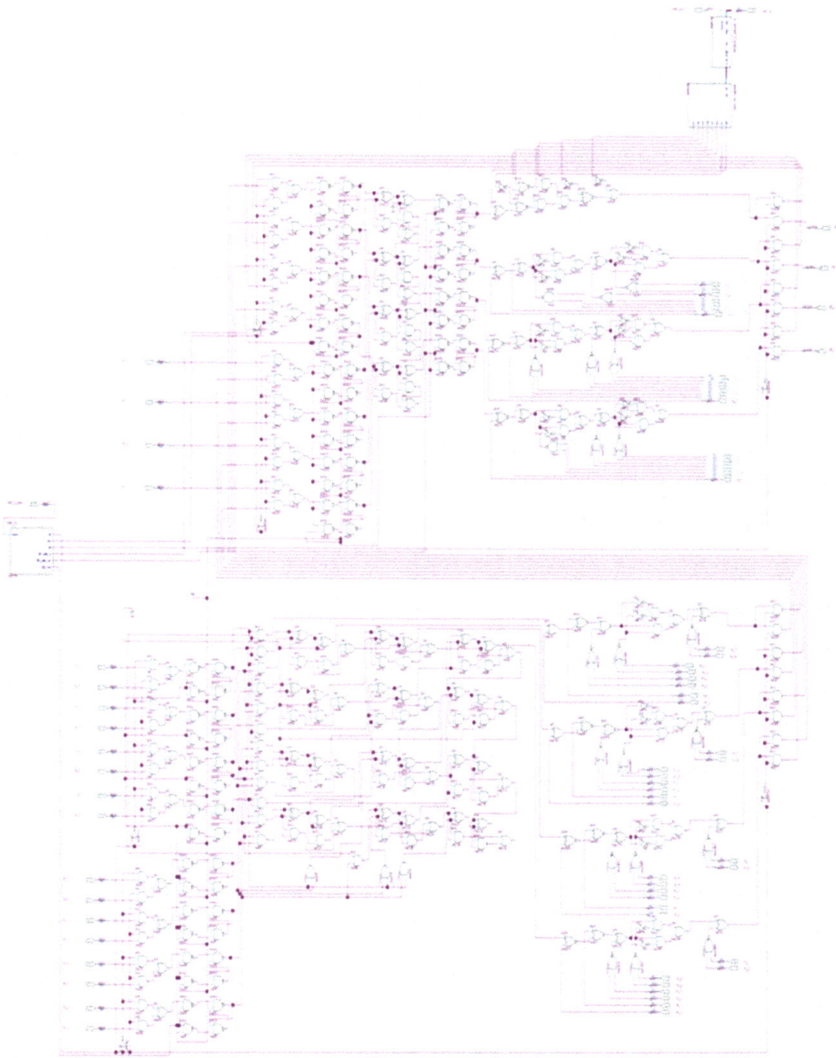

Figure 9.17 FIR obfuscated with SHA-512 (Sengupta and Rathor, 2019)

The length of n and m depend on the size and the number of AND–OR networks used in each ARL block. To decode one bit of ARL unit output, a total $n + m$ of the challenge input bits should be known in advance. Further, to know the 512-bit ARL output, an attacker needs to find $512 \times (n + m)$ challenge bits. More explicitly, for $n = 3$ and $m = 3$, the attacker needs to find 3,072 bits of the challenge input. Hence for an attacker to know the architecture of the 64 reconfigured ILBs, 512 bits of ILB keys (for $n = 3$ and $m = 3$) are required to be decoded which, in turn, requires up to 3,027 bits of challenge input to be decoded.

- In order to reconfigure 64 ILBs, only one ARL unit is integrated with a functionally obfuscated DSP design. Further, in case more ILBs are needed to be reconfigured (such as in large designs), then only the number of ARL blocks are required to be increased in the same ARL unit. Whereas in the case of AES-128 and SHA-512–based approaches, extra instances of them need to be integrated to generate more encrypted keys to fed more than 64 ILBs. *Thus the ARL unit incurs considerably lower area overhead than AES-128 and SHA-512–based approaches for reconfiguring same number of ILBs*. Therefore the security technique using an ARL unit is considered a lightweight solution against removal attack (Rathor and Sengupta, 2019).

Overview

The functional obfuscation on DSP cores is employed using ILBs to safeguard against RE attacks. However, removal attacks on ILBs nullifies the purpose of functional obfuscation. The solution to removal attack is to make ILBs un-obvious or undetectable for an attacker. This can be achieved by using reconfigured ILBs instead of standard ILB structures. The structural reconfiguration of ILBs by generating their keys from custom AES-128 and SHA-512 cryptologics has been discussed earlier in Sections 9.4.2.1 and 9.4.2.2. However, AES-128 and SHA-512 cryptologics incur substantial design overheads and provide lower security (in terms of number of input bits required to be decoded by an attacker) against removal attacks. This sub-section discusses a custom lightweight and more secure solution against removal attacks. This solution has been referred as lightweight ARL unit (Rathor and Sengupta, 2019; Sengupta *et al.*, 2019). The ARL unit generates the key-bits for ILBs of a functionally obfuscated DSP core. Based on these key-bits, ILBs get structurally reconfigured ensuring security against the removal attack.

Figure 9.18 shows the protection scenario using an ARL unit against the removal attack. An ARL unit is constructed using a number of ARL blocks. The number of ARL blocks used in an ARL unit is decided based on the number of ILBs a designer intends to reconfigure. For example, say a designer intends to reconfigure P number of ILB and each ILB activates on applying an 8-bit key. Further, since each ARL block generates only one bit key, a total of $P \times 8$ ARL blocks need to be used in the ARL unit. Since an attacker is unaware of the custom design of the ARL unit, he/she fails to find the keys of reconfigured ILBs. Thus the attacker becomes unable to derive the ILB structures based on 8-bit keys. This renders the ILBs undetectable for the attacker, hence thwarting the removal attack on ILBs. Further, as the ARL unit architecture is customizable and also not known to the attacker, thus the ARL unit itself remains secured against the removal attack. Additionally, an integrated functionally obfuscated DSP core with an ARL unit is resynthesized. Post-synthesis, the internal gates of ILBs and the ARL unit are further camouflaged in the entire design. This leads to the detection of ILBs by the

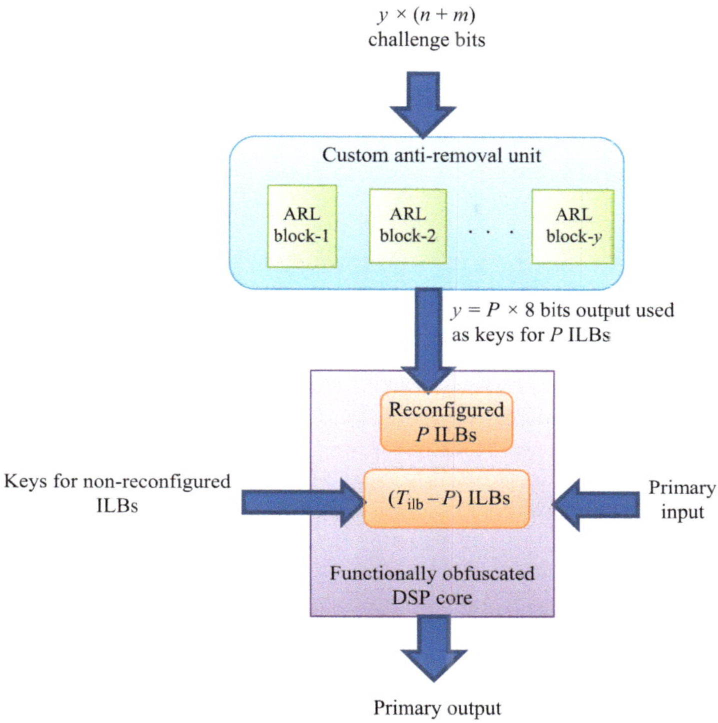

Figure 9.18 Overview of protection scenario of a functionally obfuscated DSP core against removal attack using an ARL unit (Rathor and Sengupta, 2019)

attacker being more challenging, thus enhancing the security against removal attacks (Rathor and Sengupta, 2019).

Hardware design and integration with obfuscated DSP core

Several ARL blocks jointly constitute an ARL unit, where each ARL block is responsible for producing only a single key-bit for ILBs. For producing y number of key-bits, y ARL blocks are used in an ARL unit, where $y = P \times 8$. Here P represents the number of ILBs to be reconfigured and '8' is the number of key-bits required per ILB. Further, Figure 9.19 shows the internal structure of an ARL block (Rathor and Sengupta, 2019). As shown in the figure, an ARL block comprises of several AND–OR networks responsible for producing a single bit of output. The number of AND–OR networks and size of each depend on the two sets of the challenge inputs given as follows: $S = \{S0, S1, \ldots, Sn\}$ and $C = \{C0, C1, \ldots, Cm\}$. Here the size of challenge inputs S and C has been given as n and m respectively. The size of the AND–OR network is decided by the n and number of AND–OR networks in an ARL block is determined by the m. An ARL block uses a total of 2^m AND–OR networks. Further, the size of the multiplexer used in the ARL block also

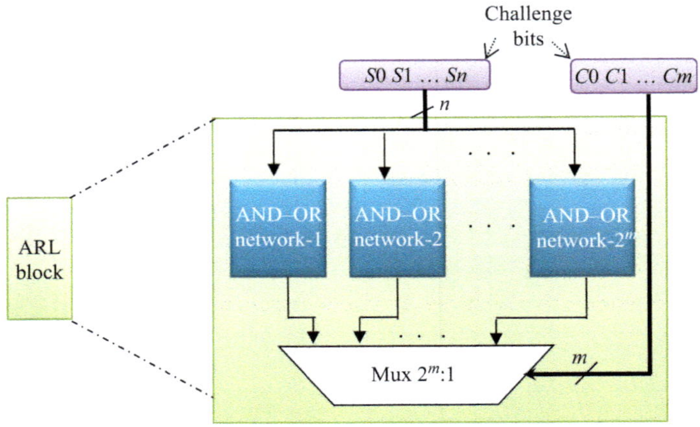

Figure 9.19 Overview of an ARL block

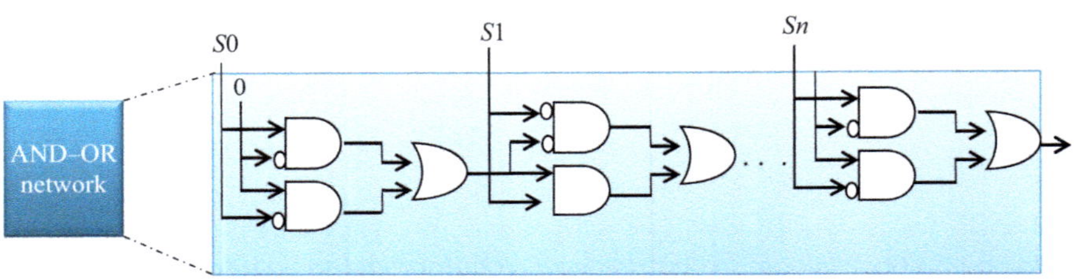

Figure 9.20 Overview of an AND–OR network used in ARL block

depends on *m*. The bit-pattern of challenge input set *S* decides the value of a key-bit generated from the selected AND–OR network. The selection of a AND–OR network depends on the challenge input set *C*. More explicitly, based on the bit-pattern of challenge input *C*, the respective AND–OR network architecture is selected to generate the one bit key. A possible architecture of an AND–OR network is shown in Figure 9.20. *Note: this architecture is flexible and swappable*. Each AND–OR network in an ARL block uses the same set of challenge input *S* (Rathor and Sengupta, 2019).

The architecture of an ARL unit depends on various factors such as the number of ARL blocks used, the size of challenge inputs *S* and *C* which, in turn, decides the structure of each ARL block and the configuration of AND–OR network. These tuning factors of ARL architecture are decided by the designer. Thus only the designer has the knowledge of the custom ARL unit. For an attacker, it is extremely hard to know the exact ARL unit architecture used and its functionality. This almost renders it infeasible for an attacker to know the ILB keys, therefore the reconfigured ILBs (based on keys generated from ARL unit) remain undetected. Hence, the removal attack on the ILBs becomes impeded. Further, the removal attack of the ARL unit itself is also impeded owing to its custom design. Moreover, the security in terms of number of challenge bits required to decode the keys of reconfigured ILBs is also high using ARL unit (from an attacker's perspective). This is because $n + m$ bits of the challenge inputs are responsible to generate the one bit of an ILB key. To reconfigure 64 ILBs, 512 key-bits are generated. This requires $512 \times (n + m)$ bits of challenge input. If we consider $n = 3$ and $m = 3$, then a total 3,072 bits of challenge input will be required to decode the 512 key-bits. On the contrary, to reconfigure 64 ILBs, 512 key-bits are generated using 4 instances of AES. To decode the 512 bits of keys, a total 1,024 bits (4×128 bits of input + 4×128 bits of fixed secret key) of challenge inputs are required. Further, in the case of the SHA-512–based approach, 1,024-bit of input should be known to the attacker to decode the 512-bit output (key-bits for ILBs). Hence the ARL unit provides higher security owing to the larger number of challenge-bits needed for decoding same number of ILB keys (or ILB structure). Furthermore, the generation of key-bits for X more ILBs using an ARL unit can be accomplished simply by adding $X \times 8$ more ARL blocks. However, the generation of key-bits for more ILBs using AES-128/SHA-512 hardware requires one extra instance to be integrated with the functionally obfuscated design. This incurs substantial area and power overheads. Hence, the ARL unit evolves to be a lightweight solution against removal attack on a functionally obfuscated DSP core (Rathor and Sengupta, 2019).

As discussed within Section 9.4.2.3, the ARL unit is integrated with the functionally obfuscated DSP core to generate ILB keys. A top-level diagram of a custom lightweight ARL unit integrated with the FIR filter design is shown in Figure 9.21 and its respective gate-level structure is shown in Figure 9.22. The implementation of the complete integrated design has been performed for cyclone-II: EP2C70F896C6 FPGA using Intel Quartus II v13.0 (Rathor and Sengupta, 2019). In the demonstrated functionally obfuscated FIR filter design, the size of multiplier and adder unit chosen is 4-bit. Hence to functionally obfuscate the FIR filter design, the total number of ILBs required is calculated to be eight (obtained from (9.1)). The output of the ARL unit has been connected to only one ILB for the sake of demonstration (shown in Figure 9.22). The internal architecture of an ARL unit (implemented in Quartus II v13.0) for generating 8-bit key for an ILB is shown in Figure 9.23. This architecture post-synthesis gets converted to the optimized gate-level representation in the obfuscated design netlist. The integration of an ARL unit eliminates the vulnerability of the functionally obfuscated FIR filter design against the removal attack.

9.4.3 Demonstration of ILB structure reconfiguration

The reconfiguration of an ILB structure is performed based on its key value, which should be unresolvable by an attacker. The randomness of the keys (for the attacker) is ensured by connecting the keys of a sub-set of ILBs with the output of AES-128 or SHA-512–based key generation hardware or lightweight ARL unit. Therefore the ILBs of a functionally obfuscated design use the random values of keys generated. These ILB keys are only known to the designer as only he/she

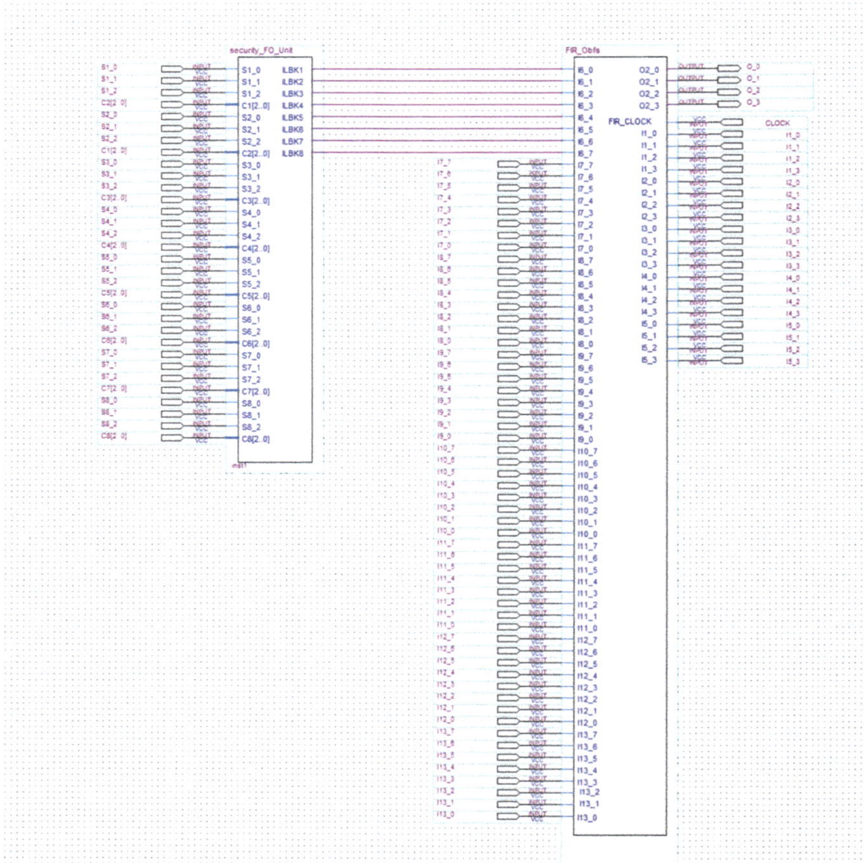

Figure 9.21 Top-level FIR obfuscated with ARL

knows the required challenge bits to generate the keys. According to these keys, the reconfiguration of ILBs structure is performed by rearranging the internal gates in such a fashion that the functionality of the ILBs do not change (i.e., it produces correct output only upon applying a valid key). For example, a reconfigured ILB based on an 8-bit key (10110000) is shown in Figure 9.24. Due to structural reconfiguration, no extra gates in the ILBs are added; rather, internal rearrangement of the gates occurs.

9.5 Analyses of case studies

This section discusses various case studies on DSP applications to analyse the effectiveness of the custom SHA-512–based ILBs-key generation hardware (Sengupta and Rathor, 2019) and custom ARL unit (Rathor and Sengupta, 2019)

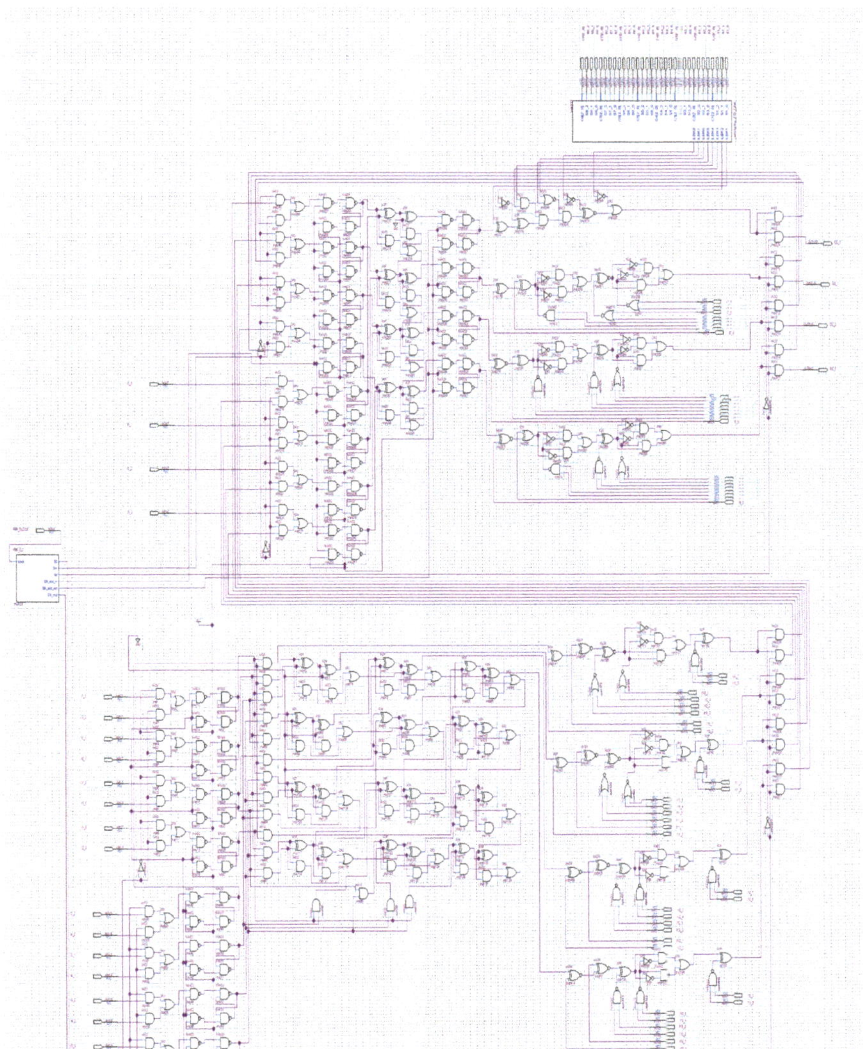

Figure 9.22 Integrated custom ARL unit with the functionally obfuscated 4-bit FIR filter (implemented in Quartus II v13.0)

against removal attacks on functionally obfuscated DSP cores. The analysis is based on assessing the security and design cost of the SHA-512–based ILBs-key generation hardware (Sengupta and Rathor, 2019) and ARL unit (Rathor and Sengupta, 2019) with respect to AES-128 (Sengupta *et al.*, 2019).

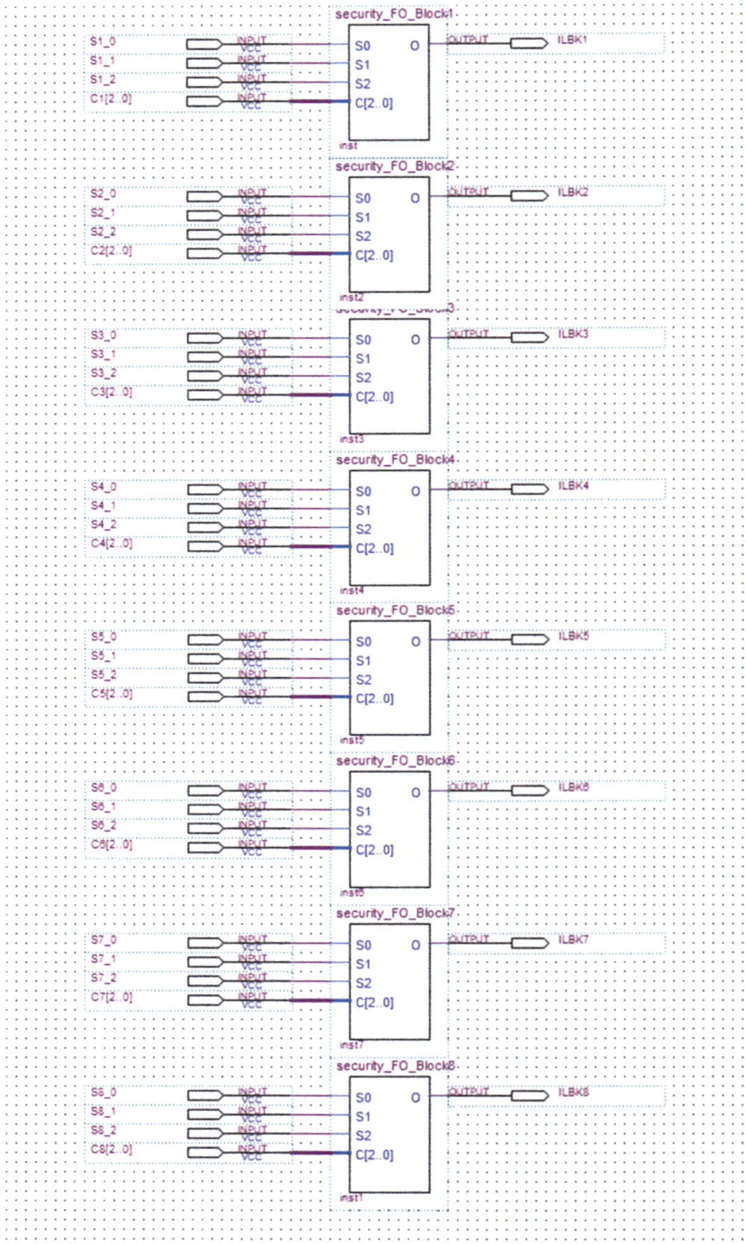

Figure 9.23 Internal architecture of an ARL unit hardware using eight ARL blocks

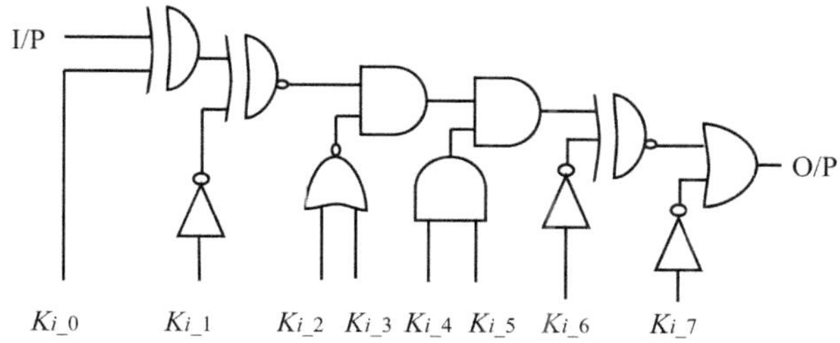

Figure 9.24 Structurally reconfigured ith ILB based on key input (10110000) (Sengupta and Rathor, 2019)

9.5.1 Security analysis

9.5.1.1 Comparative study between obfuscated DSP core with custom SHA-512–based key encryption hardware vs. obfuscated DSP core with custom AES-128–based key encryption hardware

The security analysis of the custom SHA-512–based key encryption hardware (Sengupta and Rathor, 2019) is performed on the basis of the number of encrypted key-bits generated for ILBs. If more ILB key-bits are generated, then more ILBs of a functionally obfuscated DSP design can be structurally reconfigured (resulting in enhanced security). For a comparative security analysis, the SHA-512–based key encryption logic has been compared with the AES-128 logic (Sengupta *et al.*, 2019) in terms of the number of encrypted key-bits generated for ILBs. This comparison is shown in Table 9.2. As evident from the table, the custom SHA-512–based key encryption hardware (Sengupta and Rathor, 2019) is capable to generate key-bits for a larger number of ILBs in contrast to the custom AES-128–based key encryption hardware (Sengupta *et al.*, 2019). Therefore a larger number of ILBs in an obfuscated DSP design can be structurally reconfigured using the SHA-512–based approach (Sengupta and Rathor, 2019), thus providing higher security against removal attacks. This improvement in security compared to the AES-128–based approach is achieved because single SHA-512–based key encryption hardware is capable to produce 512-bit output (that is fed as 512 key-bits for ILBs), hence up to 64 ILBs can be structurally reconfigured (assuming each ILB requires an 8-bit key) simultaneously at *gate level*. On the contrary, one AES-128 is capable to produce 128-bit output (that is fed as 128 key-bits for ILBs) resulting into structural reconfiguration of only 16 ILBs (which is much less than SHA-512) simultaneously at gate level. The *larger number of ILBs structurally reconfigured results in stronger obfuscation, thus leading to greater indistinguishability in the gate-level design netlist.*

As shown in Table 9.2, two instances of custom AES-128 have been used for four DSP applications (FIR, DWT, ARF, IIR) to structurally reconfigure 32 ILBs (requiring 256 key-bits) and three instances have been used for two larger-sized DSP applications (JPEG IDCT, MESA interpolate) to reconfigure 48 ILBs

Table 9.2 Security comparisons of the SHA-512–based approach (Sengupta and Rathor, 2019) and AES-128–based approach (Sengupta et al., 2019)

DSP applications (32-bit)	Number of key-bits generated for ILBs of a functionally obfuscated DSP design		Percentage of increase in security using Sengupta and Rathor (2019) in terms of more per cent key bits generated
	Sengupta and Rathor (2019)	Sengupta *et al.* (2019)	
FIR	512	256	50
ARF	512	256	50
DWT	512	256	50
IIR	512	256	50
MESA interpolate	512	384	25
JPEG IDCT	512	384	25

(requiring 384 key-bits). On the contrary, a single SHA-512–based key generation hardware is capable to structurally reconfigure up to 64 ILBs (requiring 512 key-bits). Hence the SHA-512–based approach achieves higher security against removal attacks. Additionally, the SHA-512–based key generation hardware is secured against side-channel attack, which is quite popular in the case of AES-128. This is because the output of SHA-512–based key encryption hardware does not use keys to produce the output.

9.5.1.2 Comparative study between obfuscated DSP core with ARL unit vs. obfuscated DSP core with AES-128–based key encryption hardware

The security analysis of the ARL unit (Rathor and Sengupta, 2019)–based approach is performed on the basis of the number of challenge bits required to decode or find the key-bits of the reconfigured ILBs. If a larger number of challenge bits is required, then it is harder to guess or find the key-bits of the reconfigured ILBs of a functionally obfuscated DSP design (resulting in enhanced security). For a comparative security analysis, the ARL unit–based approach has been compared with the AES-128–based approach (Sengupta *et al.*, 2019) in terms of the number of challenge bits required to decode or find the key-bits of ILBs (keeping the number of key-bits of ILBs the same for both approaches). This comparison has been shown in Table 9.3. As evident from the table, the ARL unit requires more challenge bits in contrast to AES-128 to generate the same number of key-bits for the ILBs of a functionally obfuscated DSP core. Therefore with the ARL unit, more challenge bits are needed to be decoded for the same number of key-bits. Thus the ARL unit provides enhanced security against removal attacks owing to the increase in the complexity of finding the challenge bits in contrast to AES-128 (from an attacker's perspective).

Table 9.3 Security comparison of ARL unit–based approach (Rathor and Sengupta, 2019) with AES-128–based approach (Sengupta et al., 2019) in terms of number of challenge bits required to decode the key-bits of the reconfigured ILBs

DSP applications (32-bit)	Number of challenge bits		Percentage of increase in security using ARL
	Rathor and Sengupta (2019)	Sengupta *et al.* (2019)	
FIR	1,536	512	66.67
ARF	1,536	512	66.67
DWT	1,536	512	66.67
IIR	1,536	512	66.67
MESA interpolate	2,304	768	66.67
JPEG IDCT	2,304	768	66.67

As shown in Table 9.3, the ARL unit requires 1,536 challenge bits to produce 256 key-bits for reconfiguring 32 ILBs (in case of FIR, DWT, ARF, IIR) and 2,304 challenge bits to produce 384 key bits for reconfiguring 48 ILBs (in case of larger design such as JPEG IDCT, MESA interpolate). On the contrary, two AES-128 units jointly require only 512 challenge bits (2×128 input bits $+ 2 \times 128$ key bits) to produce 256 key-bits for reconfiguring 32 ILBs and three AES-128 units jointly require only 768 challenge bits (3×128 input bits $+ 3 \times 128$ key-bits) to produce 384 key-bits for reconfiguring 48 ILBs (for a justified comparison, the same number of key-bits have been used for both approaches). As with the ARL unit, a larger number of input bits is required to decode the same number of key-bits, thus providing enhanced security in contrast to the AES-128–based approach.

9.5.2 Design cost analysis: comparative study between obfuscated DSP core with SHA-512 vs. obfuscated DSP core with AES-128

The design cost of an integrated SHA-512–based key generation hardware with a functionally obfuscated DSP core is evaluated using the cost function:

$$C_d(R_i) = w_1 \frac{L_d}{L_{\max}} + w_2 \frac{P_d}{P_{\max}} \tag{9.2}$$

where $C_d(R_i)$ denotes the obfuscated design cost with resource configuration R_i. Here the design cost is evaluated in terms of the design latency and power consumption denoted by L_d and P_d, respectively. Further, $L_{\max}$ and $P_{\max}$ denote the maximum execution latency and power, respectively, and w_1, w_2 represent the user-specified weights, both fixed at 0.5 to assign equal preference. Both the power and latency of an IP core design are calculated using NanGate library based on 15 nm technology scale (http://www.nangate.com/?pageid=2328).

Table 9.4 Comparative study of obfuscated design cost of the SHA-512–based approach (Sengupta and Rathor, 2019) and AES-128–based approach (Sengupta et al., 2019)

DSP applications (32-bit)	Obfuscated design cost		Percentage of reduction in cost using Sengupta and Rathor (2019)
	Sengupta and Rathor (2019)	Sengupta *et al.* (2019)	
FIR	0.604	0.754	19.9
ARF	0.619	0.775	20.1
DWT	0.819	0.981	9.2
IIR	0.827	0.987	16.2
MESA interpolate	0.319	0.519	38.5
JPEG IDCT	0.419	0.619	32.4

Note: To calculate the challenge bits of an ARL unit, the size of challenge bit sets 'S' and 'C' have been considered as $n = 3$ and $m = 3$. To generate the single key-bit of an ILB, a total $n + m = 6$ challenge bits are required. Therefore $256 \times 6 = 1{,}536$ challenge bits are required for generating 256 key-bits. Similarly, $384 \times 6 = 2{,}304$ challenge bits are required for generating 384 key-bits.

Table 9.5 Comparative study of the gate count of SHA-512–based obfuscation approach (Sengupta and Rathor, 2019) and AES-128–based obfuscation approach (Sengupta et al., 2019)

DSP applications (32-bit)	Gate count of obfuscation approach		Percentage of reduction in gate count in obfuscated design using Sengupta and Rathor (2019)
	Sengupta and Rathor (2019)	Sengupta *et al.* (2019)	
FIR	76,512	104,320	26.66
ARF	74,880	102,688	27.08
DWT	67,552	95,360	29.16
IIR	62,336	90,144	30.84
MESA interpolate	138,880	206,256	32.67
JPEG IDCT	150,816	218,192	30.88

Table 9.4 shows a comparison of the design cost of the SHA-512–based key generation hardware integrated with a functionally obfuscated DSP core (Sengupta and Rathor, 2019) with the AES-128–based counterpart (Sengupta *et al.*, 2019). As shown in the table, Sengupta and Rathor (2019) incurs lower obfuscated design cost in contrast to Sengupta *et al.* (2019). Additionally, the gate count comparison of Sengupta and Rathor (2019) and Sengupta *et al.* (2019) is shown in Table 9.5. As shown in the table, Sengupta and Rathor (2019) require less number of gates than Sengupta *et al.* (2019). The explanation behind this reduction in design cost and gate count is that a single AES-128 can produce key-bits for only 16 ILBs simultaneously. Therefore in order to produce key-bits for more ILBs, more than one

AES-128 unit needs to be integrated with a functionally obfuscated design. Further, for large designs such as JPEG and MESA interpolate, three instances of AES-128 units are required to produce key-bits for reconfiguring a significant number of ILBs (e.g., 48 ILBs used in the results discussed in Tables 9.4 and 9.5). However, single SHA-512–based key generation hardware is capable of producing key-bits for 64 ILBs simultaneously.

9.5.3 Resource usage analysis: comparative study between obfuscated DSP cores with AES vs. obfuscated DSP cores with ARL unit

An AES-128 unit can produce key-bits for only a limited number of ILBs (maximum 16). In order to produce more key-bits for more ILBs, more than one instance of AES units are needed to be integrated with the functionally obfuscated DSP design. On the contrary, to produce the key-bits for X more ILBs using an ARL unit, we only need to increase the $X \times 8$ ARL blocks inside the ARL unit. Hence the ARL unit emerges as a lightweight solution to combat removal attack. Table 9.6 shows the gate count comparison of the ARL unit (Rathor and Sengupta, 2019) and AES-128 unit (Sengupta *et al.*, 2019). For four DSP applications (viz. FIR, DWT, ARF, IIR), 32 ILBs have been structurally reconfigured and for two large DSP applications (viz. JPEG IDCT, MESA interpolate), 48 ILB have been structurally reconfigured. As shown in Table 9.6, the ARL unit results into a lower gate count than the AES-128 unit for same number of ILB reconfigurations (for a specific DSP application). Further, the post-synthesis resource usage summary of both approaches (implemented in Quartus II v13.0 for EP2C70F896C6 FPGA) has also been shown in Table 9.7. As evident from the table, the ARL unit–based obfuscation approach consumes significantly lower logic elements and dedicated logic registers compared to the AES-128–based obfuscation approach.

Table 9.6 Comparative study of the gate count of ARL unit–based obfuscation approach (Rathor and Sengupta, 2019) and AES-128–based obfuscation approach (Sengupta et al., 2019)

DSP applications (32-bit)	**Gate count**		**Percentage of reduction in gate count using ARL unit**
	Rathor and Sengupta (2019)	**Sengupta *et al.* (2019)**	
FIR	58,976	104,320	43.47
ARF	57,344	102,688	44.15
DWT	50,016	95,360	47.55
IIR	44,800	90,144	50.30
MESA interpolate	138,240	206,256	32.98
JPEG IDCT	150,176	218,192	31.17

Table 9.7 Resource utilization summary comparison

Resource utilization	Rathor and Sengupta (2019)	Sengupta *et al.* (2019)
Total logic elements (%)	0.2	0.8
Dedicated logic registers (%)	0.02	0.05

9.6 Conclusion

This chapter highlights the various hardware threats for functionally obfuscated DSP cores. Among all the potential attacks, removal attack has been emphasized for detailed discussion. As a solution to removal attacks on the functionally obfuscated DSP cores, three recent security techniques (custom AES-128, custom SHA-512 and custom ARL unit–based approaches) have been discussed. These security techniques offer security against removal attacks by generating the key-bits for a subset of ILBs (locking logic) to be reconfigured. Based on the generated key-bits, ILBs of a functionally obfuscated DSP core are structurally reconfigured (modified). The strength of a particular security technique depends on two factors, namely (1) number of ILBs that resulted into reconfiguration and (2) the number of challenge (input) bits required to decode the key-bits for the reconfigured ILBs. The security technique against the removal attack should be capable of providing higher security (capable to reconfigure a larger subset of ILBs and complexity to find challenge bits to decode the key-bits) at lower design cost. To assess the effectiveness of the security techniques against removal attack, they have been analysed based on security and design costs.

At the end of this chapter, the reader is able to understand the following concepts:

- Attack scenario for functionally obfuscated DSP cores.
- Removal attack on functionally obfuscated DSP cores.
- Solution to removal attack using custom AES-128–based approach.
- Solution to removal attack using custom SHA-512–based key encryption hardware.
- Solution to removal attack using customized lightweight ARL unit.
- Discussion on different forms of threats that use RE.
- Different security properties of locking logic used in obfuscation.
- Comparative analyses of different approaches used for securing functionally obfuscated DSP cores.

9.7 Questions and exercises

1. What is an ILB?
2. Draw a sample ILB structure.
3. What is multi-pairwise security?
4. What is run of key gates?
5. What is key-gate isolation?

6. What is SAT attack?
7. How is a key-sensitization attack launched?
8. What is a removal attack?
9. What is an ARL unit?
10. How does an ARL unit protect against removal attacks on ILBs?
11. What are the challenge bits in ARL?
12. What is S-Box, and state its properties?
13. What is the diffusion property in AES?
14. Which hardware logic could perform diffusion in AES?
15. What is modular reduction in AES?
16. What is ILB key selection logic in SHA-512 hardware?
17. What is AND–OR network in ARL unit?
18. What is majority function in SHA?
19. What is the role of RFC logic in SHA?
20. Explain G-function in AES.
21. How are the total number of ILBs estimated before insertion in a DSP obfuscated gate-level design?
22. What is muting of key-gates?
23. What are the different forms of threats on an IP core?
24. What is a design netlist?

References

E. Castillo, U. Meyer-Baese, A. Garcia, L. Parilla, and A. Lloris, "IPP@HDL: Efficient intellectual property protection scheme for IP cores," IEEE Transactions on Very Large Scale Integration (VLSI) Systems, vol. 15, no. 5, Pages: 578–590, 2007.

R. S. Chakraborty and S. Bhunia, "HARPOON: An obfuscation-based SoC design methodology for hardware protection," IEEE Transactions on Computer-Aided Design of Integrated Circuits and Systems, vol. 28, no. 10, Pages: 1493–1502, 2009.

Y. Lao and K. K. Parhi, "Obfuscating DSP circuits via high-level transformations," IEEE Transactions on Very Large Scale Integration (VLSI) Systems, vol. 23, no. 5, Pages: 819–830, 2015.

S. M. Plaza and I. L. Markov, "Solving the third-shift problem in IC piracy with test-aware logic locking," IEEE Transactions on Computer-Aided Design of Integrated Circuits and Systems, vol. 34, no. 6, Pages: 951–971, 2015.

M. Rathor and A. Sengupta, "Low-cost robust anti-removal logic for protecting functionally obfuscated DSP core against removal attack," IET Electronics Letters, vol. 55, no. 7, Pages: 374–376, 2019.

J. A. Roy, F. Koushanfar, and I. L. Markov, "EPIC: Ending piracy of integrated circuits," in Proc. Design Autom. Test Europe, Pages: 1069–1074, 2008.

R. Schneiderman, "DSPs evolving in consumer electronics applications," IEEE Signal Processing Magazine, vol. 27, no. 3, Pages: 6–10, 2010.

A. Sengupta, "Intellectual property cores: Protection designs for CE products," IEEE Consumer Electronics Magazine, vol. 5, no. 1, Pages: 83–88, 2016.

A. Sengupta, "Hardware security of CE devices [hardware matters]," IEEE Consumer Electronics Magazine, vol. 6, no. 1, Pages: 130–133, 2017.

A. Sengupta, S. P. Mohanty, F. Pescador, and P. Corcoran, "Multi-phase obfuscation of fault secured DSP designs with enhanced security feature," IEEE Transactions on Consumer Electronics, vol. 64, no. 3, Pages: 356–364, 2018.

A. Sengupta and S. Mohanty, "Advanced Encryption Standard (AES) and Its Hardware Watermarking for Ownership Protection," IP Core Protection and Hardware-Assisted Security for Consumer Electronics, Pages: 123–170, 2019, e-ISBN: 9781785618000.

A. Sengupta, D. Roy, S. P. Mohanty, and P. Corcoran, "DSP design protection in CE through algorithmic transformation based structural obfuscation," IEEE Transactions on Consumer Electronics, vol. 63, no. 4, Pages: 467–476, 2017a.

A. Sengupta, D. Kachave, and D. Roy, "Low cost functional obfuscation of reusable IP cores used in CE hardware through robust locking," IEEE Transactions on Computer-Aided Design of Integrated Circuits and Systems, vol. 38, no. 4, Pages: 604–616, 2019.

A. Sengupta and M. Rathor, "Security of functionally obfuscated DSP core against removal attack using SHA-512 based key encryption hardware," IEEE Access, vol. 7, Pages: 4598–4610, 2019.

A. Sengupta, S. Bhadauria, and S. P. Mohanty, "TL-HLS: Methodology for low cost hardware Trojan security aware scheduling with optimal loop unrolling factor during high level synthesis," IEEE Transactions on Computer-Aided Design of Integrated Circuits and Systems, vol. 36, no. 4, Pages: 655–668, 2017b.

A. Sengupta and S. P. Mohanty, "Functional obfuscation of DSP cores using robust logic locking and encryption," in IEEE Computer Society Annual Symposium on VLSI (ISVLSI), Hong Kong, Pages: 709–713, 2018.

P. Subramanyan, S. Ray, and S. Malik, "Evaluating the security of logic encryption algorithms," in IEEE International Symposium on Hardware Oriented Security and Trust (HOST), Pages: 137–143, 2015.

R. Torrance and D. James, "The state-of-the-art in IC reverse engineering," Canadian Healthcare Engineering Society, vol. 5747, Pages: 363–381, 2009.

Y. Xie and A. Srivastava, "Anti-SAT: Mitigating SAT attack on logic locking," IEEE Transactions on Computer-Aided Design of Integrated Circuits and Systems, vol. 38, no. 2, Pages: 199–207, 2019.

M. Yasin, J. J. Rajendran, O. Sinanoglu, and R. Karri, "On improving the security of logic locking," IEEE Transactions on Computer-Aided Design of Integrated Circuits and Systems, vol. 35, no. 9, Pages: 1411–1424, 2016.

M. Yasin, B. Mazumdar, O. Sinanoglu, and J. Rajendran, "Removal attacks on logic locking and camouflaging techniques," IEEE Transactions on Emerging Topics in Computing, 2017, doi: 10.1109/TETC.2017.2740364.

J. Zhang, "A practical logic obfuscation technique for hardware security," IEEE Transactions on Very Large Scale Integration (VLSI) Systems, vol. 24, no. 3, Pages: 1193–1197, 2015.

Index